# wonders of the HUMAN BODY

## Volume 1

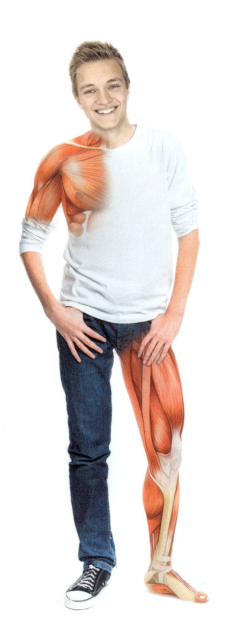

First printing: March 2022
Second printing: September 2023

Copyright © 2015, 2016, 2017 by Tommy Mitchell and Master Books®. All rights reserved. No part of this book may be used or reproduced in any manner whatsoever without written permission of the publisher, except in the case of brief quotations in articles and reviews. For information write:
Master Books
P.O. Box 726
Green Forest, AR 72638

Master Books® is a division of the New Leaf Publishing Group, LLC.

ISBN: 978-1-68344-277-6
ISBN: 978-1-61458-811-5 (digital)
Library of Congress Numbers: 2015930890, 2016935317, and 2017908538

Cover by Diana Bogardus
Interior by Jennifer Bauer

Unless otherwise noted, Scripture taken from the New King James Version®. Copyright © 1982 by Thomas Nelson. Used by permission. All rights reserved.

Please consider requesting that a copy of this volume be purchased by your local library system.

**Printed in the United States**

Please visit our website for other great titles:
www.masterbooks.com

For information regarding promotional opportunities, please contact the publicity department at pr@nlpg.com.

### Dedication
*For my beloved wife, Elizabeth*

*Cross section of glandular ducts*

# TABLE OF CONTENTS

## Unit 1: The Musculoskeletal System

### Foundations .................................................................................. 6
1. Introduction ............................................................................ 8
2. Cells ..................................................................................... 12
3. Tissues ................................................................................. 30
4. Organs & Organ Systems ...................................................... 36
5. Homeostasis ......................................................................... 40

### The Skeletal System ........................................................................ 42
1. Functions of the Skeletal System ........................................... 44
2. Bones ................................................................................... 46
3. The Skeleton ........................................................................ 62

### The Muscular System ...................................................................... 80
1. Muscle Basics ....................................................................... 82
2. Muscles…How They Move Me! ............................................... 94

## Unit 2: Cardiovascular and Respiratory Systems

### The Cardiovascular System ............................................................ 106
1. Introduction ........................................................................ 108
2. The Heart ............................................................................ 110
3. Two Kinds of Hearts ............................................................. 140
4. Blood Vessels ...................................................................... 142
5. Physiology of Circulation ..................................................... 148
6. The Circulatory System ....................................................... 156

### The Respiratory System ................................................................. 162
1. Breathing — No Big Deal? .................................................... 164
2. How We Breathe .................................................................. 184
3. Is This "Design" Just an Accident? ....................................... 196

# Unit 3: The Nervous System

1. Introduction .................................................................................. 200
2. Structure of Nervous Tissue ......................................................... 206
   - Neurons .................................................................................. 207
   - Neuroglia ................................................................................ 211
3. Nerve Signals ............................................................................... 218
   - The Resting Membrane Potential ............................................ 220
   - The Action Potential ............................................................... 221
   - The Synapse ........................................................................... 227
4. The Central Nervous System ....................................................... 232
   - The Brain ................................................................................ 233
   - The Spinal Cord ...................................................................... 257
5. The Peripheral Nervous System .................................................. 262
   - Cranial Nerves ........................................................................ 263
   - Spinal Nerves and Their Distribution ..................................... 266
   - The Autonomic Nervous System ............................................ 273
6. Special Senses ............................................................................. 278
   - Smell ....................................................................................... 280
   - Taste ....................................................................................... 282
   - Hearing ................................................................................... 284
   - Sight ....................................................................................... 292
7. The Gospel ................................................................................... 300

Glossary ........................................................................................... 302

Index ................................................................................................ 315

About the Author ............................................................................ 324

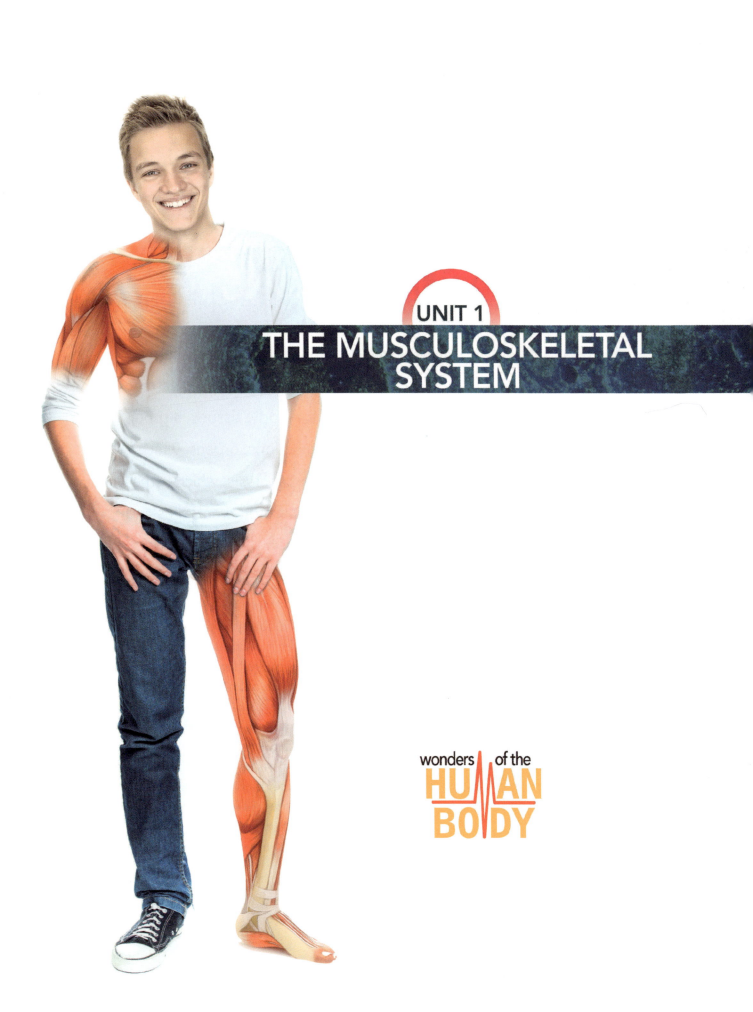

# UNIT 1
# THE MUSCULOSKELETAL SYSTEM

wonders of the
HUMAN BODY

# SECTION 1
# FOUNDATIONS

This book is introductory to the anatomy and physiology series, so you've come to the perfect starting off place! You'll be learning about cells, the basic building blocks of the body; tissues, a group of cells that perform similar or related functions; organs, which are tissues that function together; and homeostasis, which is the balance found among all your body functions. God created us and so He knows best how to care for us through all He has made!

"Where were you when I laid the foundations of the earth?" God asked this questions of Job (38:4), referring to the creation of the world. A foundation is the starting place for building something, and a good foundation is needed to create something that will last. Just as God created the heavens and the earth, He created us as well, and knows us more deeply and intricately than science will ever be able to grasp.

*For You formed my inward parts; You covered me in my mother's womb. I will praise You, for I am fearfully and wonderfully made; Marvelous are Your works, And that my soul knows very well. My frame was not hidden from You, When I was made in secret, And skillfully wrought in the lowest parts of the earth. Your eyes saw my substance, being yet unformed. And in Your book they all were written, The days fashioned for me, When as yet there were none of them (Psalm 139:13–16).*

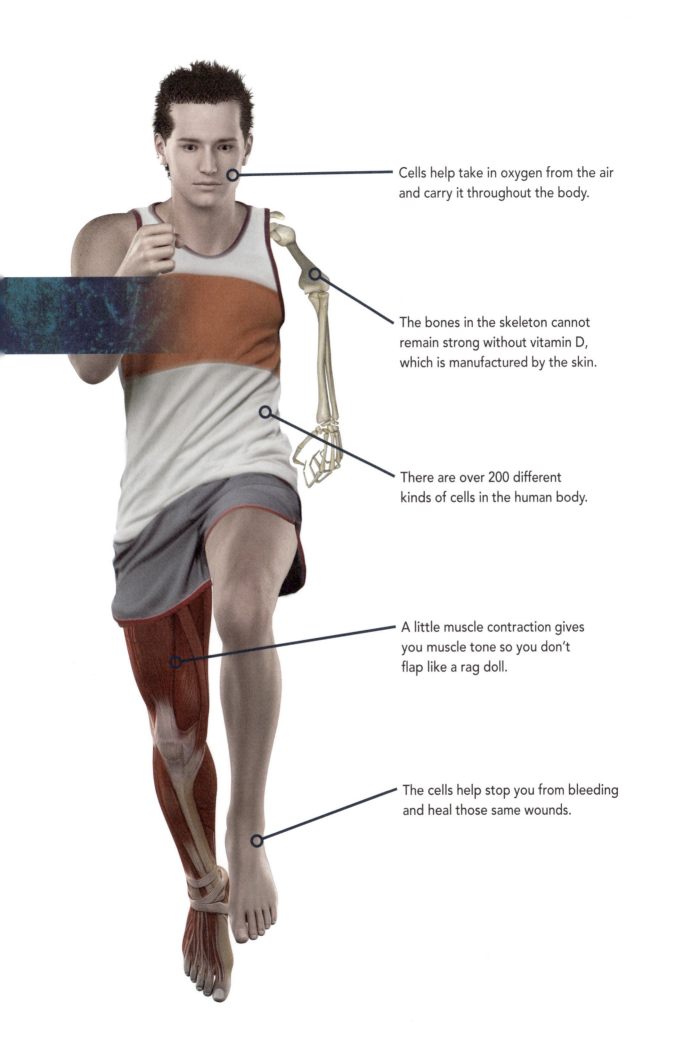

- Cells help take in oxygen from the air and carry it throughout the body.
- The bones in the skeleton cannot remain strong without vitamin D, which is manufactured by the skin.
- There are over 200 different kinds of cells in the human body.
- A little muscle contraction gives you muscle tone so you don't flap like a rag doll.
- The cells help stop you from bleeding and heal those same wounds.

# INTRODUCTION

The human body — two arms and two legs, a head, chest, and tummy — seems simple on the surface. In reality, however, it is an incredibly designed orchestra of parts perfectly created to work together. Your body's parts — from the largest bones and organs to the smallest molecules and cells — are put together with a precision no engineer could design. Your body is able to do a remarkable array of things. And it must do many of them nonstop without your attention. This series of books will take you on a guided tour of your own body, giving you a peek into its secrets, large and small, and showing you how it works and how all its parts are designed to work together. Think of this as an owner's manual to the first birthday gift you ever received — the body with which you were born.

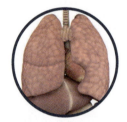
Respiratory System

Skeletal System

Muscular System

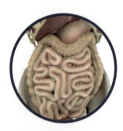
Digestive System

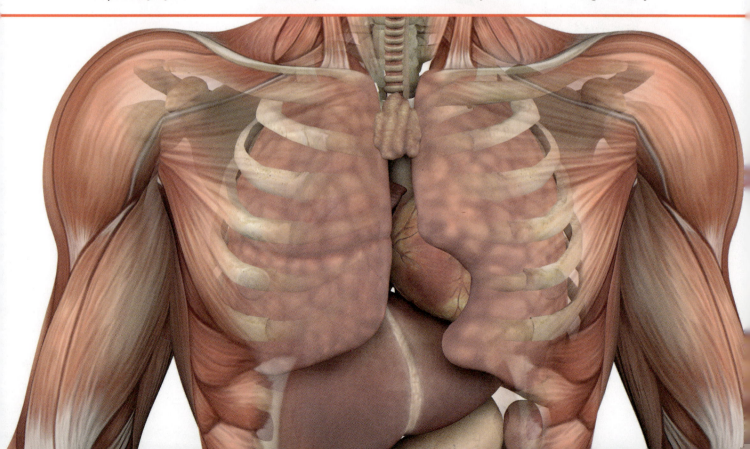

The same body that can use a hammer to drive a nail can hold a feather without crushing it. The same eyes can take in the expanse of the Grand Canyon yet detect faint light from stars many light years away. Your lungs enable you to talk and sing without forgetting to take in oxygen and release carbon dioxide. If you had to think about breathing, you could never go to sleep, for you would die within minutes.

Your digestive system breaks down food into the chemicals you need for energy. Your heart pumps blood through miles of blood vessels to deliver oxygen and nutrients to the most unseen places. Directed by your brain, you walk, run, throw a ball, play a video game, read a book, paint a picture, or even play an instrument. However, the same brain also controls many essential activities inside your body that you are probably not aware of.

## What Is Evolution?

In the Bible we read, "In the beginning God created the heavens and the earth" (Genesis 1:1). This theme is repeated time and time again. God is called Creator. In John 1:2 it states, "All things were made through him."

However, there are many people, Christians and non-Christians alike, who do not acknowledge Him as the One who created all things. Instead, they believe in something called evolution.

Rather than accepting God as Creator, they believe that billions of years ago, matter just appeared out of nowhere. Then, they believe, there was a rapid expansion of the universe called the big bang. As this process progressed, the stars and galaxies formed themselves. Then about five billion years ago, our sun was formed and later all the planets in our solar system.

They believe our planet was at first a hot molten blob that eventually cooled and became covered with water. Then all the substances necessary for living things appeared. The first simple life form then sprang into existence by accident.

Then over the next few billion years, living things became more and more complex until at last man appeared on the scene. In this view, man is no more than a cosmic accident, the result of chemicals bumping together over millions of years. Basically, we are then just a highly evolved animal. What folly is a belief such as this!

As we go through our study of the human body, you will see how absurd it is to think that these marvelously complex systems assembled themselves by accident. We are not an accident. We are made in the image of our wonderful Creator God. God tells us in the Bible where we came from: "Then God said, 'Let Us make man in Our image, according to Our likeness' " (Genesis 1:26).

Your bones give your body shape and support even while they grow, your skin protects you from the outside world, your blood carries oxygen and the tiny tools to fight off harmful germs, your liver manufactures chemicals you need while it breaks down toxins, and your kidneys rid your blood of many waste products and help control the amount of water in your body. The amazing list of unseen and unceasing processes that must go on simultaneously for your body to work properly goes on and on.

Actually, the amazing thing is that many people think that the human body is a product of chance, that we are merely an accident. They believe that our bodies developed on their own as a result of chemical reactions occurring over billions of years through a process called evolution.

Imagine that . . . chemicals combined themselves to become alive, and then the human body, so complicated, so intricate, just assembled itself! Hard to believe, isn't it?

The truth is that we are not an accident. The human body was designed by a Master Designer, the Creator God of the universe. The human body He designed is simple enough that a newborn baby instantly begins making it work and quickly learns the most complex skills. He designed it with sufficient well-orchestrated complexity and built-in control systems so that even before a baby is born the cells in her body are performing chemical reactions she would need a college degree to even begin to understand, and many that the smartest scientists of all are only beginning to discover.

You are marvelously and wonderfully made. Psalm 139:14, says, "I will praise You, for I am fearfully and wonderfully made; marvelous are Your works." As you explore the incredible features of the human body, remember to praise the Creator God who designed the human body and gave one to you.

You will hear from many people that you are nothing more than an animal, a highly evolved one, but an animal nonetheless. This idea comes from people who wrongly believe that life evolved through random processes and all by itself produced increasingly complex animals until humans appeared. They believe that humans are just animals and not special at all. God's Word tells us otherwise.

In Genesis 1:26 it says, "Then God said, 'Let Us make man in Our image, according to Our likeness.'" Therefore, we are not merely animals. God made human beings in His own image.

But you might say, "Well, we look similar to some animals, so aren't we just animals, too?" No, we are not. Humans and some animals do share many similar features, but that does not mean that animals are our ancestors. All it means is that animals and humans were all designed by the same God. We share a common designer, not a common ancestor! A wise master designer would naturally use variations of many common designs in the living things He made.

Just as words are built of letters and books are built from words, so your body is built of organs and tissues, and all the organs and tissues are made of cells. Cells are even called the building blocks of life. We are going to begin exploring the amazing designs of the human body by finding out how the cells of the major organs work.

# Anatomy

Anatomy is the study of the body's parts and how they are put together. Anatomy includes how the organs look, where they are, and how everything is connected. Anatomy is the structure of the body. For example, an anatomical study of the circulatory system includes a study of the heart itself and all the blood vessels and their connections to all the other organs.

Anatomy includes not just the things you can see with your eyes, like lungs and kidneys. Anatomy includes the microscopic structures — the cells and the tissues (collections of cells) that make up the larger parts. This study of microscopic anatomy is called histology.

# Physiology

Physiology is the study of how the parts of the body function. Physiology is the study of how everything in the body works. For example, physiology of the circulatory system focuses on how the heart works, what controls blood circulation and blood pressure, how oxygen gets into the blood, and how blood delivers oxygen to the tissues and organs.

If we didn't understand some physiology, learning about the human body would just be the dull business of memorizing the names and locations of organs and bones. But when we find out how each part works and interacts with other parts, the study of the human body really does come alive!

# Cells and Tissues and Organs, Oh My!

The best way to understand the human body is not just to memorize its parts but to begin with its building blocks and then to see how they form the more complex structures and how they work. We will begin with the most basic building block of the body, the cell. Cells are small but not simple. While cells have a simple list of parts — like nucleus, cytoplasm, and cell membrane — these are subdivided into a dizzying list of smaller parts. Many cells are like tiny factories. There are many types of cells in the body. For example, there are liver cells, muscle cells, kidney cells, nerve cells, skin cells, blood cells — well . . . you get the idea.

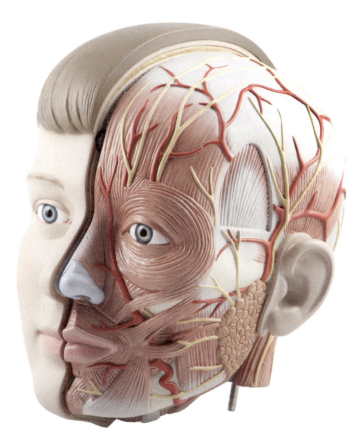

Groups of cells form tissues. Each kind of tissue can be thought of as one of four basic tissue types — epithelial, connective, muscle, and nervous. These tissues cover, connect, move, and communicate. More on that later.

Tissues combine to form more complex structures called organs. Organs are groups of tissues that have a particular function. Lungs, bones, and the brain are examples of organs.

Lastly, we will deal with organ systems. This is where we will "put things together" by exploring groups of organs that work together to do specific things in the body. For example, the bones are all connected together as the skeletal system. And all the parts that process your food — from your mouth and stomach to your liver and intestines — are part of the digestive system.

Before we can put it all together we need to go back to basics, so let's get going with the cell.

# CELLS

The cell is not only the basic building block of the body but also the basic "functional unit." What does that mean? Well, your body does a lot of things — some things you see and some that you don't. It moves. It grows. It digests food, turning some of it into energy, storing some of it, and discarding the leftovers. It manufactures many kinds of complex chemicals. It tastes, smells, sees, hears, touches, senses temperature, and feels pain. It takes in oxygen from the air and carries it all over your body. It fights infection and protects you from most germs. It stops you from bleeding when you get a cut, and later it heals the cut. All these "functions" are really performed by or inside cells. That's why we say the cell is the smallest "functional unit" of the body. The cell is where the action is!

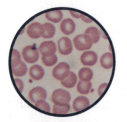
Blood Cell

Liver Cell

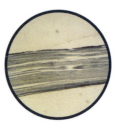
Muscle Cell

Nerve Cell

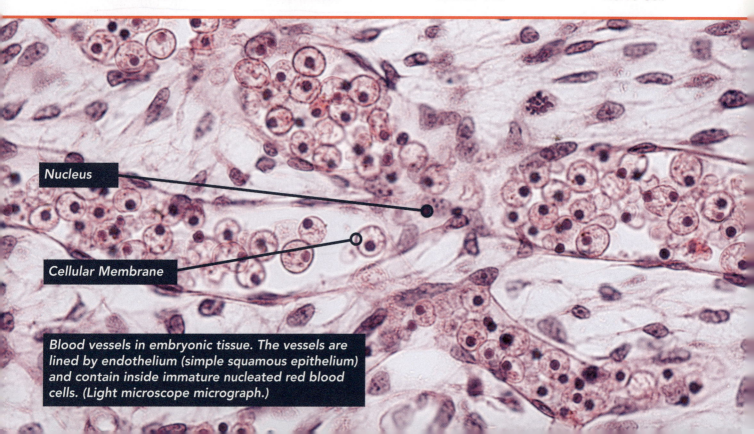

*Nucleus*

*Cellular Membrane*

*Blood vessels in embryonic tissue. The vessels are lined by endothelium (simple squamous epithelium) and contain inside immature nucleated red blood cells. (Light microscope micrograph.)*

Each cell is like a factory designed to carry out a specific function. There are over 200 different kinds of cells in the human body, and they come in all shapes and sizes. Most cells have three basic parts — a *nucleus* that directs most of the action, a *cell membrane* that forms the cell's outer border, and *cytoplasm* where most of the cell's work gets done. Most kinds of cells have many *organelles* that perform the various jobs in the cell.

*Erythrocytes* are red blood cells. Their main job is to carry oxygen. Red blood cells are packed with a red oxygen-carrying molecule (hemoglobin), which is why they are red. Erythrocytes are comparatively simple cells. The erythrocytes circulating throughout your body don't even have a nucleus or organelles.

In contrast, liver cells are much more complex. Liver cells process and store nutrients, manufacture important substances, and rid the body of some toxic chemicals. Because liver cells are involved in more complex activities than red blood cells, their structure is more complex.

Each muscle cell is designed to contract, and you can move because muscle cells work together. Certain cells in the pancreas produce *insulin* that controls the amount of sugar in your blood, because either too much or too little is bad for you. Nerve cells transmit nerve impulses so that one part of your body can communicate with another. Otherwise, your hand would not "know" that your brain told it to move. And the list goes on. Each cell has an important job to do.

For all their many differences in structure and function, most cells have a lot of things in common. Here we'll learn about a "typical" cell. Then in our journey through the human body, we will examine specific cell types in more detail.

**SO SIMPLE YET SO COMPLEX — Designed by the Master**

## Human Cells and Plant Cells

You will soon learn about many different kinds of cells found in the human body. Plants are also made of cells. Plant cells have many things in common with our cells. Plant cells have nuclei containing chromosomes that direct the cellular activities. They have mitochondria and the other organelles we have. And plant cells also have cell membranes.

But plant cells have two things our cells lack — cell walls and chloroplasts. Plant cell membranes are surrounded by a tough cell wall made of cellulose. Humans cannot make cellulose. The cell wall provides a sturdy support for plant cells and helps maintain their shape. Plant cells, unlike our cells, are also able to capture energy directly from sunlight and use it to manufacture sugar. This process is called photosynthesis. Photosynthesis takes place in special organelles called chloroplasts. The chloroplasts in plant cells contain the green pigment chlorophyll, which captures the sun's energy. God designed plant cells to produce sugars and other important foods for humans and animals to eat.

*Plant cell*

*Human cell*

# Basic Cell Structure

Regardless of size, shape, or complexity, most human cells have, as we mentioned, three main parts. The *cell membrane*, also called the *plasma membrane*, encloses the cell, forming the boundary with its *extracellular* surroundings. One could look at the plasma membrane as the bag or sack that holds all the other parts. This is no ordinary "bag" though. Even the membrane surrounding the cell is specially designed to perform a lot of vital jobs. The cell membrane keeps some things in and keeps other things out, while letting some things travel across it and actively helping other things to pass through. The cell membrane is like the ultimate doorkeeper, and then some!

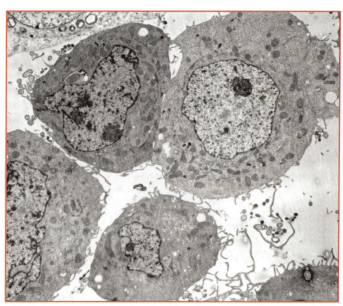

*Electron microscopic view of cells*

The control center of the cell is the *nucleus*. It directs the activities of the cell. The nucleus stores all the instructions the cell needs to function. These instructions are in code. The code is written into the structure of DNA, long chain-like molecules that are stored in the nucleus. The blueprint for making each protein the cell it is supposed to make is written in a *gene* in this DNA. Except for mature red blood cells, all cells in the body have at least one nucleus. Some have several *nuclei*.

In between the cell membrane and the nucleus, or nuclei, is the *cytoplasm*. All the parts of the cell that are not part of the nucleus or cell membrane are part of the cytoplasm. Many little "workstations" called *organelles* float in the *cytosol*, which is the cytoplasm's fluid. Dissolved in the cytosol are also many substances like sugars and electrolytes. (*Electrolytes* include sodium ions, potassium ions, calcium ions, and so forth. *Ions* are charged chemicals, and we'll learn later that the way they move into and out of cells is very important.) Large molecules such as enzymes also float around in the cytosol, each doing an important job.

### TAKING A CLOSER LOOK
## Human Cell Structure

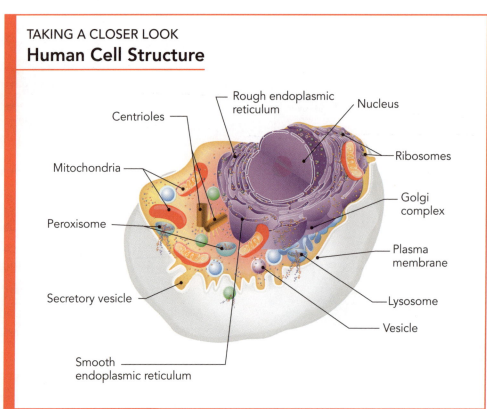

# The Plasma Membrane

The plasma membrane is the envelope that contains the other components of the cell. Within it is the cytoplasm, its organelles, and the nucleus. Without the plasma membrane, the cell would have no form or structure. The plasma membrane holds the cell together.

However, the plasma membrane is far more than just a container. It helps separate the two major fluid compartments of the body, the *intracellular fluid* — fluid inside cells — from the *extracellular fluid* — fluid that is outside cells. The plasma membrane is also involved in moving fluid, nutrients, and other substances into and out of the cell while forming a barrier to things that should stay out.

Most of the intracellular fluid and most of the extracellular fluid is water, but the concentration of the chemicals dissolved in them makes them very, very different. The chemicals dissolved in these fluids are "water soluble," which means they can dissolve in water. You probably already know that sugar and salt dissolve in water, and oil does not. Well, sugar molecules are water-soluble. Salts are made of ions, like sodium ions and potassium ions and chloride ions, and such salts are also water-soluble. Fats and oils, however, are not water-soluble: they do not dissolve in water. Another name for a fat is *lipid*.

## Its Structure

The plasma membrane is actually made up of two layers of molecules. These molecules are called

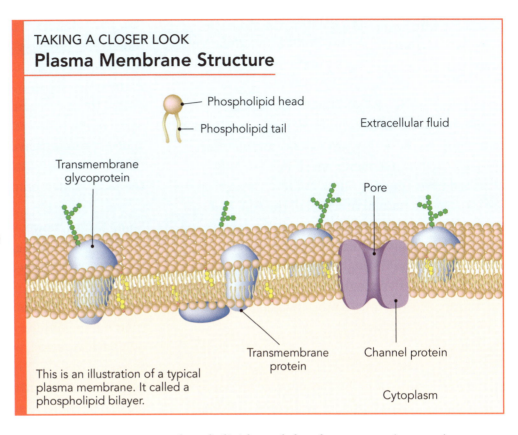

**TAKING A CLOSER LOOK**
**Plasma Membrane Structure**

This is an illustration of a typical plasma membrane. It called a phospholipid bilayer.

*phospholipids*, and they have a very interesting shape, as you can see in the illustration.

These molecules have what can be described as a "head" and two "tails." The "head" of the molecule is charged. This portion of the molecule is water-soluble (known as *hydrophilic*, a word that literally means "water-loving") and is therefore attracted to water. The tail portion is uncharged and avoids water (known as *hydrophobic*, a word that literally means "water-fearing"). These characteristics of phospholipids are important not only in the structure of the plasma membrane, but also for its function.

The plasma membrane is composed of these two layers of phospholipids, creatively called a *phospholipid bilayer*, which means "two layers of phospholipids." The phospholipid molecules are lying with the heads facing the outer and inner surface of the plasma membrane and the tails pointing to the interior of the membrane. The hydrophilic (water-loving) heads of the molecules are in contact with the watery fluid inside and outside the cells. The hydrophobic (water "fearing") tails are pointing toward each other, as far from the watery fluids as possible. This helps maintain the integrity of the membrane.

In addition to the phospholipids, the plasma membrane has a lot of protein molecules embedded in it. Some of these proteins extend completely through the plasma membrane. Some are only attached to its inner or outer surface. These proteins are vital to the normal function of the cell. Some of them ferry certain substances across the membrane. Some form a doorway allowing particular sorts of molecules to pass through. Some of them are like name tags that identify the cell to other cells. Some even form attachments to other neighboring cells.

## Its Function

So beyond just holding the contents of the cell in one container, what is the function of the plasma membrane? Well, among other things, it helps regulate what goes into and out of the cell.

Some substances, like water and certain lipid (fat) molecules, can pass directly through the plasma membrane and get into or out of the cell. However, many other substances cannot easily get into cells. Often, these can gain access to the cell by means of some of the proteins in the plasma membrane. These special proteins have a channel in them to allow things into a cell that could not pass directly through the plasma membrane.

Some things, however, are too large even for protein channels. So in the case of the largest molecules, there is a special mechanism called *endocytosis*. In this case, a portion of the plasma membrane folds into the cell, surrounds the molecules needed, and then the membrane pinches off, forming a small bubble-like *vesicle*, which is then processed inside the cell. Occasionally this process is reversed and vesicles formed within the cell merge with the plasma membrane and release products made by the cell. The process of releasing material from inside the cell is called *exocytosis*.

Further, the plasma membrane is able to respond to cellular signals because of some of the proteins on its outer surface. These proteins bind to certain molecules that cause the cell to react in a specific way.

There are also special proteins on the outer surface of the plasma membrane that help identify the cell. In other words, these proteins are like an identification tag for the cell, so the body itself can know which cells are which. When we study the immune system, you will see this in action. So the plasma membrane isn't just any old bag, is it?

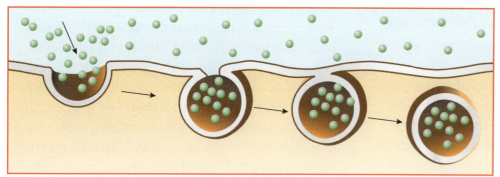
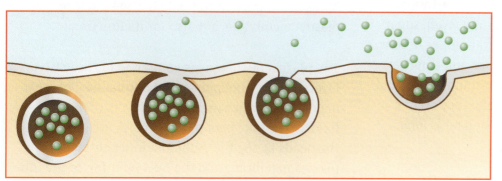

Vesicles can transport material into and out of cells. During endocytosis, shown on top, material is transported into a cell by packaging it into a vesicle. Exocytosis is shown in the bottom illustration. There, a vesicle merges with the cell membrane and the material it contains is released.

## Cell Markers

*So Simple Yet Designed by the Master So Complex*

The plasma membrane contains some special proteins called glycoproteins. These proteins have carbohydrate (sugar) groups attached that protrude into the extracellular fluid. These carbohydrate groups along with other special molecules called glycolipids form a coating on the cell surface known as the glycocalyx.

The pattern of the glycocalyx varies from cell to cell. It is distinctive enough that it forms a molecular "signature" for a cell. This is one way that cells can recognize one another.

## Cytosol

Cytosol is the liquid found inside the cell. It surrounds the organelles and the nucleus. The *cytosol* plus the *organelles* make up the *cytoplasm*.

The cytosol is mostly water. Water makes up 70 to 75 percent of the volume of the cell. The cytosol contains many substances, and the cell works hard to maintain the appropriate balance of the substances found there.

There are lots of ions (charged atoms or molecules) in the cytosol, mostly potassium, sodium, chloride, and bicarbonate ions. These ions help maintain the electrical balance between the inside and outside of the cell (called the *membrane potential*, as we will explore later), as well as help maintain the correct water concentration inside the cell.

The cytosol also contains lots of proteins and *amino acids*. (Amino acids are the building blocks of proteins; we'll get more into that later.) These proteins and amino acids provide the raw materials for many of the activities of the cell.

### Endoplasmic Reticulum

The *endoplasmic reticulum* is a network of tubes and membranes that is connected to the nuclear membrane. The endoplasmic reticulum, or ER, comes in two forms, *rough ER* and *smooth ER*.

Rough ER is bumpy because it is covered with *ribosomes*. Ribosomes are little factories for making protein. Rough ER is primarily involved with protein production. Proteins that are made in the ribosomes can be modified by the endoplasmic reticulum to fit them for their particular jobs. The particular proteins and lipids that make up the plasma membrane are made in the rough ER.

### TAKING A CLOSER LOOK
**Endoplasmic Reticulum**

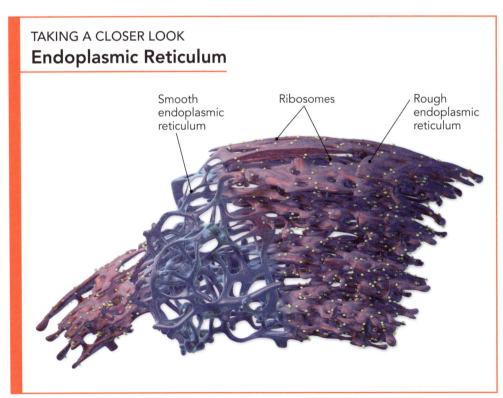

Smooth endoplasmic reticulum — Ribosomes — Rough endoplasmic reticulum

Smooth ER is more tube-like in appearance and is not covered with ribosomes. It is more involved with production of fats, certain hormones, and the breakdown of some toxins that enter the cell.

## Golgi Apparatus

The Golgi apparatus is a collection of small flattened sacs that stack on one another. They tend to be flatter in the middle and more rounded on the ends.

Cells produce lots of things, especially fats and proteins. The Golgi apparatus helps the cell transport these products to where they are needed. It does this by forming little sacs, or vesicles, around the needed items. These vesicles pinch off from the Golgi apparatus and travel to their destination. Sometimes this is within the cell itself. Sometimes the vesicle moves to the plasma membrane and releases its contents outside the cell via *exocytosis*.

The Golgi apparatus is an exquisitely designed delivery system. Without it, the cell could not function.

## Lysosomes

*Lysosomes* are small vesicles containing enzymes that can digest many kinds of molecules and debris. This may seem surprising. After all, aren't these types of substances dangerous to the cell itself? Yes, they can be, but they are still very necessary.

Lysosomes break down worn-out organelles, bacteria, and toxic substances. For example, white blood cells contain a large number of lysosomes. That is how they are able to help rid the body of invading bacteria.

Lysosomes also aid the cell by breaking down substances the cell needs for nutrition, particularly large molecules the cell takes in. In fact, the

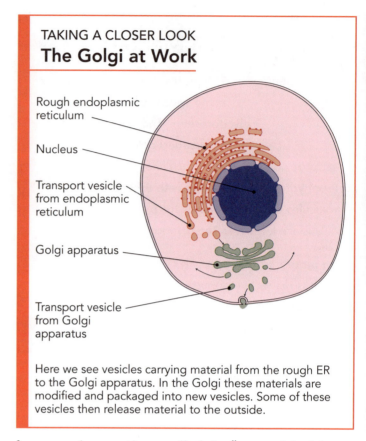

**TAKING A CLOSER LOOK**
**The Golgi at Work**

Here we see vesicles carrying material from the rough ER to the Golgi apparatus. In the Golgi these materials are modified and packaged into new vesicles. Some of these vesicles then release material to the outside.

lysosome is sometimes called the "stomach" of the cell. And by breaking down organelles that are worn out or no longer needed, the lysosomes recycle valuable materials.

## Ribosomes

Ribosomes are found floating in the cytoplasm and attached to the rough endoplasmic reticulum. These are little structures, but they have a very big job. *Ribosomes* are where proteins are made. Let's consider where a ribosome gets its protein-building instructions.

You may remember that the nucleus of a cell directs the cell's activities. The instructions for what the cell is supposed to do are stored in the nucleus. The "blueprints" for how to build the proteins a cell is supposed to build are mostly stored in the nucleus. These "blueprints" or "recipes" for building proteins are called genes.

## TAKING A CLOSER LOOK
### Ribosomes

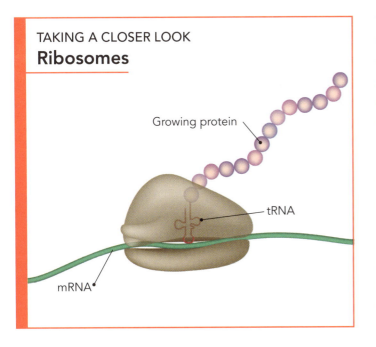

Genes with protein-building instructions are in the nucleus, but the protein-making ribosomes are located in the cytoplasm. How can the ribosomes get their instructions? Well, copies of the instructions, called *messenger RNA*, are made in the nucleus. Those messages move from the nucleus into the cytoplasm. There, ribosomes read the messenger RNA's instructions and build the protein described, stitching together a string of *amino acids*, which are the building blocks of proteins. The ribosome follows the "recipe" stored in the nucleus and copied onto messenger RNA.

## Mitochondria

The *mitochondria* are often called the "powerhouses" of the cell. They are called that because they generate and store energy. Mitochondria are like super battery chargers.

These are elongated bean-shaped structures with lots of folded membranes inside. Unlike the other organelles in the cell, mitochondria even contain some genes used to reproduce themselves! (Remember, all the rest of the genes in your body's cells are stored in the nuceli.)

The mitochondria are responsible for producing high-energy molecules. Those high-energy molecules are like batteries: they store energy until the cell needs the energy for something. One of the most important high-energy molecules is ATP (which stands for *adenosine triphosphate*, if you want to show off to your friends . . .). This molecule stores energy needed to fuel cellular activities.

ATP is actually built from ADP, *adenosine diphosphate*. ADP is like a battery that needs to be recharged. And ATP is like a fully charged battery. As you might guess from the names *triphosphate* and *biphosphate*, ATP contains three "phosphates" and ADP contains two "phosphates." The bonds that hold phosphate onto ADP and ATP store a lot of energy,

## TAKING A CLOSER LOOK
### Mitochondria

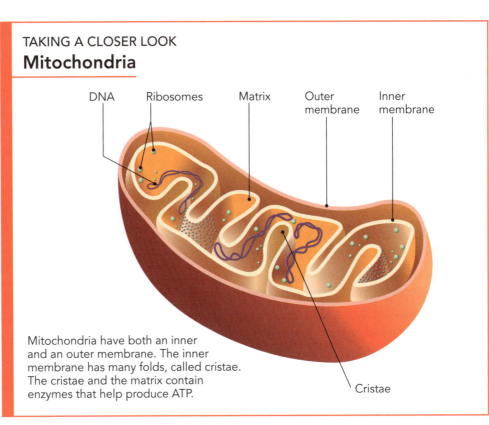

Mitochondria have both an inner and an outer membrane. The inner membrane has many folds, called cristae. The cristae and the matrix contain enzymes that help produce ATP.

## THE MUSCULOSKETETAL SYSTEM

*SO SIMPLE YET SO COMPLEX — Designed by the Master*

### Making Mitochondria

The nucleus is not the only place that DNA is found in the cell. Mitochondria have multiple copies of their own DNA. This DNA exists as a circular molecule containing 37 genes. Interestingly enough, mitochondrial DNA is inherited only from the mother. In addition, mitochondria contain RNA and ribosomes. During times of increased energy needs, the mitochondria can reproduce themselves to increase their number. They grow and divide by pinching in half.

much like a battery stores energy until it is needed. When energy is needed, a high-energy bond in ATP (or in other similar high-energy molecules) is broken and the energy released from it is used to power whatever the cell needs to do.

But where does the mitochondria get the energy to charge these chemical batteries? After all, you've learned before that energy cannot be created or destroyed but only transformed from one form to another. The fuel that provides the energy for the mitochondria's charging operation comes from sugar.

The process of providing energy to the cell is kind of like putting wood in a stove or putting gasoline in a car. Wood and gasoline are both fuels. The wood in the stove burns to make heat that can be used to cook food or heat your home. The gasoline in a car is burned by the engine and provides energy to make the car move. It is not all that different to make energy for a cell. The cell's favorite fuel is not wood or gasoline but the sugar *glucose*. The energy produced when it is *metabolized* — a sort of very controlled way of "burning" the fuel — must be captured and stored in chemical "batteries" like ATP.

Remember, think of ATP and ADP like rechargeable batteries. The primary fuel for cells is the sugar glucose. Glucose is taken into the mitochondria through a series of chemical reactions, and the molecule ATP is produced by recharging ADP with energy from glucose. Just as burning wood or gasoline depends on oxygen, this chain of chemical reactions in the mitochondria also requires oxygen (so thank your lungs here!).

The number of mitochondria in a cell depends on the energy needs of the cell. Liver cells, for example, are involved in making proteins, making cholesterol and other lipids (fats), making and secreting bile, and many other things. So you may well imagine that it takes lots of energy to perform all these functions. In fact, a liver cell can have as many as 2,000 mitochondria!

### Centrosome and Cytoskeleton

You might ask yourself, "What keeps all this stuff in place?" Well, there is an answer! The cell has a sort

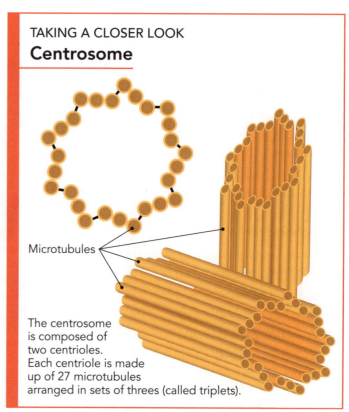

**TAKING A CLOSER LOOK**
**Centrosome**

Microtubules

The centrosome is composed of two centrioles. Each centriole is made up of 27 microtubules arranged in sets of threes (called triplets).

of skeleton, called a cytoskeleton, that helps with that task. This cytoskeleton is composed of a network of tubes and filaments that run throughout the cell. Though not pictured in most diagrams of cells, these fine tubes and filaments provide support for the organelles.

But this support system does more than just hold things in one place. Along with the cytoskeleton, there are special motor proteins that help organelles move around. Mitochondria, lysosomes, and vesicles all move around the cell with the help of these amazing structures.

Another very special organelle, called the *centrosome*, is necessary for cellular reproduction. After all, most kinds of cells wear out and must therefore reproduce, or duplicate, themselves. We'll go into the complex process of how a cell divides in two later.

Sometimes it seems like all the action is in the nucleus when we talk about cell division. But if it weren't for the centrosome, which is located outside the nucleus in the cytoplasm, cellular reproduction would be a disorganized chaotic mess. Nothing would end up in the right place!

The centrosome is an L-shaped structure made up of two barrel-shaped *centrioles*. These centrioles are responsible for helping form a complex of *microtubules*, called the *mitotic spindle*, which guides the cell's chromosomes during cell division.

## Nucleus

The nucleus is the control center of the cell. Stored in the DNA (deoxyribonucleic acid) in each cell's nucleus are the genetic instructions needed to make all the proteins in the body. The genes — the little recipes for building proteins — and even the regulations that determine how and when those genes are to be used are part of the DNA. The nucleus regulates the types of proteins made by its cell and their

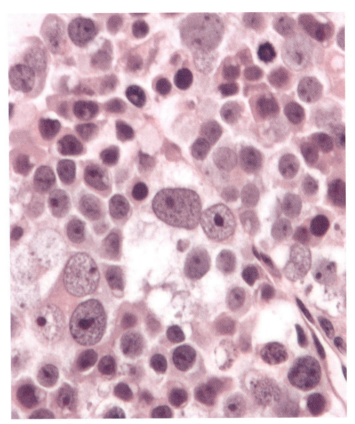

*Micrograph of a spermatocytic seminoma*

amounts. Even though the nucleus contains a copy of your entire *genome*, only the information needed by each cell type is ever turned on and used.

The majority of cells have one nucleus. However, there are exceptions. Skeletal muscle cells (and a few other cell types) have more than one nucleus, and mature human red blood cells have none.

Just as the cell has a cell membrane, so the nucleus has a *nuclear membrane*. You recall that messages — in the form of messenger RNA — must pass from the nucleus into the cytoplasm to deliver instructions to the ribosomes. Did you wonder how the message gets through? Well, the outer part of the nuclear membrane connects to rough endoplasmic reticulum. Through tiny pores in the nuclear membrane, substances can pass from the nucleus into the cytoplasm. That way the instructions from the nucleus can reach the cytoplasm where they can be implemented.

## DNA

DNA — deoxyribonucleic acid — is one of the most amazing molecules in the universe. In your DNA is contained all the information needed to make your body!

DNA is a big molecule made up of two long strings of smaller molecules called *nucleotides*. There are four different kinds of nucleotides present in DNA. These four nucleotides are the building blocks of DNA. Two long strands of nucleotides are attracted to each other and form a structure that looks like a twisted ladder. That structure is called a *double helix*.

So what is so amazing about long strings of chemicals?

Well, it turns out that the order in which the nucleotides are found in DNA is very, very important. Those four nucleotides in DNA aren't just DNA's building blocks. They are the "letters" in a code — the genetic code of life that is used not only in the human body but in all the living things God designed!

You see, DNA is not just a string of chemicals. It is a very complex system of information! For decades now, scientists have studied the "letters" and "words" in the DNA and how they work.

Imagine each nucleotide as a "letter." Three "letters" form a "word." And a group of "words" can give coded instructions for building a protein or even for regulating how those instructions are carried out. The DNA in a human cell contains over 3 billion nucleotides. The instructions coded in your DNA determine which proteins can be made.

Each section of DNA that has the information for a particular protein is called a "gene." Another way of looking at this is to think of a certain group of nucleotide "words" combining to make up a genetic "book." Other sets of nucleotide words make up other books, and so on.

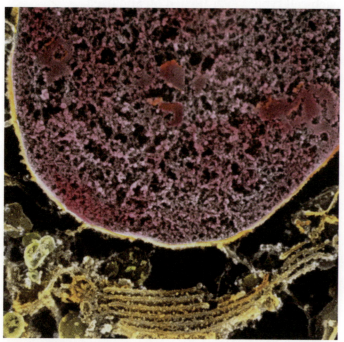

*Colored high resolution scanning electron micrograph of the nucleus and rough endoplasmic reticulum of a primordial testis germ cell.*

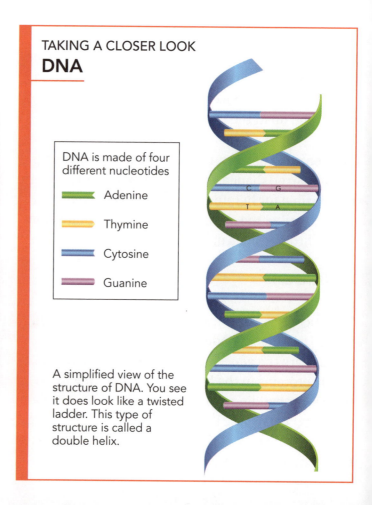

### TAKING A CLOSER LOOK
### DNA

DNA is made of four different nucleotides
- Adenine
- Thymine
- Cytosine
- Guanine

A simplified view of the structure of DNA. You see it does look like a twisted ladder. This type of structure is called a double helix.

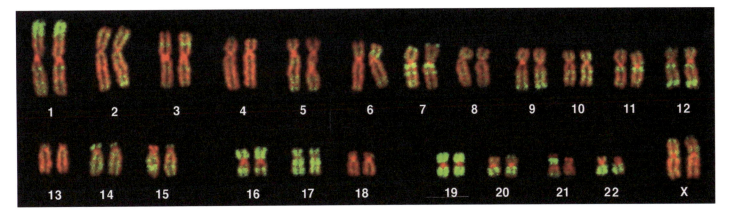

This is a picture of a person's chromosomes. As you can see, humans have 23 pairs of chromosomes. The autosomal chromosomes are numbered according to their length. Number one is the longest and number 22 is the shortest. The remaining pair are the sex chromosomes.

However, DNA also contains coded instructions for other things, like the directions for what kind of cell each cell is supposed to be or how busy it is supposed to be. Some scientists have claimed that the DNA that didn't code for proteins was leftover evolutionary junk with no purpose. Bible-believing scientists know that evolution did not create life, DNA, or the human body. Therefore, these scientists predicted that none of our DNA was evolutionary "junk." Now, scientists have begun finding that "junk" DNA really does have a purpose. The double-helix structure of DNA was discovered in 1953, but scientists are just beginning to figure out how much coded information is contained in each molecule of DNA.

So each strand of DNA is made up of many, many genes. Each gene gives the instructions for building a protein. Proteins are built out of amino acids. Proteins are a kind of biological molecule, and they do much of the "work" in your body. Lots of molecules you may have heard of are proteins. *Enzymes* that perform all the chemical reactions in your cells, *antibodies* that fight infectious invaders, *taste receptors* in your tongue, *collagen* that holds much of your body together, the *actin* and *myosin* molecules that make your muscles contract, the *clotting factors* that make your blood clot, and the transport proteins and identification proteins embedded in your cell

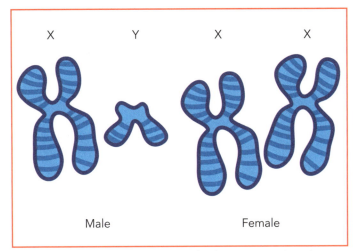

The sex chromosomes determine whether a person is male or female. A person who has an XY pair is male. Those who have an XX pair are female.

### TAKING A CLOSER LOOK
## Genes and Chromosomes

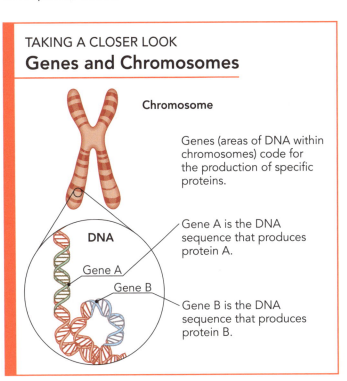

Chromosome

Genes (areas of DNA within chromosomes) code for the production of specific proteins.

Gene A is the DNA sequence that produces protein A.

Gene B is the DNA sequence that produces protein B.

membranes are all proteins. Each and every protein molecule must be built in a cell, following the instructions from the nucleus.

Each double helix molecule of DNA is carefully organized and packaged into a chromosome. Each *chromosome* is like a section of a huge library where lots of books are stored. A chromosome consists of DNA — like the books — and special proteins that help package it and take care of it — like "shelves." Human beings have 46 chromosomes in each of the body cells.

The DNA in one of your cells would be about 6 feet long if it were stretched out. In just this tiny strand of DNA is contained enough information to fill hundreds of books, and the DNA in just one of your cells contains the coded information to build your whole body!

## DNA at Work

So what exactly does DNA do?

### DNA Can Make Proteins

We said that DNA was more than just a string of molecules. It is a complex system of information. This information is used primarily to make the proteins in our body.

Proteins are one of the most important substances in the body. Proteins are made up of long chains of molecules called *amino acids*. For proteins to function properly, the order of these amino acid building blocks must be correct. So there must be a very precise process to make proteins.

TAKING A CLOSER LOOK
### Building a Protein

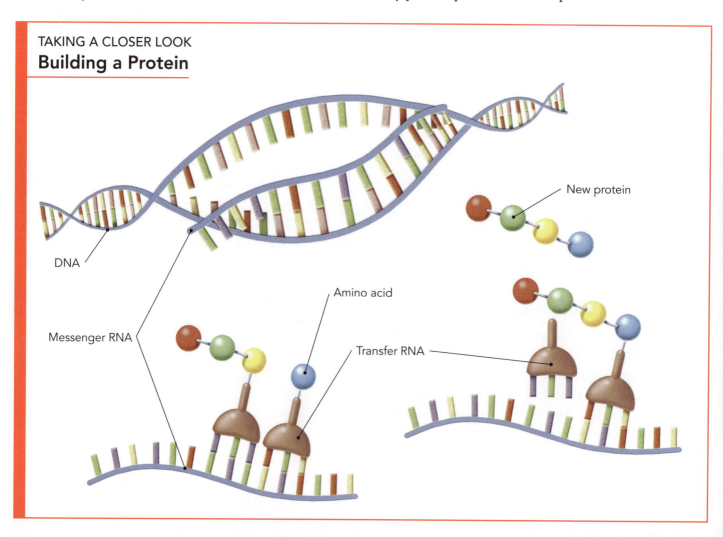

The process of protein making is incredible. DNA uncoils and exposes the gene that contains the necessary instructions. Then the particular segment of the DNA "ladder" that contains the information about the protein splits to expose its nucleotides. (Remember the "words" and "letters"!) Then these nucleotide "words" are read and a special molecule is made. This special molecule is called messenger RNA (mRNA). The mRNA takes the information from the DNA and leaves the nucleus through the pores in the nuclear membrane. Outside the nucleus, the mRNA connects to ribosomes.

Once they're on a ribosome, the mRNA is "read" by another type of RNA, called *transfer RNA* (tRNA). Each kind of tRNA carries the code for a particular amino acid and an attachment for that amino acid. As each segment of the mRNA is read, the tRNA brings the correct amino acid, in the correct order, and the protein is assembled. The ribosome stitches together each protein, folding it carefully so that it will work just right.

## What Is "Junk DNA"?

Only a small portion of our DNA actually contains the information that codes for proteins. So what, then, is the purpose of the rest of our DNA?

Many scientists over the last few decades have felt that if any portion of DNA did not actually code for proteins, it had no purpose. For that reason, many scientists began to refer to this part of our DNA as "junk DNA." They felt that these useless regions of DNA were merely left over from our evolutionary past.

However, in recent years, it has been shown that junk DNA is not junk at all. These regions of our DNA are quite active and serve many functions, such as helping switch genes on and off. Every day, researchers are discovering more about how "junk DNA" actually works!

You see, our Master Designer does not make "junk!"

## Do Humans and Chimps Have Similar DNA?

It is often said that the DNA of humans and chimps are 98 percent alike. This popular notion has been repeated and repeated so often that most people believe it to be true. Many scientists promote this idea to support their mistaken idea that humans and chimps evolved from a common ancestor a few million years ago. This supposed similarity in DNA is used as "proof" of an evolutionary link between humans and chimps.

Actually, when you really examine the data, you find that the similarity between human and chimp DNA is more like 70 percent. It is nowhere near the 98 percent that some people claim.

Even though there is a 70 percent similarity, that 30 percent difference means an awful lot. Between humans and chimps there are millions and millions of sequences in the DNA that are different. That is obvious as humans and chimps are distinctly different creatures.

So how can we explain the 70 percent of our DNA that is similar to the chimp's? This is simple for the Christian. We understand that all living things have a common Designer, not a common ancestor, as evolution would suggest. This amazing Designer would allow for many design similarities in the creatures he created. These similar features would be reflected in similarities in our DNA.

This process occurs thousand of times each second, and countless proteins are made in our cells each day.

If that were all DNA could do, it would be amazing. But there's more....

**DNA Can Make DNA**

Well, DNA is able (with the help of a series of proteins and enzymes) to reproduce itself. By doing this, the information contained in the DNA can be passed on when the cell divides.

It works like this.

> **TAKING A CLOSER LOOK**
> **DNA Replication**
>
> During replication, one parent strand of DNA becomes two daughter strands of DNA. This process requires the presence of several enzymes that perform critical functions such as uncoiling the parent strand or correctly assembling nucleotides to make the daughter strands.

You have seen that DNA looks sort of like a twisted ladder. When it is time for a cell to divide, the membrane around the nucleus temporarily dissolves and the DNA duplicates itself.

First, the DNA in each chromosome uncoils. Then it splits into two strands (almost as if the rungs of the ladder were split in two). With the help of a special set of enzymes, each strand of DNA is copied. When the process is finished, there are two complete sets of chromosomes where there was one set before. Each set of chromosomes is then placed in the newly formed nucleus of a new "daughter cell."

# How Cells Divide

While we are on the subject, let's take a closer look at how cells divide. After all, we continually need more cells as we grow and worn-out cells need to be replaced. How does this happen? Let's explore the cell cycle and see how this works! The *cell cycle* is sort of like the life cycle of a cell. There is a time for a cell to focus on its job, whatever that happens to be. And then for most kinds of cells there is a time for it to copy itself and become two "daughter cells."

The part of the cell cycle when a cell is not actually splitting into two cells is called *interphase*. That's when a cell simply does its job, or jobs. During this time, most of the protein-making activity of the cell occurs. The substances that the cell makes for the body's use are manufactured during interphase. Also, during interphase more organelles are made so that there are enough to supply both daughter cells after division. Near the end of interphase, the cell prepares to divide. The DNA in the nucleus duplicates during this part of interphase. For a short period of time, then, the cell has twice its normal amount of DNA — 46 *pairs* of chromosomes rather than just 46 chromosomes! Because these duplicated chromosomes are stuck together, we often use another name to describe them here — a *chromatid*. A chromosome and its copy, stuck together, is called *a pair of sister chromatids*. Remember, the DNA gets duplicated during interphase so that it is all ready to be split between the two new cells that will be formed during cell division.

The part of the cell cycle that is directly involved with dividing the cell into two daughter cells is called *mitosis*. So the working phase of a cell's cycle is interphase, and the dividing phase of a cell's cycle is mitosis. Mitosis can be broken down into four steps, called phases (wouldn't you just know it . . .). We will examine each in turn.

The first phase of mitosis is called *prophase*. Remember that the DNA gets duplicated before interphase is over. That DNA is a tangled mess like spaghetti, however, and it must be sorted out before the chromosomes can be assigned to each daughter cell. During prophase, the DNA coils and tightens, or *condenses*, so that the chromosomes are dark enough to be visible under a microscope.

Remember the centrioles in the cytoplasm? Well the membrane around the nucleus dissolves, allowing

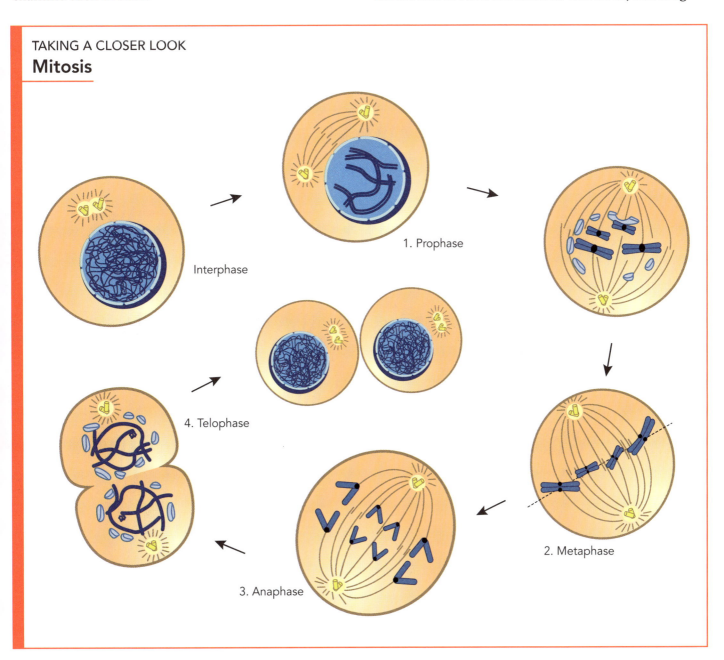

TAKING A CLOSER LOOK
**Mitosis**

Interphase
1. Prophase
2. Metaphase
3. Anaphase
4. Telophase

# 28
## THE MUSCULOSKETETAL SYSTEM

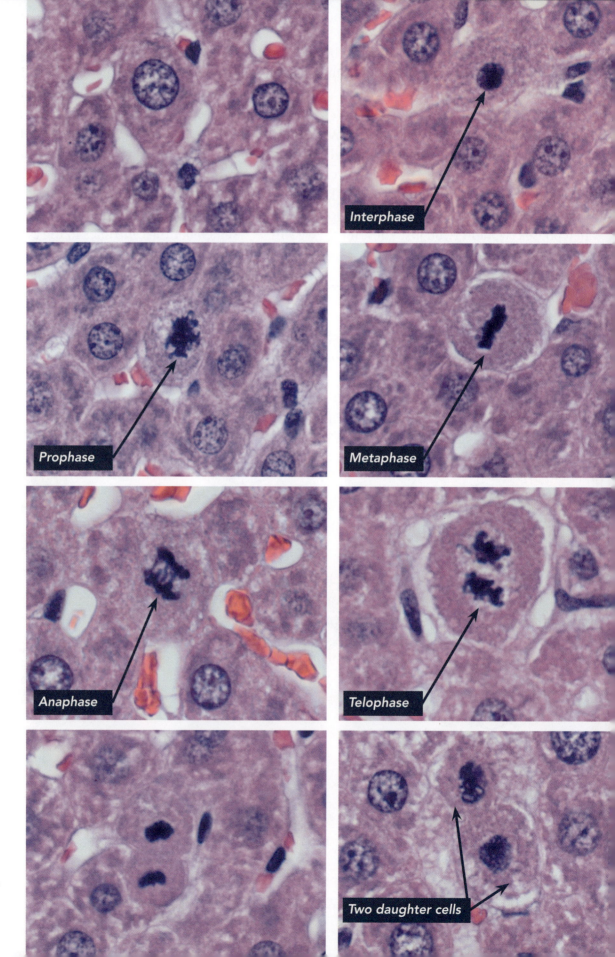

Hepatocytes (liver cells) undergoing mitosis

the centrioles to build a scaffold on which the chromosomes can be organized. The centrioles separate, moving to opposite ends of the cell. A series of microtubules form and anchor to the centrioles. These microtubules attach to the duplicated chromosome pairs — *the sister chromatids* — and begin moving them to the center of the cell.

When all the chromosomes, traveling along the microtubules strung between the centrioles, arrive at the center of the cell, the cell is in *metaphase*. Metaphase is the second phase of mitosis.

The third phase is called *anaphase*. This is the shortest phase of mitosis. At this time, the sister chromatids are pulled apart, each chromatid, or chromosome, moving to opposite ends of the cell. Anaphase ends when the chromosomes reach opposite poles of the cell. Now there is a complete complement of genetic material, enough for one new cell, gathered together. Because the sister chromatids stayed attached to each other until they were all lined up in the middle, and then were pulled apart in opposite directions, each daughter cell should contain identical copies of the genes in the original "parent" cell.

The final phase of mitosis is called *telophase*. During this phase, the chromosomes uncoil and become much less visible. New nuclear membranes form at each end of the cell, encircling the group of chromosomes. Thus, for a brief time the cell has two nuclei (each identical to the original nucleus in the parent cell). Then the cell pinches off in the center, forming two daughter cells.

# Moving On . . .

So we have taken a look at the cell and its parts. It is difficult to imagine how people can call the cell "simple," because it certainly isn't. As we continue our journey exploring the human body, there will be many examples of the complex functions performed by cells.

### Is DNA Just an Accident?

Many people think so. One common evolutionary belief is that millions of years ago, DNA just formed itself from chemicals, building the complex DNA molecule itself as well as the complex coded messages in it.

You see, many people believe that millions of years ago there was no life on earth. They believe earth's oceans were full of chemicals that, all by themselves, formed the nucleotides from which DNA is made. Then they believe DNA assembled itself from the nucleotides. Yep, they believe that strands of DNA, millions of nucleotides long, just came together . . . in exactly the right order . . . by chance.

But even if that could happen — and nothing in science has ever discovered any way that it could — the evolutionary story still wouldn't make sense. After all, DNA is not just a string of chemicals; it is a very complex information system. So even if DNA could have assembled itself, where did the coded language contained in the DNA come from? Without a source of information and a language code to record that information, the nucleotides in DNA really would just be a string of nonsensical chemicals. You see, information does not come from matter. Information only comes from a higher source of information. And who is the highest source of information?

DNA is not the result of random chance processes. It is another testimony to the magnificent Creator God, the source of all information.

# TISSUES

As we have seen, cells are marvelously complex. They can do amazing things. However, one cell cannot do the work of many cells. A group of cells that perform similar or related functions is called a tissue. Tissues have many functions, but there are really only four basic categories of tissues. All tissues in the body belong to one of these four groups. These tissue types are epithelial tissue, muscle tissue, connective tissue, and nervous tissue. In a nutshell, they cover, move, connect, and communicate, but that description barely scratches the surface. Let's explore each type in turn.

Epithelial Tissue

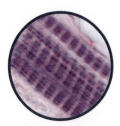
Muscle Tissue

Connective Tissue

Nervous Tissue

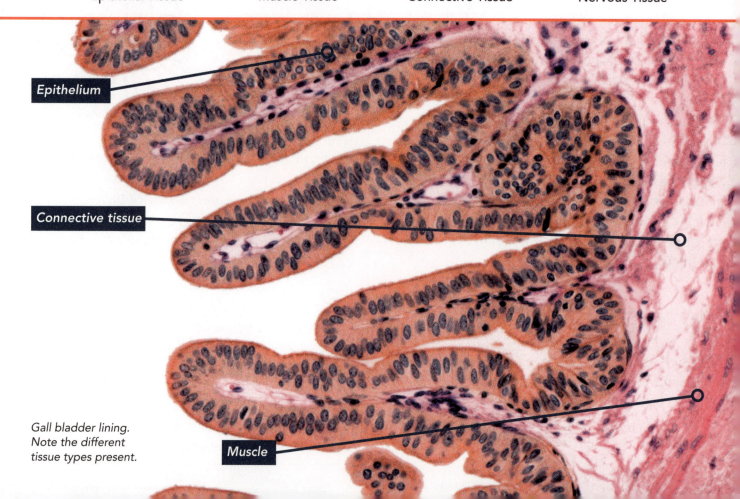

Gall bladder lining. Note the different tissue types present.

# TISSUES

*Epithelial tissue* (or *epithelium*) lines your body cavities or covers surfaces. For example, you may have heard that your skin has layers of cells. Well, the outer layer of skin is an epithelial layer, called *epidermis*. The sheets of cells that line the stomach and intestines, as well as the cells that line the heart, blood vessels, and the lungs are also kinds of epithelial tissue. As you might imagine, the epithelial tissues in each of these locations differ a great deal, but they cover surfaces or line cavities or tubes. They are all in the category called *epithelium*.

Epithelial tissue is specialized to perform many activities. It can act as a barrier for protection. For example, your skin keeps out bacteria, and it is waterproof! The lining of the digestive tract is designed to absorb water and many kinds of molecules, then digestive enzymes break down your food. The epithelium of the kidneys helps rid the body of waste products. Your hair follicles are a specialized kind of epithelium that grows hair!

Another very special type of epithelial tissue is called *glandular epithelium*. This tissue forms the glands of the body. Glands are groups of cells that produce and secrete certain substances. (Some of

## TAKING A CLOSER LOOK
### Types of Epithelial Tissue Cells

| Cells | Location | Function |
|---|---|---|
| *Simple squamous* | Air sacs of lungs and the lining of the heart, blood vessels, and lymphatic vessels. | Allows materials to pass through by diffusion and filtration, and secretes lubricating substance. |
| *Stratified squamous* | Lines the esophagus, mouth, and vagina. | Protects against abrasion. |
| *Simple cuboidal* | In ducts and secretory portions of small glands and in kidney tubules. | Secretes and absorbs. |
| *Transitional* | Lines the bladder, uretha, and the ureters. | Allows the urinary organs to expand and stretch. |
| *Simple columnar* | Ciliated tissues are in bronchi, uterine tubes, and uterus; smooth (nonciliated tissues) are in the digestive tract, bladder. | Absorbs; it also secretes mucous and enzymes. |
| *Pseudostratified columnar* | Ciliated tissue lines the trachea and much of the upper respiratory tract. | Secretes mucus; ciliated tissue moves mucus. |

these are used in the body, and some are discharged outside the body or even into the cavities and tubes that they line.) We will learn much about glands as we study the various body systems.

*Muscle tissue* is responsible for movement. Muscles move large parts of your body, like your legs and your fingers. Muscle tissue also moves your stomach's walls to churn your food, and muscle tissue moves to guide food through your digestive tract. Another kind of muscle tissue moves to keep the heart pumping, and still other muscles work together to help your lungs draw in air!

Even though there are several kinds of muscle tissue, each is made of muscle cells. Muscle cells contain structures called *myofilaments* that allow the cells to contract. A little muscle contraction gives you

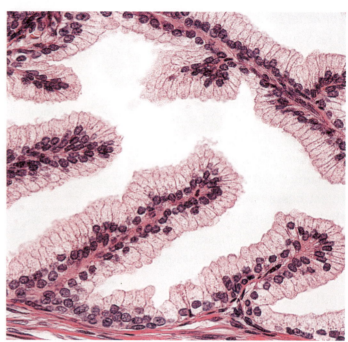
*The rows of tall cells you see in this photomicrograph are epithelial tissue.*

## Tissue Types

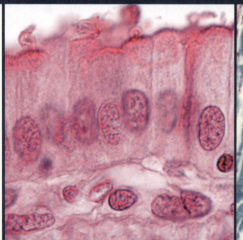

### Epithelial Tissue

Epithelial tissue (or epithelium) lines body cavities or covers surfaces. For example, the outer layer of skin is epithelium. The sheet of cells that line the stomach and intestines, as well as the cells that line the heart, blood vessels, and the lungs, is epithelial tissue.

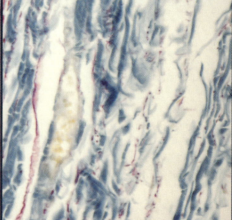

### Connective Tissue

Connective tissue helps provide a framework for the body. It also helps connect and support other organs in the body. Further, it helps insulate the body, and it even helps transport substances throughout the body. This tissue can be hard or soft. Some connective tissue stretches. One type is even fluid. Connective tissue is comprised of three parts: cells, fibers, and ground substance.

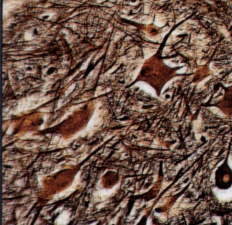

### Nerve Tissue

Nervous tissue is the primary component of the nervous system. The nervous system regulates and controls bodily functions.

Nerve cells are incredible. They are able to receive signals or input from other cells, generate a nerve impulse, and transmit a signal to other nerve cells or organs.

muscle tone so you don't flop like a rag doll. A little more muscle contraction produces movement.

There are three types of muscle tissue.

Skeletal muscle is attached to the bones of the skeleton. When it contracts, it allows us to move our arms and legs, or grasp something with our hands, or smile when we're happy. This type of muscle contracts when we want it to; it is under our conscious control. That is why it is often referred to as voluntary muscle. The distinctively striped cellular structure of skeletal muscle, easily seen under the microscope, differs from that of other types of muscle. The skeletal muscle cells are usually arranged neatly side by side so that they can all pull in the same direction.

### SO SIMPLE YET SO COMPLEX — Designed by the Master

Did you know that you are constantly shedding your skin? Your skin contains many layers of cells. The bottom layer of cells multiplies rapidly. Crowded out by newer cells, older cell layers are pushed upward. Meanwhile, many of these cells produce a protein called keratin. Your hair and nails are also made of keratin produced by the cells in your hair follicles and nail beds, so you can imagine that keratin molecules are tough. Keratin-producing cells in your skin gradually fill with keratin as they age and are pushed closer and closer to the surface of your skin. By the time they near the surface, these skin cells die and leave behind the keratin they contained as part of a waterproof protective layer on the surface of your skin. This layer is constantly rubbed away by your day-to-day activities but is quickly replaced.

## Muscle Tissue Types

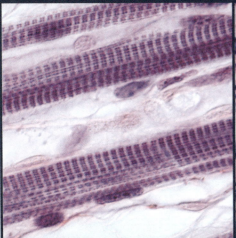

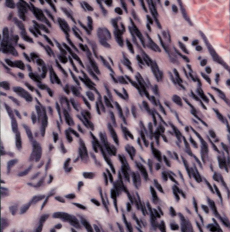

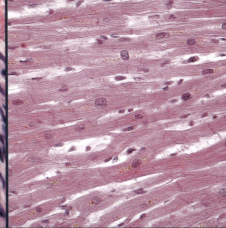

### Skeletal Muscle Tissue
Skeletal muscle is attached to the bones of the skeleton. When it contracts, it allows us to move our arms and legs, or grasp something with our hands, or smile when we're happy. It has a structure that is distinct from other types of muscle as we will see.

### Smooth Muscle Tissue
Smooth muscle is found in the walls of most of the hollow organs of the body. For example, it is found in the walls of our digestive tract where it helps push our food as it is digested. Smooth muscle is found in blood vessels, the urinary tract, the respiratory tract, the prostate, among other places. Smooth muscle is not under our direct control, and is sometimes referred to as involuntary muscle.

### Cardiac Muscle
The third type of smooth muscle is cardiac muscle. It is found only in the walls of the heart. This type of muscle is also an involuntary muscle.

Smooth muscle is found in the walls of most of the hollow organs of the body. In the walls of our digestive tract, smooth muscle squeezes the food in our stomach and then pushes it through our intestines as it is digested. Smooth muscle is found in blood vessels, in the urinary tract, and in the air passages in the lungs. Smooth muscle is not under our direct control, and is sometimes referred to as *involuntary muscle*. Under the microscope, smooth muscle is not striped, and the muscle cells are woven and crisscrossed, not lined up in neat rows.

The third type of smooth muscle is *cardiac muscle*. It is found only in the walls of the heart. This type of muscle is also an involuntary muscle. That is good, for if we had to remember to tell our heart to beat, we wouldn't have time to think about anything else, much less sleep! Under the microscope, cardiac muscle cells look striped, like skeletal muscles, but they are arranged differently.

*Connective tissue* helps provide a framework for the body. It also helps connect and support other organs in the body. Connective tissue helps insulate the body, and it even helps transport substances throughout the body.

This tissue can be hard or soft. Some connective tissue stretches. One type is even fluid. If this seems like a riddle — after all, what kind of tissue could do and be *all* these things? — remember that "connective tissue" is a category that includes a lot of different tissues.

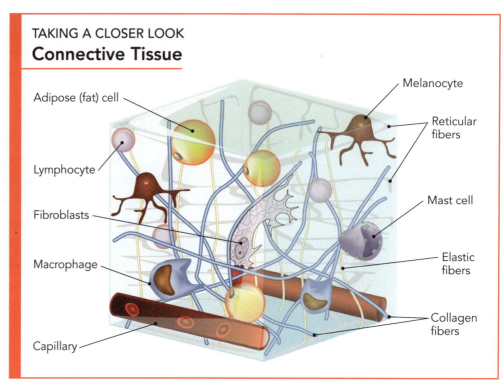

TAKING A CLOSER LOOK
**Connective Tissue**

- Adipose (fat) cell
- Lymphocyte
- Fibroblasts
- Macrophage
- Capillary
- Melanocyte
- Reticular fibers
- Mast cell
- Elastic fibers
- Collagen fibers

### The Fibroblast

Connective tissue provides support and helps hold things together. These important functions would not be possible without a cell known as a *fibroblast*. The fibroblast is one of the most important cells found in connective tissue. The fibroblast is the cell most responsible for the production of *collagen fibers*. Collagen fibers help give connective tissue its strength. Fibroblasts also secrete the *ground substance* (or *matrix*) that fills in the space surrounding the fibers and cells of connective tissue. Fibroblasts also play a very important role in wound healing. When tissue is injured, fibroblasts migrate to the damaged area and help begin the healing process by making new collagen.

No wonder that fibroblasts are the one of the most common cells found in connective tissue. Without the fibroblast, we would fall apart!

There are many varieties of connective tissue. The fibrous capsules that cover many organs are connective tissue. The elastic fibers in the walls of arteries are connective tissue. The cartilage found throughout the body is connective tissue. Ligaments that bind bone to bone are connective tissue. Tendons that hook muscles to bones are connective tissue. Hard bone and soft *adipose* (fat) tissue are examples of connective tissue. Amazingly, even blood is considered connective tissue.

What do they all have in common? Well, for starters, connective tissues are each composed of three parts: cells, fibers, and ground substance.

*Ground substance* (also known as *matrix*) is the material that fills the space between the cells. It is composed mainly of fluids and protein.

The fibers of connective tissue are located in the ground substance. The most common type of fiber is made of long, stringy protein molecules called *collagen*. Collagen fibers are very tough and durable. They are coiled and intertwined and cross-linked to each other. In fact, ounce for ounce, a common type of collagen fiber is stronger than steel!

The cells must manufacture the fibers and the ground substance that make up connective tissue.

Without connective tissue we would fall apart.

*Nervous tissue* is the primary component of the nervous system. The nervous system regulates and controls bodily functions.

Nerve cells are called *neurons*. Neurons are able to receive signals from other cells, generate a signal, or *nerve impulse*, and transmit a signal to other nerve cells or organs.

Nervous tissue is found in the central nervous system and the peripheral nervous system. The central nervous system has only two parts: the brain and the spinal cord. The peripheral nervous system consists of all the other nerves outside your body; in other words, all the nervous tissue outside the brain and spinal cord.

### TAKING A CLOSER LOOK
## Types of Neurons

**Multipolar neuron**
- Motor neuron
- Pyramidal neuron
- Purkinje cell

**Bipolar neuron**
- Retinal neuron
- Olfactory neuron

**Unipolar neuron**
(touch and pain sensory neuron)

**Anaxonic neuron**
(Amacrine cell)

# ORGANS & ORGAN SYSTEMS

An organ is a collection of various types of tissues that work together to perform a function. The heart is an organ. The kidney is an organ. A lung is an organ. Each of these organs is designed to do certain things. The heart pumps blood, the kidney rids the body of waste products and extra water, and the lungs get oxygen into the blood and carbon dioxide out. Other organs are the brain, liver, stomach, gall bladder, small intestine, large intestine, pancreas, spleen . . . you get the idea.

Heart

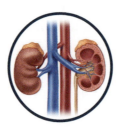
Kidney

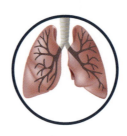
Lung

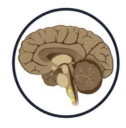
Brain

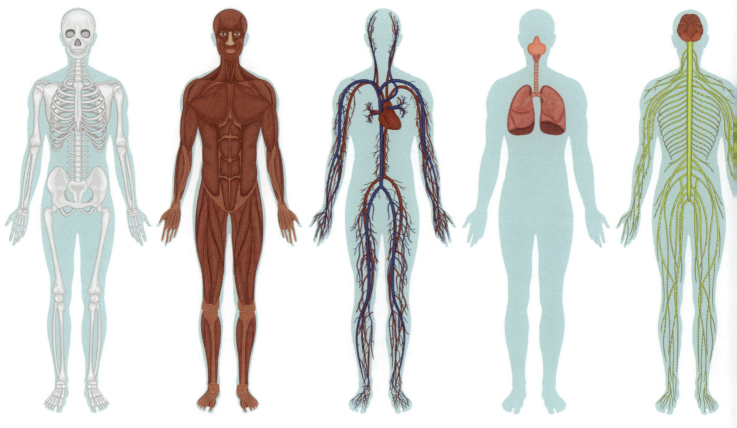

Skeletal System　　Muscular System　　Cardiovascular System　　Respiratory System　　Nervous System

However, even as important and complex as each organ is, none can do their job alone. They require other organs and structures to assist them. The heart cannot function without the veins and arteries. The kidney cannot function without the ureters and the bladder. The lungs would be useless without the nose, the trachea, and the bronchial tubes to bring air to them and the diaphragm to help the chest cavity to expand and draw air in. No part of the digestive tract would do you any good if the other parts weren't there for food to travel through as it gets digested.

These collections of organs and structures are called organ systems. The heart, veins, and arteries, for instance, are parts of the circulatory system. The kidneys, ureters (tubes that drain the kidneys), and the bladder are the urinary (or excretory) system. The lungs are part of the respiratory system. It is logical

### SO SIMPLE YET SO COMPLEX — Designed by the Master

## Programmed Cell Death

Did you know that some cells are designed to self-destruct? This process is called *programmed cell death*. Our Creator designed this process to eliminate worn out cells, but that's not all! Programmed cell death also makes it possible to shape a developing baby's delicate body parts. For instance, long before birth, a baby's fingers are webbed. Once the fingers reach a certain stage of growth, the cells in the webs die away and leave the individual fingers. In an adult, programmed cell death serves to keep the right number of healthy cells in many tissues and organs so that they maintain the correct shape and don't grow too large. When a cell receives a self-destruct signal, enzymes that chop its largest molecules into pieces are activated. The cell shrinks, becomes a misshapen blob, and disintegrates.

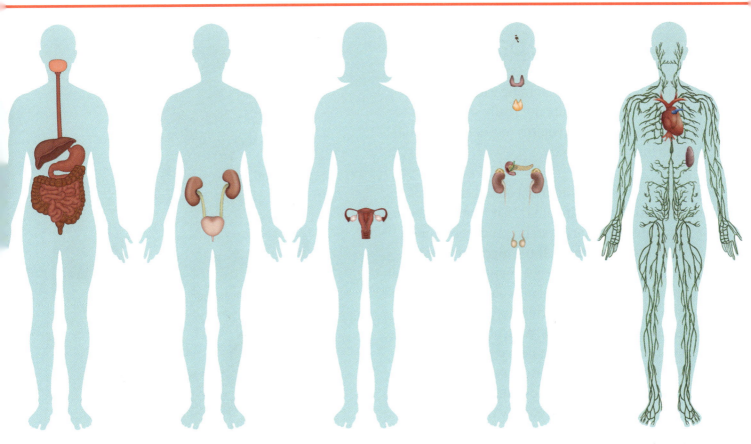

Digestive System     Urinary System     Reproductive (female) System     Endocrine (male) System     Lymphatic System

# THE MUSCULOSKETETAL SYSTEM

to explore the enormous complexity of the human body by breaking it down into the various organ systems.

| Body System | Organs Included |
|---|---|
| Skeletal System | Bones and joints |
| Muscular System | Muscles |
| Cardiovascular System | Heart and blood vessels |
| Respiratory System | Upper airway (nose, pharynx, larynx), trachea, and lungs |
| Nervous System | Brain, spinal cord, and nerves |
| Digestive System | Mouth, esophagus, stomach, intestines, liver, gall bladder, and pancreas |
| Urinary System | Kidneys, ureters, and bladder |
| Reproductive System | (Male) Testes, genital ducts, and prostate<br>(Female) Ovaries, uterus, fallopian tubes, and breasts |
| Integumentary System | Skin, nails, and hair |
| Endocrine System | Pituitary gland, hypothalamus, thyroid gland, parathyroid glands, pancreas, adrenal glands, testes (male), and ovaries (female) |
| Lymphatic System | Lymph nodes, lymph vessels, thymus, tonsils, and spleen |

## But They Are Not Really Separate

Even though we will be examining each organ system separately, they are not really separate or independent. Each one requires one or more of the others to function correctly.

For example, the bones in the skeleton cannot function without the vitamin D that is provided when the skin (integumentary system) produces vitamin D in response to sunshine. If the digestive system did not break down food to get energy, then no other system could operate. The same could be said of the circulatory system that provides oxygen to tissues and removes carbon dioxide. And, of course, the circulatory system could not deliver oxygen to the rest of the body if the respiratory system didn't bring oxygen into contact with blood pumped to the lungs by the heart. *Hormones* secreted by the endocrine system help regulate the action of the kidneys. The muscles in the legs (muscular system) compress

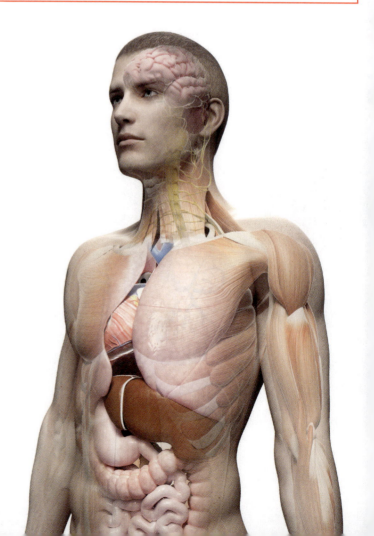

veins and help return blood to the heart (circulatory system). The list could go on and on.

# Directions on the Map of Your Body

A very important aspect of the study of anatomy is knowing where certain things are in relation to other things. That being the case, there are special terms that anatomists use to help navigate around the body. This all begins with a reference point known as the *anatomical position*.

The anatomical position is defined as the body in an upright posture with the feet spread slightly apart. The arms are down to each side with the palms of the hands turned forward with the thumbs pointing away from the body. This is the starting point from which we describe where one part is in relation to another.

The most common terms used to describe the location or position of body parts are as follows:

*Anterior* and *posterior* — These describe structures at the front (anterior) or the back (posterior) of the body. If an organ is closer to the front of the body than another, it is said to be anterior to the other organ. Your belly button is anterior to your backbone.

*Proximal* and *distal* — These describe whether something is closer (proximal) or farther away (distal) from the middle of the body. For example, the knee is proximal to the foot, and the hand is distal to the elbow.

*Medial* and *lateral* — These describe whether something is closer (medial) or farther away (lateral) from the midline, or center line, of the body. For example, the ears are lateral to the eyes.

*Superior* and *inferior* — These describe whether something is above (superior) or below (inferior) something else. For example, the knee is inferior to the hip.

There are, of course, many other anatomic terms. We will occasionally be introducing these through this series.

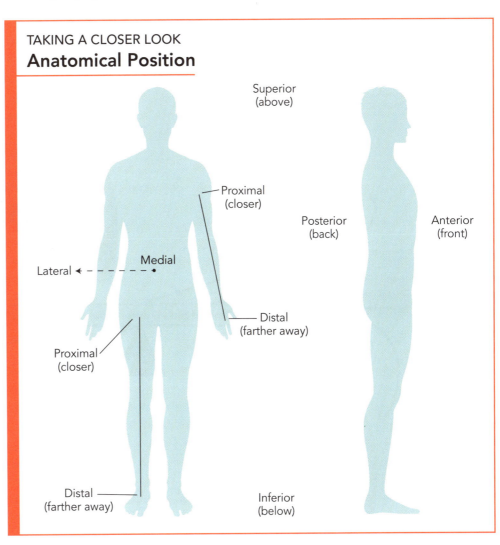

TAKING A CLOSER LOOK
**Anatomical Position**

# HOMEOSTASIS

Before we proceed further there is a basic concept in physiology that you need to understand. That concept is called homeostasis. This means the body has many mechanisms to help maintain a balance or "equilibrium" among its many systems. The body functions best within certain limits. You need to have plenty of water in your body, but not too much. Body temperature can vary a little, but it normally stays not too high and not too low. The minerals in your body are very important, but it is also important that you have neither too much nor too little of them circulating in your bloodstream.

Body Temperature    Minerals    Water    Sugar (glucose)

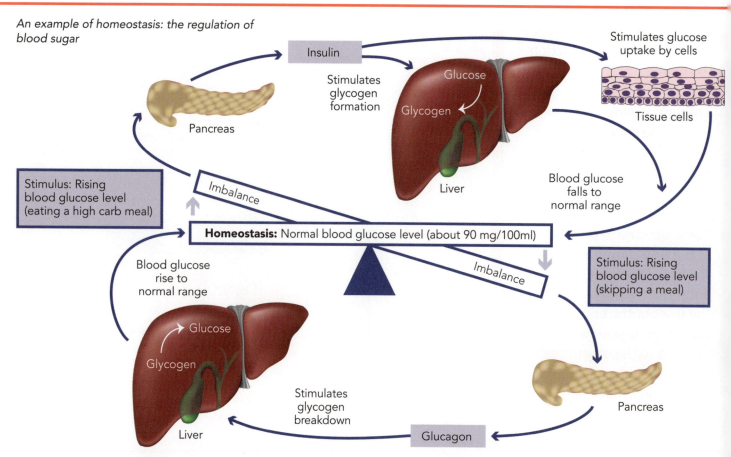

An example of homeostasis: the regulation of blood sugar

There are trillions of cells, multiple types of tissues, and many organs and organ systems in our bodies. All try to continue functioning as they were designed. All the cells, tissues, and organs must work together to achieve this goal.

However, the internal environment of the body must be kept within strict ranges in order for these systems to operate correctly. There are many, many control systems in our bodies designed to help maintain the necessary balance.

For example, our body temperature must be kept within a very narrow range. If our body temperature gets too high or too low, cells cannot work properly. You will see that there are control systems to monitor and help maintain correct body temperature.

The amount of sugar in our blood must be kept within certain limits. You will see that multiple systems play a role in controlling our blood sugar. The same could be said about the amount of calcium or potassium in our blood, the volume of fluid in our blood vessels, or the level of acids in our body.

Even though the human body contains countless little factories in about a dozen organ systems working simultaneously day and night, drawing on resources and producing waste products like acids, carbon dioxide, and excess heat, the human body is designed to maintain homeostasis, a constant set of conditions under which all the tissues work best. We will examine many of the feedback systems that help achieve this balance, and we will also see instances of what happens when things go wrong.

## Is It All an Accident?

The number of intricate relationships that exist between organ systems is truly mind-boggling. Amazingly, many people think that all these organs and systems and the manner in which they interact is merely a cosmic accident, a product of time and chance. They attribute all this to chemicals banging together over millions of years. Their view is that given enough time, anything can happen.

As we proceed, it will be apparent that the human body is not just an accident. Nothing this complex can happen by chance. Many of the body's systems cannot work unless others are already in place and working properly. This "irreducible complexity" leaves no wiggle room for the random processes of evolution. There is a design and a purpose to how the body works. We are truly "fearfully and wonderfully made." The human body is a testament to the power and majesty of our Creator God.

# SECTION 2

# THE SKELETAL SYSTEM

Just as a house needs a solid foundation, so it must have a strong framework to support it. Wooden or metal beams are connected by structural steel to form the skeletal structure of the home. All this is done before the brick or other siding covers the outside, and wood, plaster, or other materials cover the inside walls. God created us with a brilliantly designed skeleton that can move with us, support us, and help get us wherever we need to go!

Thousands of years ago, there was much misunderstanding about the human body and how everything worked together so intricately. However, God has always known, and those who followed Him gained insights into the mastery of the Creator. It was God who called forth light from the vast darkness. It was also God who intricately formed Adam from the dust. The brilliant arrangement of the human body was not lost on the prophet Ezekiel when he had a vision of dry bones in a valley:

*The hand of the Lord came upon me and brought me out in the Spirit of the Lord, and set me down in the midst of the valley; and it was full of bones… And He said to me, "Son of man, can these bones live?" So I answered, "O Lord God, You know." Again He said to me, "Prophesy to these bones…" So I prophesied as I was commanded; and as I prophesied, there was a noise, and suddenly a rattling; and the bones came together, bone to bone. Indeed, as I looked, the sinews and the flesh came upon them, and the skin covered them over; but there was no breath in them (Ezekiel 37:1-8).*

Bone marrow helps to create red and white blood cells, and these help us in many ways, including fighting bacterial infections.

Both giraffes and humans have seven bones in their necks.

The vertebral column is made up of 26 bones.

Your bones are about six times stronger than steel, if measured by weight.

# FUNCTIONS OF THE SKELETAL SYSTEM

The most recognizable organ system in the human body is probably the skeletal system. Skeletons are popular props in movies, and most people have seen a picture of one or even a plastic or cardboard model of a human skeleton somewhere. But as important as all those big bones are, the skeletal system contains many small, less familiar bones as well as the ligaments that hold the bones together, the cartilage that cushions many of their ends, and the joints that allow them to move with stability and purpose.

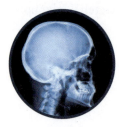
Skull

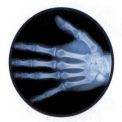
Hand

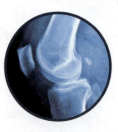
Knee Joint

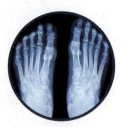
Feet

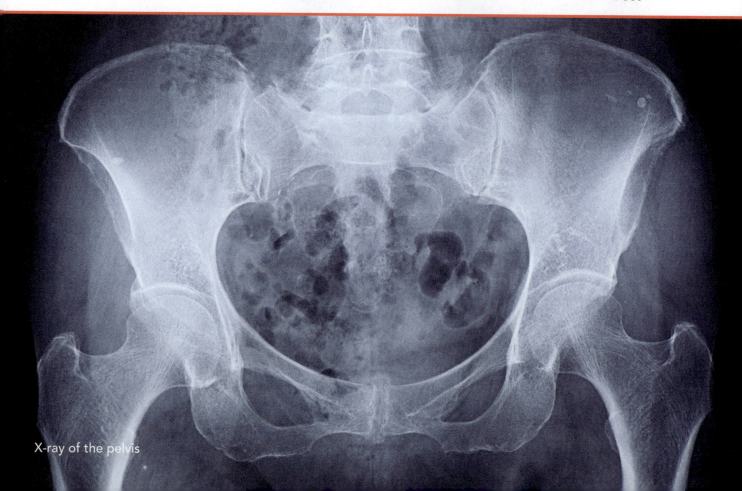
X-ray of the pelvis

# FUNCTIONS OF THE SKELETAL SYSTEM

Bones make up about 20 percent of the mass of the body (say about 30 pounds in a 150-pound person), and they consist mostly of connective tissues. And while bones obviously hold you up, they actually do a lot of other less dramatic but equally important things.

The first, and most obvious, function of the skeletal system is indeed to provide *support* for the body — to hold you up. We could not stand up, move around, pick things up, or even breathe properly if not for the framework of the skeleton. The bones provide a structure to bear weight so we can walk. They provide attachments for the muscles to help keep us upright. Without the bones inside us, we would all be just big lumps of tissue lying on the ground!

The second function of the skeletal system is to provide *protection* for our other organ systems. The skull protects the brain. Without the skull's cranial cavity to cradle your brain, it could be damaged by simply combing your hair or laying your head on your pillow, and bumping your head would be disastrous.

The rib cage is a protective shield for the heart and lungs and provides a sturdy space into which your lungs can expand. God designed your bones to protect these three very important organs as well as many other less obvious ones. The strength of the bones themselves help keep muscles, blood vessels, and several abdominal organs from being crushed when we fall or are struck by some object.

The bones of the skeleton do not move themselves, but without them we could not move. Why? Because the bones provide places where muscles can attach. The muscles then contract, and bones to which they are attached move. Bend your knee. A muscle behind your thigh pulled on a bone below your knee. Make a fist. Several muscles in your forearm pulled on the bones in your fingers. You get the idea!

The skeletal system is also a great warehouse for storing many things the body uses regularly. Bones can store certain types of fats that the body needs to hang onto. Bones also store certain minerals, such as calcium, magnesium, and phosphate. These minerals are constantly needed by the body, and it is important that the body have a reservoir of them to draw on so that the amounts of these minerals in the blood can be kept at a safe and fairly constant level. As the body uses these minerals, their concentration in the blood decreases. When this decrease is detected, the minerals stored in the bones are released to return the mineral levels in the blood to normal. At the same time, the mineral stores in the bones are resupplied from the minerals we take in from our diet.

Finally, your bones are where most of your *blood cells* are made! Most of our blood cells are formed (by a process called *hematopoiesis*) in the marrow cavities of the many bones of the body. We would be in quite a fix without a constant resupply of blood.

Blood cells die rapidly. Some blood cells live for only a matter of days. Some can live a few months. When they die, they need to be replaced. If we could not constantly make new blood cells, we could not survive.

So you can see our bones do more than just hold us up. They do much more!

### TAKING A CLOSER LOOK
### Blood Cell Production

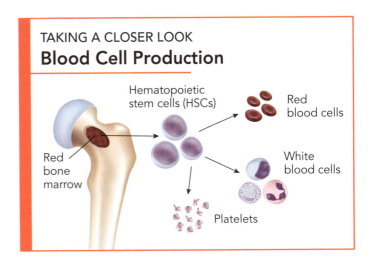

# 2

# BONES

If you have ever spent time looking at a skeleton, you will see that bones come in many shapes and sizes. Anatomists have sorted bones into certain classes based on their shape.

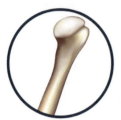
Long Bones

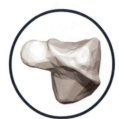
Short Bones

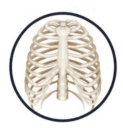
Flat Bones

Irregular Bones

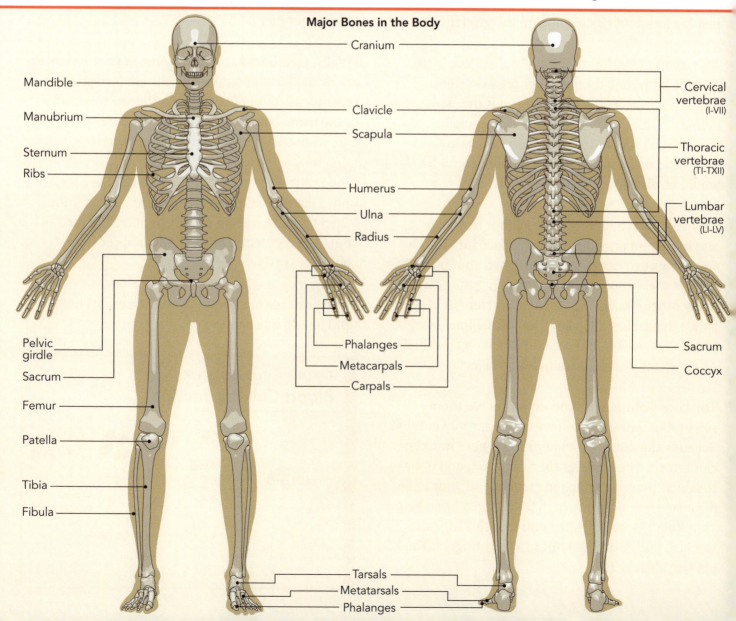

**Major Bones in the Body**

Front labels: Cranium, Mandible, Manubrium, Sternum, Ribs, Clavicle, Scapula, Humerus, Ulna, Radius, Pelvic girdle, Sacrum, Phalanges, Metacarpals, Carpals, Femur, Patella, Tibia, Fibula, Tarsals, Metatarsals, Phalanges

Back labels: Cervical vertebrae (I-VII), Thoracic vertebrae (TI-TXII), Lumbar vertebrae (LI-LV), Sacrum, Coccyx

First, there are the *long bones*. These bones are longer than they are wide. Examples of long bones are the humerus (the upper arm), the femur (thigh bone), and the tibia (the shin bone). Look at the picture of the skeleton. Find these bones and see if you can pick out some other long bones.

When you looked at the skeleton to find the long bones, you probably saw the other bone in the lower leg (the fibula) and the two bones in the forearm (the radius and the ulna) right away. (By the way, the end of the ulna is what you hit when you hit your "funny bone." The radius is the forearm bone nearest your thumb.) Did you find some smaller long bones? Even though they are not that long, the bones in the fingers are considered long bones! They are longer than they are wide.

Short bones are, incredibly enough, called *short bones*. These are bones that are about as wide as they are long. The small bones in the wrist, called carpal bones, are short bones.

Bones that are thin are called *flat bones*. These bones can be either long or short, and they are usually curved. The bones in the skull are flat bones, as are the ribs. The skull's flat bones are fused together.

Bones that are small and round are called *sesamoid bones*. These bones are embedded inside tendons, and are found where tendons pass over joints. The most prominent sesamoid bone is your kneecap. Smaller sesamoid bones are found in the hands and feet.

Bones that cannot be easily placed in any of the categories are called *irregular bones*. The bones in the vertebral column (backbone) are irregular bones.

# Gross Anatomy of Bone

Gross anatomy examines the body at a macroscopic level. Let's examine the structure of a typical long bone from the outside in.

Remember, long bones are longer than they are wide. They are comprised of the diaphysis, the epiphyses, and the periosteum.

The main shaft or midsection of the bone is known as the *diaphysis* (don't let these terms spook you, you'll have them down in no time). It is made up primarily of thick *compact bone* and has a central cavity, called the *medullary cavity*. ("Medullary" means "in the middle.") Blood cell-making bone marrow or fat is stored in the medullary cavity of long bones.

### TAKING A CLOSER LOOK
### Long Bone Anatomy

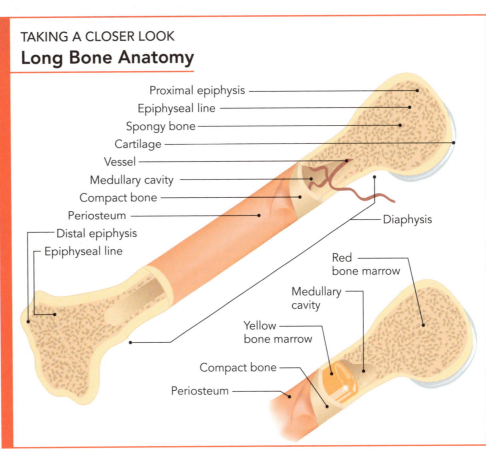

The rounded, broader end of a long bone (the joint end) is called the *epiphysis*. Joint surfaces are covered with a layer of cartilage, which makes the joints move smoothly. The joint cartilage also absorbs shock and cushions the joint during movement.

Now let's look more closely....

The outermost layer of bone is a thin, fibrous membrane called the *periosteum*. It is a tough, durable structure that covers most of the outer surface (all except the joint surfaces) of all bones. The periosteum contains special nerve endings that sense pain, which is no surprise at all to anyone who has had a broken bone! Man, does THAT hurt! The periosteum also has blood vessels that help provide nutrients to the bone.

That seems quite a lot for a little membrane, but there's more. There are special cells in the periosteum that build new bone. These cells are very important as bone grows or when a bone is broken. We will see how later.

Finally, the periosteum helps provide a place for the tendons of muscles to attach to bones. Without those attachments, we would not be able to move!

Now let's look at the inner structure of a long bone.

If a bone is cut in cross section (see below), you can see that beneath the periosteum is a dense, thick layer that is called *compact bone*. As you might expect, compact bone is very durable and strong. It is primarily responsible for the support of the body. Because it is so dense, between 75 and 80 percent of the weight of the skeleton is due to compact bone.

The innermost part of the bone is not as dense. It is almost like a sponge or a honeycomb in its

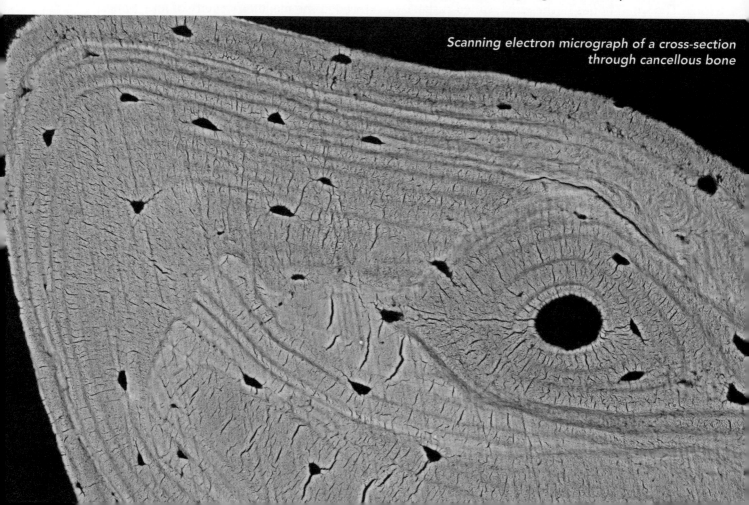

*Scanning electron micrograph of a cross-section through cancellous bone*

appearance. In fact, it is often referred to as *spongy bone*. (It is also called *cancellous bone*.)

In a typical long bone, the spongy bone is found primarily near the ends of the bones, close to the joints. In the center of the diaphysis (remember, the mid shaft of a long bone), is found the medullary cavity. This cavity contains either red bone marrow, where blood cells are made, or yellow bone marrow, where fats are stored.

## What Goes On Inside

The inside of a bone is actually a very busy place. However, looks can be deceiving. Bones are one of the most active organs in the body.

They are constantly building and rebuilding, not only to grow or repair injuries, but even to remodel themselves in response to the stresses our activities put on them. And for the most part, that stress is actually *good* for our bones. Why? Because bones respond to this by becoming even stronger.

Bones also respond to signals from other parts of the body. For instance, the bones are the warehouses from which important minerals like calcium or phosphorous can be quickly accessed and released. And as key blood-making centers, bones rapidly respond to infection or blood loss by producing more of the right kinds of blood cells.

In order to truly understand how amazing bones are, we need to take a much closer look. We will start with three tiny cells . . . three kinds of bone cells.

## Bone Cells

Bone activity is controlled primarily by the actions of three different cell types: the osteoblast, the osteoclast, and the osteocyte. These all sound very much

### Tips for Healthy Bone

There are steps you can take to keep your bones healthy!

Your bones must get the calcium and phosphorus they need from your diet. Phosphorus is plentiful in most foods, so it is unusual to run low on this mineral. However, it is important to make sure you eat and drink plenty of calcium. Good sources of calcium are milk, dairy products, and leafy green vegetables. The more, the better!

Also, getting enough sunlight (but don't get sunburned!) is important. You see, when our skin is exposed to sunlight, vitamin D is produced. Vitamin D helps your intestines take in calcium from the food you eat.

Another thing that keeps bones healthy is exercise. The stresses that are placed on bones as you run or play sports, lift weights, walk, or run help keep them strong. Because your bones are designed to detect how strong they need to be, weight-bearing exercise (like running, but not swimming) actually cause the osteoblasts to make your bones stronger.

alike, and there is a reason for this. All three names begin with the prefix "osteo," which is the Greek word for "bone." (Clever, huh?)

The term "cyte" means cell. "Blast" means "immature or precursor." "Clast" means "something that breaks." You can now guess what each cell does.

An *osteocyte* is a mature bone cell. An *osteoblast* is an immature cell or a bone-building cell. *Osteoclasts* break down bone. Thanks to the combined actions of these cells, bone is like a construction zone in which the project is never finished.

## Building Bone — Osteoblast

The function of the osteoblast is simple: it builds new bone. Whether in a bone that is getting bigger as a person gets older, a bone that is healing after a fracture, or simply a bone that is being remodeled, the osteoblast has a vital role.

Remember that bone is a connective tissue. One characteristic of connective tissue is that it has ground substance (also called *matrix*). The osteoblast makes the ground substance in bone. The bony matrix consists of collagen protein fibers made strong by minerals that are added to them.

The osteoblasts make collagen, a strong fibrous protein. The long collagen fibers are laid down outside the osteoblast cells. Collagen gives bone its *tensile strength*. (That just means "strength under tension.") It keeps bone from pulling apart and makes it somewhat flexible.

After making the collagen fibers, the osteoblasts deposit mineral salts on them. Those mineral salts contain calcium and phosphorus. Mineralization makes the bones strong and hard.

## Resorbing Bone — Osteoclast

The osteoclast is a cell that resorbs or breaks down bone. This might seem odd that there is a cell that breaks down bone. After all, aren't bones better if they are stronger? Why would a cell take away bone? Well, there is a good reason for this.

As has been mentioned, bone is a very dynamic tissue. You will see this in how bone constantly remodels itself. When bone wears out, the osteoclasts help remove it so it can be rebuilt.

Osteoclasts are large cells with multiple nuclei. When seen under the microscope, the osteoclast is often said to look "foamy." This is because the

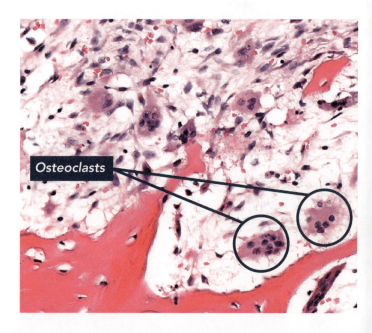

cytoplasm of the osteoclast contains lots of vesicles containing the substances needed to help break down bone tissue.

## Mature Bone — Osteocyte

When osteoblasts become trapped in the ground substance, they become osteocytes. Osteocytes are mature bone cells that have become walled off within the bone they have made.

In compact bone, osteocytes are located in spaces called lacunae. These cells have long extensions, called "processes" (sometimes called "legs"), that reach through small channels in the bone and come in contact with processes from other osteocytes. In this way, even though they are walled off from other cells, the osteocytes can communicate with one another.

There was a time when osteocytes were thought to have little function, but it is now known that these cells are more than just inactive osteoblasts. Osteocytes can act as sensory cells that help control the activity of osteoblasts and osteocytes. They can help detect how strong your bones need to be and let the other cells know. Through contact with other osteocytes via these "processes" they can transmit signals to other cells.

*Osteocyte*

## Bone Structure

Let's take a look at how bones work. We will start with compact bone.

The basic unit of compact bone is called an *osteon*. An osteon looks like a long tube or cylinder made up of layers. These layers are called *lamellae*. In the center is a canal that contains blood vessels and nerve fibers. (You might not think so by its external appearance, but bone has a very rich blood supply.) Surrounding this central canal are layers of ground substance. Between the lamellae are found the small chambers, or *lacunae*, where the osteocytes are found. The processes of the osteocytes extend through the lamellae to communicate with other bone cells.

One way to visualize the osteon is to think of the rings in a tree trunk. The rings get bigger as you go out from the center. So it is with the layers of the osteon: they get larger as you move from the center.

Multiple osteons are packed together with extra ground substance to hold them together. This extra material can be thought of as glue that holds the osteons in place.

The outer layer of compact bone is made up of very large rings of ground substance and osteocytes. These rings are called *circumferential lamellae*, because they encase all the osteons in the circumference of the bone. The fibrous outer covering of the bone, the periosteum, covers the circumferential lamellae.

Spongy bone is somewhat different from compact bone in that it is not packed as tightly in appearance. It looks much like a sponge, and at first glance it seems quite random in its layout. However, our Master Designer is not that haphazard in His design. As it turns out, the struts of spongy bone are laid out precisely to manage the weight and stress placed on our bones! It's not random at all.

## TAKING A CLOSER LOOK
### Osteon

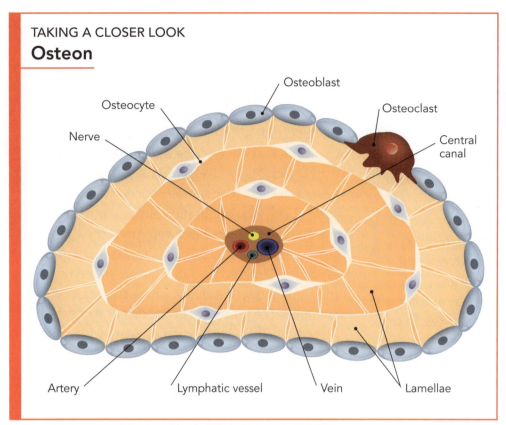

When you examine spongy bones you will see the struts, called *trabeculae*, most prominently. Under a microscope, you will not see osteons in spongy bone like in compact bone. Rather, you will note that the trabeculae are made up of irregularly shaped layers of bone. These are most commonly only a few layers thick. There are osteocytes in these layers.

In spongy bone, blood vessels are not found in the layers. Rather, the blood supply is located in the spaces between the trabeculae. (These spaces look like the air spaces in a sponge.)

# Bone Growth

As we grow, our bones must grow also. This seems obvious, right? After all, we have very small twig-like leg bones as a newborn baby, but we have large strong leg bones as a teenager or adult. Our skeleton must grow along with the rest of our body.

Let's take a look at how this happens.

Bones do not grow by merely adding new bone tissue at the end. As it happens, bones could not grow like that. You see, most bones have joint surfaces at each end. A joint surface is covered by a layer of cartilage, which is a special type of connective tissue that helps protect the joint from stress and allows the joint to move smoothly. So obviously, bones cannot grow by putting new bone at the ends.

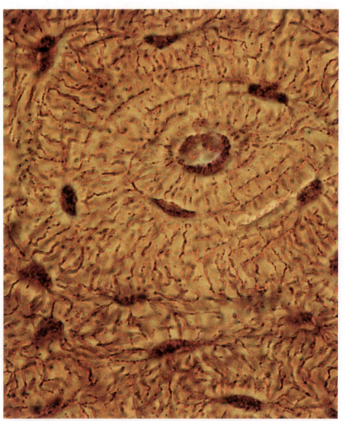

*This is an osteon as seen through a light microscope. Compare it to the diagram above. Can you find the central canal surrounded by lamellae? How many osteocytes do you see?*

How bones grow is another testament to our wonderful Designer who created a special process by which bones can grow. Bones grow at an area called the *epiphyseal plate* (sometimes known as the growth plate).

Recall that a long bone has a main shaft (diaphysis) and rounded ends (each called an epiphysis). At the junction between the main shaft and a rounded end is a thin *epiphyseal plate* made of cartilage. On an X-ray, this plate looks like a gap in the bone. Growth of a long bone happens at the epiphyseal plate.

This is how a long bone grows longer. Cartilage is made by cells called *chondrocytes*, so there are chondrocytes in the thin cartilage of the epiphyseal plate. Along the side of the epiphyseal plate that faces the epiphysis, chondrocytes make new cartilage. However, the cartilaginous plate doesn't get thicker. Instead of the plate getting bigger, something interesting happens.

### TAKING A CLOSER LOOK
## Bone Growth

**Growth**
Bone grows in length because:
- Cartilage grows here
- Cartilage replaced by bone here
- Cartilage grows here
- Cartilage replaced by bone here

**Remodeling**
Growing shaft is remodeled by:
- Bone resorbed here
- Bone added by appositional growth here
- Bone resorbed here

## Growth Hormone

Human growth hormone (hGH) helps regulate bone growth. If a child's pituitary gland secretes too much hGH, *gigantism* can result, leading to a height of seven to nine feet! If the pituitary secretes too little hGH, the child is shorter than normal.

After the epiphyseal plates calcify with adulthood, bones cannot grow longer. Therefore excessive hGH in an adult causes bones in the hands, feet, and jaw to enlarge. This is called *acromegaly*.

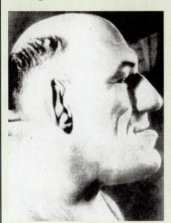

Along the other side of the epiphyseal plate — the side next to the diaphysis — the chondrocytes die. The cartilage in this area of the plate begins to accumulate calcium. Then osteoblasts begin to make new bone there.

So, as the epiphyseal cartilage grows on the side of the epiphysis, new bone is formed on the side of the diaphysis. This is how the diaphysis (the main shaft of the bone) gets longer without disturbing the joint surfaces at the rounded ends.

If long bones did not grow like this, the joints would have to stop working while growing. Aren't you glad God designed your bones to grow longer from the epiphyseal plates so you can keep moving while your bones grow!

Ok, so that's how bones get longer, but don't bones get BIGGER? Yes, they do. They must, in order to withstand your weight as you grow. The long bones in a tiny baby are like little twigs and could never carry your weight. A long, very thin bone isn't of much use, is it?

Remember, we mentioned the membrane that covers the bone? That membrane is called the *periosteum*. Well, there are osteoblasts along the inner surface of the periosteum, and these cells make new bone. In this way the bone becomes thicker. But that can be a problem. When more bone is made it becomes heavier. It does not necessarily become stronger.

This process is kept under control by osteoclasts that are located in the middle (inside the medullary cavity) of the bone. The cells remove bone and make the medullary cavity bigger. In this fashion the bone gets larger but does not get too heavy.

Remember that all this activity is exquisitely regulated to keep too little bone from being made or too much bone from being broken down. Then end result is that we get bigger stronger bones.

## Rickets

Rickets is a bone disease in children that is the result of a deficiency of vitamin D. The signs of rickets include bowed legs, defects in the spine or pelvis, and, on occasion, muscle weakness. Other signs include low blood calcium levels and an increased tendency for fractures.

The primary cause of rickets is a deficiency of vitamin D. Vitamin D aids in the absorption of calcium in the digestive tract. When vitamin D levels are low, less calcium is obtained from the diet. This lack of calcium leads to the skeletal defects characteristics of rickets.

The obvious treatment for rickets is to increase the levels of vitamin D. This can be accomplished by increased exposure to sunlight and increasing the intake of foods rich in vitamin D, such as eggs yolks and fish oils.

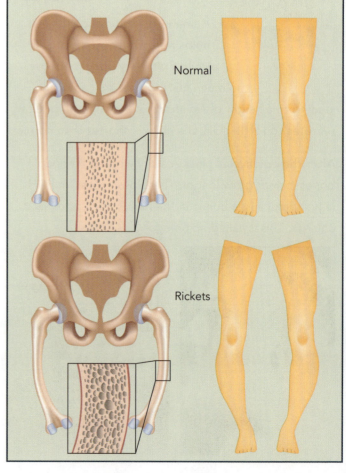

The average person grows in height until around age 17 or 18. Due to hormonal changes, there is no significant change in height after that age. At that point, the epiphyseal plate becomes calcified, and is then called the epiphyseal line.

## Bone Remodeling

Though bone does not look like it, it is, in fact, quite active even after a person has finished growing taller. Bone must continually build up and break down in order to remain healthy and strong. You see, as bone gets older, changes occur in the calcium compounds in the bone, causing the bones to become brittle. A brittle bone is more easily broken, so that could present a problem. Right? By getting rid of bone before it gets old and brittle and replacing it with fresh new bone, the bone stays strong and hard, yet just flexible enough to not be brittle.

God designed osteoclasts to break down and recycle older bone and osteoblasts to build new bone. We recycle about 5 percent of our bone mass every week!

In order to stay truly healthy, bones need to bear weight. Staying active therefore helps your bones remain strong. You have heard the phrase, "Use it or lose it!" People who exercise more tend to have stronger bones.

When weight is put on bones, they respond to this good kind of stress by making more bone tissue. Osteoblasts secrete more bony matrix when they are under stress. For example, football players who lift lots of weights have thicker bones at the points where the muscles attach to the bones. Tennis players tend to have stronger bones in their dominant arm. (Which arm depends on whether they are right or left handed.)

On the other hand, people who do not put much stress on their bones tend to have weaker bones.

For example, people who are sick and require long periods of bed rest lose bone strength. Astronauts who stay in space for long periods of time can also develop weaker bones, and therefore they must use special exercise equipment during long missions. These men and women have to exercise regularly; otherwise, the weightless environment in space would cause them to lose bone mass.

## Bone and the Body

When we briefly described the various organ systems, it was said that even though each system could be

*Without the stress of gravity during long missions in space, astronauts' bones lose calcium and strength without the proper exercise. Here astronaut Sunni Williams runs on the first treadmill installed on the International Space Station.*

studied separately, they all truly work together. In this way, the entire body can function correctly.

It is important to remember this when we study the skeletal system. Our bones do so much more than just hold us up.

Let's look at one of the ways our bones interact with other systems in the body.

Bones store certain important materials that are needed for the body to function. One of the most important is calcium.

Calcium is obviously important to help us develop and maintain strong bones and teeth. In fact, about 98 percent of the body's supply of calcium is found in bones and teeth. But calcium is used for more than building bones and teeth. That other 2 percent is just as important. In fact, you couldn't live without it.

Calcium is vital for the operation of many other organ systems. Without calcium, blood could not clot, nerves could not transmit signals, muscles could not contract, and many enzymes could not work. *Enzymes* are special proteins that speed up and control chemical reactions in the body. So you see that calcium does more than make bones hard. It is needed in the proper concentration for many things.

Thyroid gland

Parathyroid gland

Having either too much or too little calcium in the blood is dangerous.

So what does all this have to do with calcium being stored in bones? Well, our bones act as a sort of warehouse for calcium. When the body needs more calcium, it sends to the warehouse to get some more. How the body asks for more calcium is fascinating. It asks by using a hormone called PTH (parathyroid hormone). Hormones, remember, are chemical signals released from the brain or certain glands to tell other parts of the body what to do.

In our neck we have four small glands called parathyroid glands. These glands monitor the level of calcium in our blood. When the calcium level starts to get low, PTH is made and released into the blood. When PTH reaches the bones, osteoclasts are stimulated to act, and bone is broken down, releasing calcium into the blood. So, the calcium level in the blood goes up.

PTH can help increase the calcium level in other ways, too. PTH can signal the kidney to stop filtering out as much calcium. So if less calcium leaves the body in the urine, then more is kept in our blood. Another thing that PTH does is increase vitamin D production in the skin. Remember that vitamin D helps increase the amount of calcium we absorb from our food.

So when calcium levels go down, PTH production increases, and we get more calcium into our blood by extracting some from bone, absorbing more from our food, and losing less in our urine. Pretty neat, huh?

But are there times when the calcium level is too high? Yes, that does occur. Do we have a mechanism to correct this? Yes, we do!

There is another gland in our neck called the thyroid gland. It is a single gland, much larger than the parathyroids. The thyroid gland has several functions,

## Osteoporosis

Osteoporosis is a disease, primarily of the elderly, that results in bones that are very fragile. It is a very serious health issue for older people in our society.

Osteoporosis occurs when bone *resorption* exceeds bone deposition. In other words, more of the bone gets broken down for recycling than gets replaced. This results in a loss of bone. Thus, the density of bone decreases over time. This decrease in bone density results is a much higher risk of fractures, particularly fractures of the spine or hip.

Osteoporosis is more common in women than in men. Risk factors for osteoporosis include a diet low in calcium, a history of smoking, and a lack of exercise.

The treatment of osteoporosis includes supplementation with calcium and vitamin D. In addition, patients can be given medications that decrease the activity of osteoclasts, which, in turn, slows the resorption of bone. An important part of both treatment and prevention of osteoporosis is weight-bearing exercise. Remember — putting stress on bones can be a good thing as it stimulates bones to build and become stronger.

Osteoporosis

Normal bone

but among the hormones it makes is one called *calcitonin*. Do you notice part of the word "calcium" in the name "calcitonin"? Like the parathyroids, the thyroid gland monitors calcium blood levels. When the calcium level gets too high, calcitonin is secreted.

As you might expect, calcitonin does pretty much the opposite of PTH. Calcitonin decreases the activity of osteoclasts. Thus, less bone is broken down, and less calcium is released into the blood. Further, calcitonin decreases the amount of calcium absorbed from our food, and it stimulates the kidneys to release more calcium into the urine to be eliminated.

Thus, the blood level of calcium is maintained within very precise limits. Without types of monitoring systems such as this, our bodies would quickly be so out of balance that we could not function. This is an example of *homeostasis*, the body's way of keeping equilibrium in a system.

## Arthritis

Arthritis is inflammation of one or more joints. Arthritis affects millions of people and is a leading cause of disability in people over the age of 65.

There are many kinds of arthritis. The most common are *osteoarthritis* and *rheumatoid arthritis.*

In osteoarthritis (OA), joint cartilage gradually deteriorates. This results in pain, swelling, and restricted motion in the joints affected. Aging, obesity, trauma, or overuse can cause osteoarthritis, especially in larger joints like knees and hips.

In rheumatoid arthritis (RA), a person's own immune system attacks the body's own tissues, often attacking the joints of smaller bones like those in the hands.

# Broken Bones

Have you ever broken a bone? It wasn't very pleasant, was it? Sort of reminded you that we live in a fallen, sin-cursed world, right?

Obviously, a broken bone is not only very painful but it cannot function normally. You cannot walk properly on a broken ankle. If your arm is broken, you cannot throw a ball or lift a book. A break in a bone is called a *fracture*. We would be in quite a fix if broken bones stayed broken!

Fortunately, our Creator designed a way for fractured bones to heal and repair themselves. There are several stages to the healing process.

When a bone is broken, blood vessels inside the bone and in the periosteum are damaged. Blood leaks out of these damaged blood vessels. From this a mass of clotted blood called a *hematoma* forms at the place where the bone is broken.

Then, osteoblasts, along with two other types of connective tissue–producing cells, fibroblasts that make collagen and chondroblasts that produce cartilage matrix, continue the healing process. A cartilage-like layer of tissue known as a *callus* is formed. The callus is made inside the fracture itself and also on the external surface of the bone. The callus provides support and helps stabilize the fracture as it heals.

Within a week or so, the callus made of cartilage begins to be converted to spongy bone. New trabeculae of spongy bone are made as healing continues. Over this period, the callus made of cartilage becomes a bony callus. Usually in six to eight weeks, the initial stages of healing are complete, and the bone is functional again.

But we are not done yet. As you recall, the outer part of bone is compact bone, not spongy bone. If the break in the bone were only filled in with spongy bone, the healed bone wouldn't be very strong, would it? Spongy bone is hard, but it isn't as hard by itself as it is when it is wrapped in compact bone. So something more must happen. After the fracture is initially healed with spongy bone, more bone remodeling is needed. Over the next 8 to 12 months, the bony callus undergoes a significant transformation. Osteoclasts begin the job of slowly breaking down the spongy bone. Then osteoblasts begin to make new compact bone to replace it.

### TAKING A CLOSER LOOK
### Repair of Broken Bones

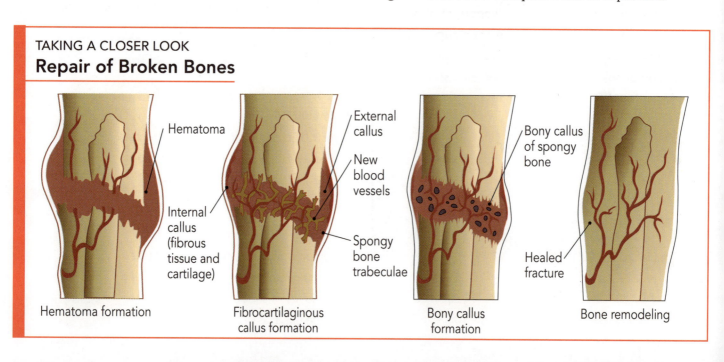

Hematoma formation | Fibrocartilaginous callus formation | Bony callus formation | Bone remodeling

The process of breaking down bone and making new bone has to be done in a very precise manner. There must be the proper balance at all times between removing old bone and making new bone.

After the bone remodeling at the fracture site is finished, there often remains a small thickening of the bone, a remnant of the callus left behind.

## Types of Fractures

There are many different types of fractures that can occur. The simplest way of looking at these is to consider four basic categories: complete, incomplete, simple, and compound. Most fractures can be described using these categories.

Fractures may be *complete* or *incomplete*. A complete fracture is one where the bone breaks into two or more separate pieces. An *incomplete* fracture is one in which the bone is cracked but not broken all the way through.

A *simple fracture* (also called a *closed fracture*) is one in which the bone is broken but does not break though the skin. A *compound fracture* (also called an *open fracture*) is where the fractured bone breaks through the skin.

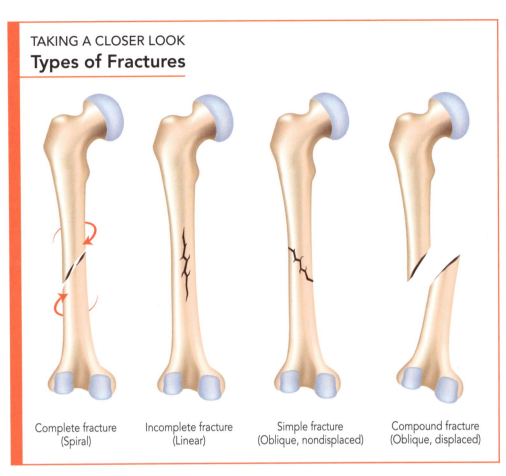

TAKING A CLOSER LOOK
**Types of Fractures**

Complete fracture (Spiral) | Incomplete fracture (Linear) | Simple fracture (Oblique, nondisplaced) | Compound fracture (Oblique, displaced)

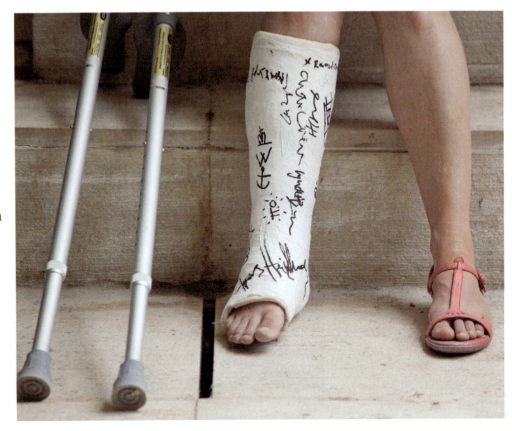

# THE MUSCULOSKETETAL SYSTEM

# Replacing Joints

Joints severely damaged by arthritis or injuries can cause chronic pain and make it very difficult to move around. Sometimes these joints need to be surgically removed and replaced with an artificial joint made of metal, plastic, or ceramic. Hips and knees are the most commonly replaced joints. Hip replacement may be performed because of ongoing pain and decreased mobility, as in the case of arthritis, or in order to repair a broken hip.

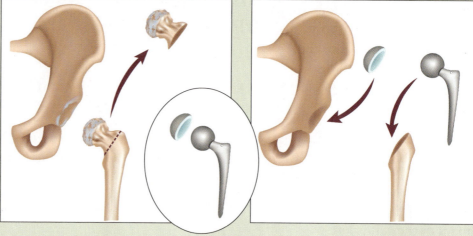

In a total hip replacement, pictured here, both the "ball" and the "socket" are being replaced. The head of the femur—the thigh bone—and the diseased cartilage that covers it are cut off. An artificial, or *prosthetic*, femoral head attached to a spike-shaped stem is inserted into the end of the femur. A cup-shaped ceramic socket is placed in the pelvic bone to cradle the new ball-shaped femoral head. During the healing process, a patient is given physical therapy to progressively increase the strength of the new joint and to encourage the patient to keep moving. With surgery like this, a person who was unable to walk or even get up out of a chair due to severe arthritis may be able to enjoy taking walks and moving more freely again.

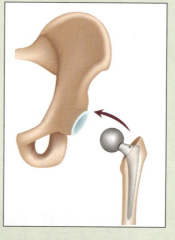
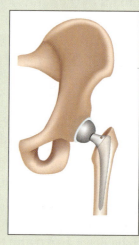

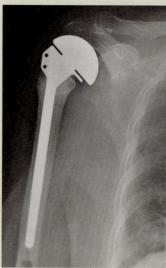
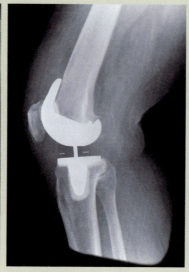
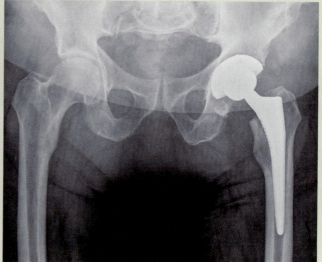

Shoulder replacement        Knee replacement        Hip replacement

# Treatment of Fractures

A simple fracture is usually *immobilized* with a cast or a splint. This supports it while the healing process begins. By immobilizing the area so that the broken pieces of bone cannot shift, the cast can decrease the pain in the early days after a broken bone. Beyond this, a cast will help keep the person from putting weight or stress on the area before it can heal properly. And, of course, the better the fracture is immobilized, the less chance there is of disrupting the fragile material being laid down in the fracture site in the early stages of healing.

More serious fractures need more intensive intervention. Sometimes the bones are not in a proper position to heal correctly. In certain of these situations, the bones must be put into alignment. This is called a *reduction*. This can be very painful and often requires the patient be heavily sedated or even put to sleep for the procedure. After the proper position of the bones is achieved it can be put in a cast until it heals.

The most serious fractures sometimes require surgery. The surgeon may use nails, surgical screws, wires, plates, or rods to fix the bones into position to allow the fracture to heal. Surgical intervention for a fracture might be preferred over more conservative treatments in certain situations. For example, with hip fractures, surgery is the preferred treatment rather than having a patient endure a prolonged period of bed rest and immobilization.

## Arthroscopy

*So Simple Yet So Complex — Designed by the Master*

To repair damaged joints, surgeons don't always have to make big incisions. Instead, joint damage sometimes can be repaired using arthroscopy. An arthrocope—a small telescope for looking into joints—is inserted through a small incision, usually only about an inch long. The joint can then be examined for any abnormality, such as torn ligaments or the effects of severe arthritis. Using other small instruments, a surgical repair can often be made with this technique.

Because the incisions are very small, the healing time for patients having arthroscopic surgery is much faster and they can soon return to their normal activities.

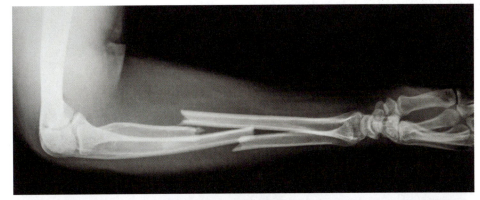

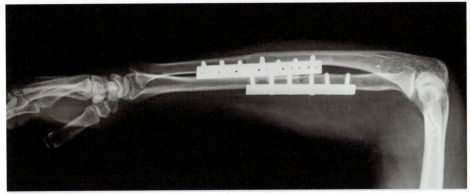

*This X-ray of the forearm shows a complete fracture of the* **radius**, *the bone behind the thumb, and the ulna, the bone beside it. The second X-ray taken after surgery shows that these bones have been realigned and plates attached to them to hold them in place for healing to occur.*

# THE SKELETON

When people see a skeleton, the first thing many think of is something dead or something scary. In reality, nothing could be further from the truth.

The human skeleton is alive! It is composed of bones, ligaments that hold the bones together, cartilage, and joints. The skeleton is intricately designed but incredibly strong.

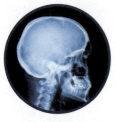
Skull

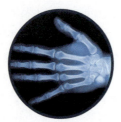
Hand

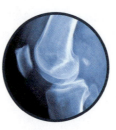
Knee Joint

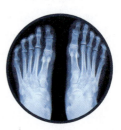
Feet

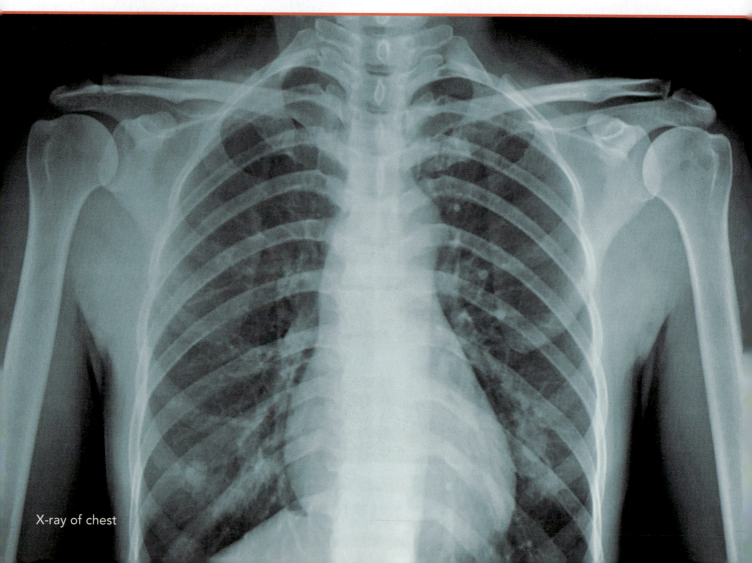
X-ray of chest

# THE SKELETON

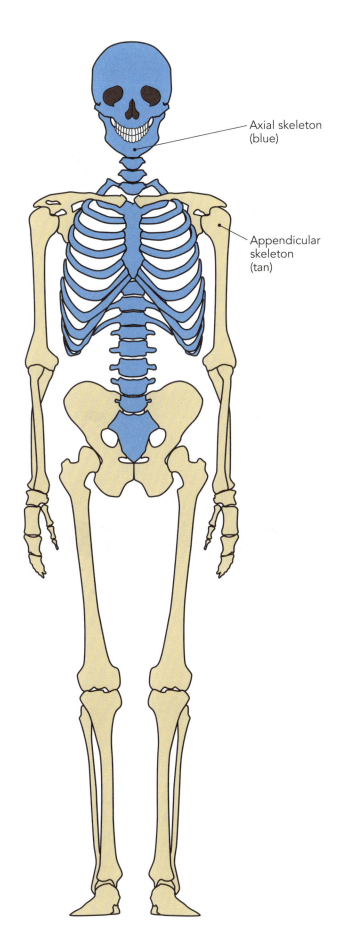

Axial skeleton (blue)

Appendicular skeleton (tan)

The human skeleton of a newborn is made of 270 bones. As we grow and mature, some bones fuse together so that as adults we have a total of 206 bones. As we mentioned earlier, our skeletons account for about 20 percent of our body weight (say about 30 pounds in a 150-pound person).

The bones of the skeleton are divided into two groups, the *axial skeleton* and the *appendicular skeleton*.

The *axial skeleton* is made up of the skull, the vertebral column (the "back bones"), and the ribs. The skull protects the brain, gives us the solid foundation for a face, and holds our teeth so we can chew. The axial skeleton's vertebral column holds us upright so we can stand, walk, and sit without flopping over. The vertebral column protects the spinal cord, but is designed to allow us to bend and twist without damaging the delicate nerves on which all our voluntary movements depend. The ribs and vertebral column protect our heart and lungs and form a cavity into which the lungs can expand as we breathe.

The *appendicular skeleton* is made up of the upper and lower limbs as well as the bones that connect them to the axial skeleton. Without an appendicular skeleton we could not walk or run. We could not reach out and touch anything. We could not pick up or throw anything.

The arms are connected to the axial skeleton by the *pectoral girdle*. Your shoulder blades and collarbones are parts of your pectoral girdles. The legs are attached to the axial skeleton by the pelvic girdle. Your hipbones make up the *pelvic girdle*. The pelvis is like a bowl, open at the bottom, made of several hip bones fused together and attached to the lower part of the vertebral column.

Neither part of the skeleton is more or less important than the other. These divisions of the skeleton merely make it easier to systematically study and understand how the skeleton functions.

# Terms of Movement

However, before we look at joints, we need to define anatomical terms of joint movement. That is, we need to understand the movement of joints themselves.

The most common joint movements are as follows:

*Flexion and extension* — Flexion means a movement that decreases the angle between two parts of the body. This is really not as complicated as it sounds. When you flex your elbow, you decrease the angle between the humerus and the ulna. When you flex your knee, you decrease the angle between the tibia and the femur. So logically, extension would be the opposite. Extension is where the angle between two parts of the body is increased. When you extend your elbow, the angle between the humerus and the ulna is increased.

*Abduction and adduction* — Abduction is movement away from the midline of the body. Adduction is movement toward the midline of the body. When you raise your arm directly away from your side, this is abduction of the shoulder joint. When you lower your arm back to your side, this is adduction of the shoulder.

*Rotation* — Rotation means moving a part around an axis. For example, when you turn your head from side to side, you are rotating the cervical vertebrae.

Other, more specific, anatomical movements will be defined later.

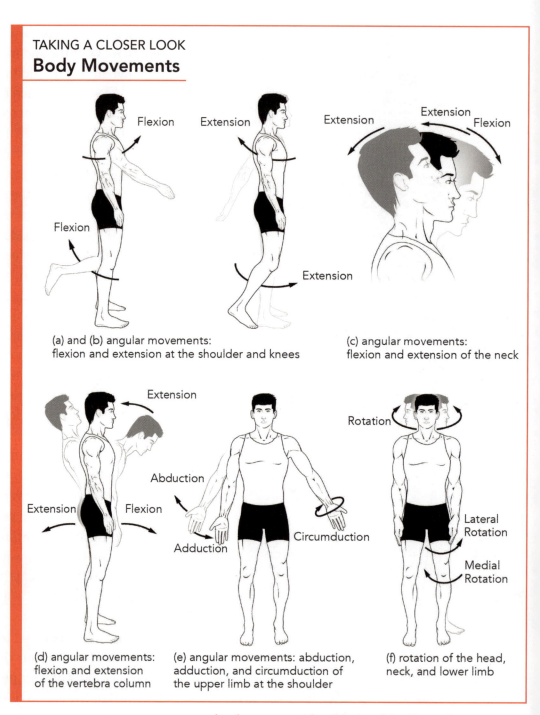

**TAKING A CLOSER LOOK**
**Body Movements**

(a) and (b) angular movements: flexion and extension at the shoulder and knees

(c) angular movements: flexion and extension of the neck

(d) angular movements: flexion and extension of the vertebra column

(e) angular movements: abduction, adduction, and circumduction of the upper limb at the shoulder

(f) rotation of the head, neck, and lower limb

# Joints

Joints are the places where two or more bones meet, and they hold the skeleton together. Some joints allow a wide range of motion. Other joints allow for much less movement. Some joints allow no movement at all.

There are three categories of joints: *fibrous* joints, *cartilaginous* joints, and *synovial* joints.

*Fibrous joints* connect bones with dense fibrous connective tissue. Generally speaking, fibrous joints permit no movement. For example, the sutures between the bones of the skull are fibrous joints. In adults, these joints are immobile.

In *cartilaginous joints*, as you might have guessed, the bones are joined by cartilage. This type of joint allows very little movement. A *symphysis* is one place where bones are joined this way. The bones that form the front part of your pelvis are joined together at the *pubic symphysis*. The *pubic symphysis* is a cartilaginous joint.

The third type of joint, the *synovial joint*, allows the most movement. Most joints are synovial joints. When you think of a joint, you probably think of the synovial joints that allow you to move so well.

The ends of the bones in a synovial joint are covered by cartilage. Because the cartilage is inside a joint, it is called *articular cartilage*. (*Articulate* means

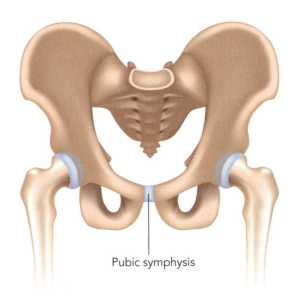

Pubic symphysis

"connected by a joint.") This cartilage provides a durable surface that allows the joint to move smoothly without damaging the bone underneath the cartilage. Imagine how uncomfortable it would be if the bare ends of the bones had to rub together, uncushioned by articular cartilage.

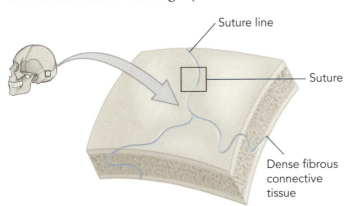

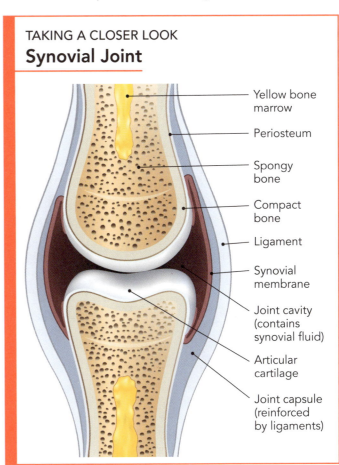

TAKING A CLOSER LOOK
**Synovial Joint**

- Yellow bone marrow
- Periosteum
- Spongy bone
- Compact bone
- Ligament
- Synovial membrane
- Joint cavity (contains synovial fluid)
- Articular cartilage
- Joint capsule (reinforced by ligaments)

## THE MUSCULOSKETETAL SYSTEM

But the cartilage cushion alone isn't enough to make the synovial joints move freely. They need to be lubricated. The joint capsule makes such lubrication possible. The joint capsule has a tough fibrous layer on the outside and a lubricant-making layer called the *synovial membrane* on the inside. The outer dense connective tissue that surrounds the joint helps hold the bones together. The inner membrane is made of less dense connective tissue and helps create a space that contains a special fluid called *synovial fluid*. Synovial fluid lubricates the joint and reduces the friction produced when the cartilage covering the bone ends rub together. Beyond this, synovial fluid bathes the articular cartilage with oxygen and nutrients. This is very important, because there is no way for blood vessels to supply these things to the cartilage inside a moving joint, for they would be ripped and torn by the movement of the joint.

There are several types of synovial joints. Each makes a particular sort of movement possible.

First of all, there are *hinge joints*. You can think of this joint as being like a door hinge. This joint can flex or extend only. A hinge joint does not twist. The knee is an example of a hinge joint.

*Saddle joints* allow more motion than hinge joints. If a hinge joint can flex and extend in one plane, a saddle joint allows this flexion-and-extension, plus movement "side to side." The thumb joint is a good example. Try moving your thumbs up and down. Now move them side to side. You see, this range of motion is more complicated than a hinge joint. Just think of all the things you could not do if your thumb only moved in one plane!

A *ball-and-socket joint* is made of a rounded end of one bone fitting into a rounded cavity in another bone. This type of joint allows the maximum range of movement. It can move up and down, and side to side, as well as rotate. Your hip joint is a good example. Stop and see how many directions you can move your hip joint. Amazing, huh? This wide range of movement allows us to be able to run, walk, and change directions with ease. You have another ball-and-socket joint in your shoulder. Thus, you can raise and lower your arm, move your arm out from your side,

### TAKING A CLOSER LOOK
### Types of Synovial Joints

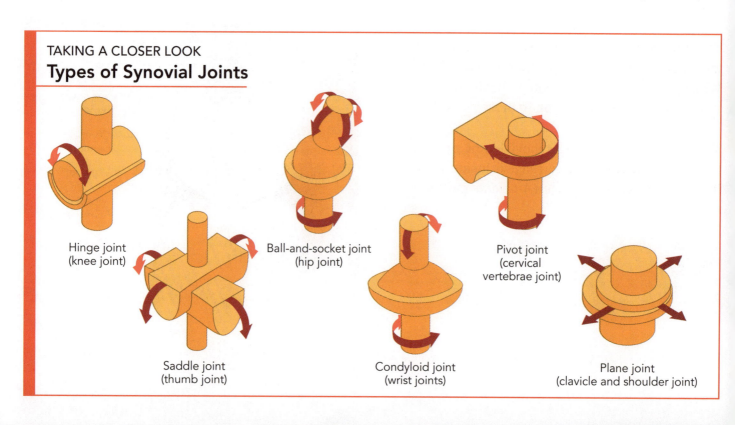

Hinge joint (knee joint)

Saddle joint (thumb joint)

Ball-and-socket joint (hip joint)

Condyloid joint (wrist joints)

Pivot joint (cervical vertebrae joint)

Plane joint (clavicle and shoulder joint)

and rotate it 360 degrees. In fact, the joint around the shoulder is called the "rotator cuff" because rotation is such an important part of the shoulder's motions.

*Condyloid joints* are similar to ball-and-socket joints, but they are more restricted in the range of motion they allow. The joints in your wrist are condyloid joints.

A *pivot joint* is a relatively simple joint in that it primarily allows rotation. Think of turning your head, and you get the idea. The joint between the first and second cervical vertebrae is a pivot joint.

A *plane joint* is made up of flat surfaces and allows only a small gliding motion. The joint between the end of the clavicle and the shoulder blade, the acromioclavicular joint, is a plane joint. It allows just a little bit of movement.

# Ligaments

You know, joints don't just stick together. If joints are to have the ability to move, then the surfaces of the joints must be able to glide or slide against one another. Ligaments make this possible.

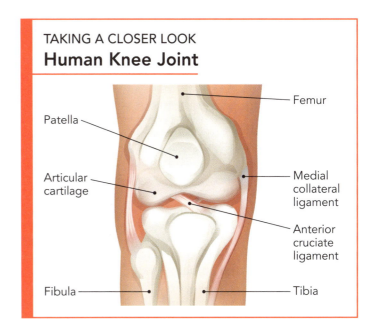

### TAKING A CLOSER LOOK
### Human Knee Joint

Labels: Patella, Articular cartilage, Fibula, Femur, Medial collateral ligament, Anterior cruciate ligament, Tibia

Ligaments are bands of dense fibrous connective tissue, primarily made of collagen fibers. They connect bone to bone to help form a joint. They are quite strong, but at the same time they are able to stretch to some degree and are, of course, very flexible. With movement of a joint, the ligament stretches. When the joint moves back, the tension is relieved and the ligament returns to its normal shape. In this way, ligaments help joints be stable while at the same time allowing joints their maximum range of movement.

Even though ligaments are strong and durable, they do have their limits. Perhaps you have heard about a football player who tore the ACL (anterior cruciate ligament) in his knee and had to have surgery to repair it. Or maybe you've heard about a pitcher in baseball who had to have surgery on his elbow to repair ligament damage. In cases like these, either through overuse or because of a sudden stress or impact, a ligament cannot withstand the force and is damaged.

Just as ligaments attach bones to other bones, so tendons are bands of connective tissue that attach muscles to bones. We'll learn more about them when we discuss muscles.

## Gout

Gout refers to recurrent bouts of painful swelling and redness in a joint, often at the base of the big toe. Like other forms of inflammatory arthritis, gout can be quite debilitating.

Gout is caused by having elevated levels of uric acid in the blood. This uric acid can crystallize, and those tiny crystals can be deposited in the kidneys or in joints and tendons. Historically gout was thought to be a "rich man's disease" because it was more common in obese people who ate a lot of meat and consumed a lot of alcohol.

# The Axial Skeleton

The axial skeleton is made up of the skull, the vertebral column, and the ribs. This division of the skeleton forms the central axis of the body. A very important function of the axial skeleton is that it protects the central nervous system (the brain and spinal cord) and the organs in the thorax (the heart and lungs). Eighty of the 206 bones in the body are in the axial skeleton.

## The Skull

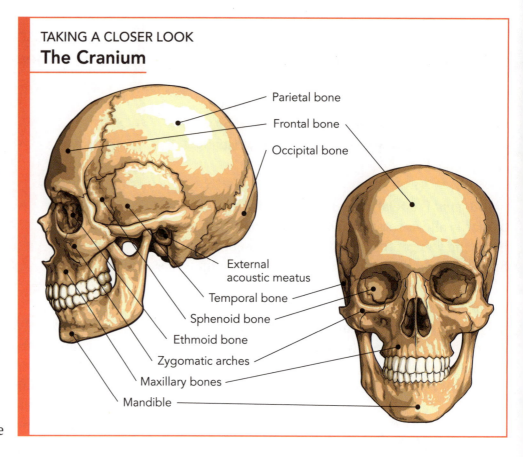

TAKING A CLOSER LOOK
**The Cranium**

- Parietal bone
- Frontal bone
- Occipital bone
- External acoustic meatus
- Temporal bone
- Sphenoid bone
- Ethmoid bone
- Zygomatic arches
- Maxillary bones
- Mandible

Perhaps the most recognizable part of the skeleton is the skull. The skull protects the brain. It has special cavities that protect our sense organs, without which we could not see, hear, smell, or taste. The skull contains the jaws and teeth that enable us to chew our food. It also is a place for our facial muscles to attach so we can smile, laugh, frown, grimace, and shape words with our mouths.

### The Cranium

The most prominent feature of the skull is the cranium. This is the rounded part of the skull at the top. While it looks like one solid upside-down bowl, the cranium is actually made up of eight bones. In an adult, these bones are fused tightly together.

On the front of the cranium is the *frontal bone*. This is the bone in your forehead. Directly behind the frontal bone are the *parietal bones* (one on each side). If you touch the top of your head and run your fingers front to back, you will run your fingers along the connection between your parietal bones. Lightly tap the bones on the top of your skull to each side. These are the parietal bones. Then on the *back* of your skull, protecting the part of your brain that enables you to see, is the *occipital bone*. (Your eyes are not on the back of your head, but they send information there!)

Looking from the side at a skull, you see one of the temporal bones. There is a *temporal bone* on each side, below the parietal bone. Your temples are named for your temporal bones. Buried within the thickest part of the temporal bone on each side is an intricately shaped labyrinthine chamber that perfectly matches the shape of your inner ear. In this little chamber, the delicate organs that enable you to keep your balance are protected. In front of the temporal bone is the *sphenoid bone*. The wing-shaped sphenoid bone forms a floor for your brain. Finally, located deep between the eyes is the *ethmoid bone*. The ethmoid bone has a lot of little holes through which the olfactory nerves pass to carry information about what you are smelling from your nose to your brain. The sinuses are formed by air cavities in some of these cranial bones.

Even though the cranial bones are fused together and therefore do not move in adults, they do have joints. The joints that hold the cranial bones together are called *sutures*. The word *suture* also means "stitch," so think of the fused cranial bones as being stitched together. These sutures are fibrous joints that hold bones together but permit no movement and so give our brains a great deal of protection.

## The Facial Bones

The skull not only has cranial bones to protect the brain but also facial bones, including our jaws. The cheekbones you can feel below your eyes are called *zygomatic arches*. Below these are the *maxillary bone*s. These form the upper jaw, and hold your upper teeth.

The most prominent of the facial bones is the *mandible*, or the lower jawbone. This is the only facial bone that moves. And it's a good thing it can. If the mandible could not move, we could not chew or speak.

Our lower teeth are anchored in the mandible. The oral cavity formed between our upper and lower jaw provides protection for our tongue and upper airway.

## Holes in Our Head?

If you examine a skull closely, you will see many, many holes. Some are large, some are small. The most obvious holes in the skull are on the front. These are, of course, our eye sockets. The eye sockets protect our eyes and also provide a passage for the *optic nerves* to connect with the brain. The optic nerves carry the impulses created by what our eyes see to our brain.

You can also see the opening of the nasal passage on the front of the skull as well as the ear canal on either side. This ear canal is also called the *external acoustic meatus* — *external* meaning "outer," *acoustic* meaning "hearing," and *meatus* meaning "opening." You see, even the big words assigned to anatomical structures usually make sense.

At the base of the skull is a very large opening called the *foramen magnum*. *Foramen* means "opening" and *magnum* means "big." Can you guess what this "big opening" is for? Of course you can. Through this opening passes the spinal cord, which is connected to the brainstem. Through the foramen magnum, then, the brain can send signals to and receive information from the body, via the spinal cord.

Beyond these larger openings there are many small holes, each also called a foramen. *Foramina* means more than one foramen. Through these little foramina small *cranial nerves* pass to control your head and neck.

Other important types of "holes" are *sinus cavities*, or *sinuses* for short. These cavities make the skull lighter. Sinuses

### Hyoid Bone

Did you know that your tongue has its own bone? The hyoid bone is not in the tongue but under it. Shaped like a horseshoe, this small bone is located in the neck between the jawbone and the thyroid gland. It is the only bone not connected to another bone. Anchored instead by muscles and ligaments to several distant points above, below, and behind, the hyoid bone provides a point of attachment for the muscles of the tongue, the floor of the mouth, the voice box, and many muscles used in swallowing.

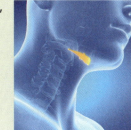

# THE MUSCULOSKETETAL SYSTEM

also heat the air we breathe, especially in the winter! Also, if the air you breathe is too dry, the moist linings of the sinuses moisten the air before it reaches the lungs.

## The Vertebral Column

The vertebral column is made up of 26 bones and provides a strong central support for our bodies. It not only holds us up, but it is vitally important because it protects the spinal cord as it extends from the base of the brain to the lower back.

This structure must then be very, very strong but also provide enough movement that we can turn our heads, twist to look behind us, bend to pick up our clothes off the floor, and shift to balance as we run up the stairs.

In the vertebral column, our Creator has provided us the means to accomplish all these things!

It is best to think of the vertebral column as being made up of five regions: cervical (neck), thoracic (chest), lumbar (lower back), sacrum, and coccyx. Each part consists of vertebrae, but the intricate shapes of the bones in each section and the way they connect allows each region to function as it should.

The cervical region is composed of 7 vertebral bones (called *vertebrae*). The thoracic region contains 12 vertebrae, and the lumbar region has 5 vertebrae. The sacrum is 1 fused bone that is formed by the fusion of 5 vertebrae. Similarly, the coccyx is 1 fused bone formed from the fusion of 3 to 5 vertebrae.

When you view the spine from the front, it appears relatively straight. However, when viewed from the side, the spine is decidedly curved. In fact, it has several curves, like an S. The curvature of the spine is vitally important to us. If our spine were not curved as it is, we could have difficulty walking with balance or carrying anything heavy. The curve of the spine

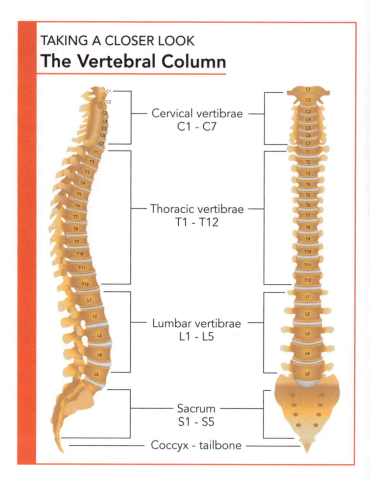

**TAKING A CLOSER LOOK**
### The Vertebral Column

- Cervical vertibrae C1 – C7
- Thoracic vertibrae T1 – T12
- Lumbar vertibrae L1 – L5
- Sacrum S1 – S5
- Coccyx – tailbone

actually helps maintain a certain degree of flexibility. This flexibility helps distribute the stresses on the spine as we walk or lift.

Let's take a look at a typical vertebral bone (or *vertebra*).

The front (anterior) part of a vertebra is called the *vertebral body*. This is the largest part of the vertebra.

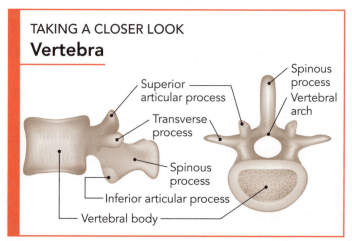

**TAKING A CLOSER LOOK**
### Vertebra

- Superior articular process
- Spinous process
- Vertebral arch
- Transverse process
- Spinous process
- Inferior articular process
- Vertebral body

It is basically oval in shape. It is the body of the vertebra that gives the spine its strength. The vertebral body provides support and bears the weight of our body.

To the rear (posterior) of the body of the vertebra is a ring of bone with a foramen (hole) in the center. This is called the vertebral arch. Along the *vertebral arch* extend several projections called *processes*. These bony processes provide attachment points for muscles and ligaments.

The vertebrae are stacked one on top of the other. As you can see from the illustration, vertebrae do increase in size somewhat as you travel down the vertebral column. Cervical vertebrae are smaller than lumbar vertebrae.

The vertebrae are separated by cushions called *intervertebral discs*. These tough, rubbery discs are made of fibrocartilage. The intervertebral discs act as shock absorbers as we walk or run.

In addition, there are very sturdy ligaments along both the front and back of the vertebral column to keep the bones in place and to provide further stability. Without these ligaments the spinal column would be no more stable than a stack of coins. These ligaments keep the stacked vertebrae from sliding off of each other as the spinal column bends and twists.

There is a large hole in each of the vertebrae, called a *vertebral foramen*. The hole is formed by the arch-like projections on the back part of each vertebra. As the vertebrae are stacked one on top of the other, these holes match up to form a long canal, called the vertebral canal. Can you guess what it is for? Of course, this is where the spinal cord is located. This long canal protects the spinal cord as it runs from the base of the brain to the lower back. Between each pair of the vertebrae are small openings called *intervertebral foramina*. (*Inter* means "between." So this word just means "holes between the vertebrae"!)

## Is the Back Poorly Designed?

Evolutionists believe that many thousands of years ago, humans walked on all fours. So it is their contention that the human spine is best suited to movement on all fours, and when humans began to walk upright, the spine was no longer ideal. For this reason you will often hear the claim that the back is an example of poor design. Thus, the evolutionist will claim that evolution must be true because there could not be an all-knowing Creator God that would be responsible for such shoddy workmanship.

Actually, nothing could be further from the truth. The human spine is a marvel of design and performs its task very, very well. The curvature of the spine allows the weight of the body to be properly distributed so that we are well-balanced while at the same time providing maximum flexibility to allow us to move efficiently in our environment.

Evolutionists say that humans have lots and lots of back problems because our spine is not designed for walking on two feet. That claim is simply incorrect. The Creator God of the universe knew exactly what He was doing when He designed the human spine.

The problem is that we all live in a fallen, cursed world. Our bodies wear out. Things begin to fail as a result of the decay that occurs in our world. Many factors contribute to the back problems that people suffer from. These factors include such things as improper or excessive lifting, obesity, smoking, and poor posture, to name but a few. Thus, back problems are not a result of poor design, but rather are the result of the effect of sin on our world.

Through these openings, nerves branch out from the spinal cord and go to various parts of the body.

# The Thoracic Cage

The axial skeleton not only protects the spinal cord, it also protects the heart and lungs. The heart and lungs are protected inside a thoracic cage. The thoracic cage is made up of the 12 thoracic vertebrae in the rear, the 12 pairs of ribs on the sides, and the *sternum* (breastbone) in the front.

## The Ribs

We have 12 pairs of ribs. That's 24 ribs in all. Each rib attaches in the rear to a vertebra. The first seven ribs are called "true" ribs. They are called this because they attach in the front to the sternum. Thus, each pair of true ribs, combined with one thoracic vertebra and a part of the sternum, forms a ring. These ribs are attached to the sternum by flexible connective tissue called *costal cartilage*. This cartilage allows the ribs to move, rising and falling with every breath. Is this important? It's actually very important if you want to breathe. You see, the rib cage expands and contracts during respiration. As we breathe in, our rib cage expands. As we exhale our rib cage contracts.

Ribs eight through ten are called "false ribs." This is kind of an unfortunate description because they really are ribs. They are called "false ribs" because they do not directly attach to the sternum in the front. Rather, in the front, these ribs attach to the costal cartilage above them. Thus, they do have an attachment and form part of the continuous expanding and contracting wall of the thoracic cage, but the attachment to the sternum is more indirect.

Ribs 11 and 12 are called "floating ribs." They don't really float, because they are attached to thoracic vertebrae in the back. The term "floating" refers to the fact they have no attachment in the front.

## The Sternum

The front portion of the thoracic cage is the sternum, or breastbone. As was noted previously, rib pairs one through seven attach to the sternum by means of costal cartilages. This bone is quite strong and helps provide protection to the heart and lungs.

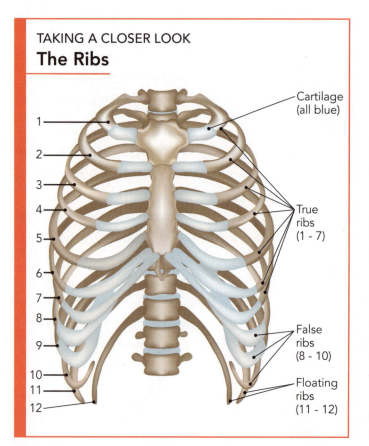

TAKING A CLOSER LOOK
**The Ribs**

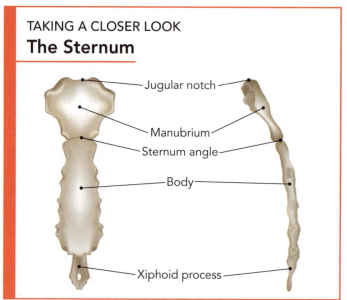

TAKING A CLOSER LOOK
**The Sternum**

# The Appendicular Skeleton

The appendicular skeleton is made up of the upper and lower limbs and the "girdles" that attach them to the body. The appendicular skeleton allows us to both move around in our environment (via our lower extremities) and interact with our environment (via our upper extremities). Of the total of 206 bones in our body, 126 bones are found in the appendicular skeleton.

## Do Men Have Fewer Ribs Than Women?

Over the years, many people have been taught that women have one more rib than men. This is due to what the Bible tells us about the creation of Eve.

"And the Lord God caused a deep sleep to fall on Adam, and he slept; and He took one of his ribs, and closed up the flesh in its place. Then the rib which the Lord God had taken from man He made into a woman, and He brought her to the man" (Genesis 2:21–22).

So if the Lord took a rib from Adam to make Eve, then women should have more ribs than men, right? Actually, no. Both men and women have 12 pairs of ribs. It is very easy to demonstrate. All you need to do is count the number of ribs in a man and in a woman, and there you have it.

It simply comes down to this. Removing a rib from Adam did not mean that his children would have one less rib. After all, in Adam's DNA is the information to code for 12 pairs of ribs, so the children should have 12 pairs of ribs.

So, as is usually the case, it's all in our DNA.

# The Upper Limb

The upper limb consists of the pectoral (or shoulder) girdle, arm, forearm, wrist, and hand.

## The Pectoral Girdle

The pair of bones that attaches each upper limb to the body is called a pectoral girdle. Some people call it the "shoulder girdle." Both terms are correct.

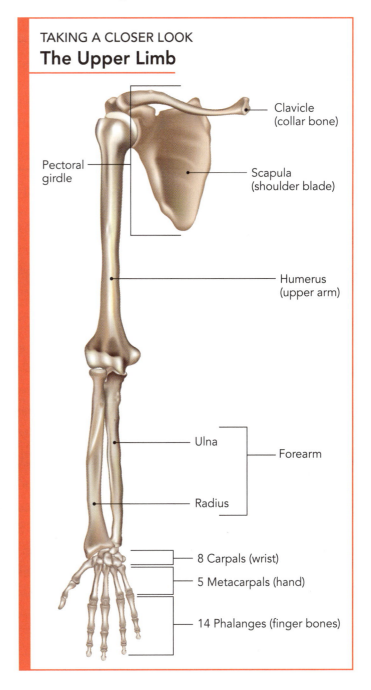

TAKING A CLOSER LOOK
**The Upper Limb**

- Clavicle (collar bone)
- Pectoral girdle
- Scapula (shoulder blade)
- Humerus (upper arm)
- Ulna
- Radius
- Forearm
- 8 Carpals (wrist)
- 5 Metacarpals (hand)
- 14 Phalanges (finger bones)

# THE MUSCULOSKETETAL SYSTEM

The pectoral girdle consists of the *clavicle* (collarbone) and the *scapula* (shoulder blade). The clavicle attaches at one end to the sternum and to the scapula at the other end. Interestingly enough, the scapula does not directly attach to anything but the clavicle. The scapula does not attach to the axial skeleton. Instead, the scapula is attached to the body by a series of muscles. This is a good thing in many ways. The main advantage of this design is that it allows the scapula to move more freely. This allows much more movement of our arms.

However, the more movement or flexibility a joint has, the less stable it is. So there is a price to pay for the amazing range of motion. The shoulder joint is not as stable as many other joints in the body.

## The Arm

You are probably accustomed to thinking of your upper limb as your "arm," but in actual anatomical terms, your *arm* is only the part between your shoulder and elbow. The part between elbow and wrist is the *forearm*. Therefore, there is only *one* bone in your arm: the *humerus*. The humerus forms a ball-and-socket joint with the shoulder on the top and a hinge joint with the forearm's *ulna* below, at the elbow.

## The Forearm

The forearm consists of two bones: the *radius* and the *ulna*. The radius is the one on the same side as your thumb. The bones attach on one end to the humerus to form the elbow and on the other end to the wrist.

## The Dislocated Shoulder

The shoulder joint has the greatest range of movement of any joint in the body. Its amazing design gives us the ability to move our arms in almost any direction. However, the trade-off for all this flexibility is that the shoulder joint is much more unstable than other joints. This makes the shoulder susceptible to injury, especially dislocation.

A shoulder dislocation is a situation where the ball-shaped upper end of the humerus is separated from the shallow, bowl-shaped depression in the scapula (called the *glenoid fossa*). (This ball-and-socket joint is called the *glenohumeral joint*.) The most common form of shoulder dislocation is where the humerus is displaced forward, or *anteriorly*, called an anterior dislocation. This is usually the result of a fall where a person lands on an outstretched arm, pushing the top of the arm bone forward and out of its socket.

When a dislocation occurs, the goal of treatment is obviously to return the shoulder to its normal position. There are a variety of techniques that can be used to coax the shoulder back into position. If this is successful, then the arm is usually placed in a sling for a period of time. Then strengthening exercises and physical therapy can help the shoulder recover its strength.

In the most severe cases, shoulder dislocations can require surgery to correct. In these more extreme cases, the recovery time is certainly much longer.

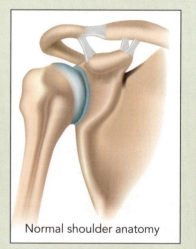

Normal shoulder anatomy

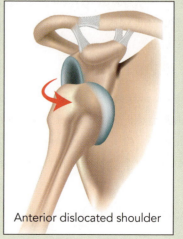

Anterior dislocated shoulder

At the elbow, the *proximal* end of the ulna and the *distal* end of the humerus form a hinge joint. This part of the elbow is primarily concerned with flexion/extension movement. At the elbow, the proximal end of the radius forms a pivot joint rather than a hinge joint. This joint allows the rotation of the forearm.

Try this. Put your finger on the tip of your elbow. This is the proximal end of the ulna and is called the *olecranon*. Now rotate your forearm. You will notice the olecranon does not move. This is because this rotation movement involves the radius and not the ulna. (When you whack the nerve that runs close to this bone, it hurts in a particular way that most of us don't like, and we say we hit our "funny bone.")

## The Wrist and Hand

The human hand is an engineering marvel. Its movements are so varied that we can grip a bat, put spin on a ball, or play the violin. Our hands are strong enough that we can grip a hammer to drive a nail, yet they are so delicate that we can hold a baby bird without crushing it or pick up a tiny pin. Only a wise Designer could create something like our hands!

But here is something to think about. On each side, in the wrist and hand, are 27 bones. Just think of that! In combination, the wrists and hands have a total of 54 bones out of the 206 bones in our bodies! Considering that, they must be pretty important, right? You bet they are! How could we do all the incredible things humans do without our hands!

In the wrist are 8 bones called *carpal bones*. These are stacked in two rows containing four bones each. Just move your wrist around to see how this arrangement allows so much movement!

Next we have five bones called *metacarpals*, one for each digit in our hand (clever, huh?). These are the bones in the palm of our hand. These allow some movement, but not nearly as much as our fingers. Finally, we have the bones of the thumb and fingers. These are called phalanges; each finger has three phalanges, the thumb has only two.

Wow, what an intricate collection of bones and joints! Just think of how incredible your arms (oops, I mean, your upper limbs) are.

## The Lower Limb

The lower limb consists of the pelvic girdle (or just the "pelvis"), thigh, leg, ankle, and foot. You probably thought the leg started at your hip, but in anatomic terms it doesn't. The *thigh* — with only one bone — connects the hip and the knee. The *leg* — with two bones — begins at the knee and goes to the ankle.

### The Pelvic Girdle

The pelvic girdle is the structure that attaches the lower limbs to the body. The pelvis consists of two hipbones and the sacrum, which is part of the vertebral column. The hipbones have a large

---

TAKING A CLOSER LOOK
**Bones of the Wrist and Hand**

- Phalanges: Distal, Middle, Proximal
- Metacarpal bones
- Carpal bones

# THE MUSCULOSKETETAL SYSTEM

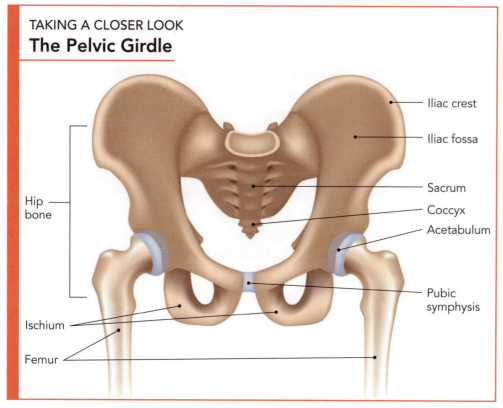

**TAKING A CLOSER LOOK**
**The Pelvic Girdle**

There is only one bone in the thigh, and it is called the *femur*. The femur is the longest and strongest bone in the body. It has to be to endure the stress that is placed on it. This bone has to withstand the stress that comes with all the running and jumping and lifting that we all do each day.

The femur attaches to the pelvis at the hip joint. The ball of the femur fits into a special cavity on the pelvis known as the *acetabulum*. A complex set of ligaments holds this joint in place. Can you guess what type of joint this is? Sure you can, it is a ball-and-socket joint!

wing-like prominence on each side, called the iliac crest. You can feel this just below your waist on each side. The hipbones are connected to the sacrum in the back. They are connected to each other in front with a fibrocartilage joint called the *pubic symphysis*. (Remember, this is a cartilaginous joint as we described earlier.) Below the sacrum is the tip of the vertebral column, the *coccyx* (sometimes called the tailbone).

This structure is thus attached to the axial skeleton, and it connects the lower limbs by means of the hip joint. The pelvis is like a large, strong bowl, open in the bottom. It is a sturdy ring that bears the weight of the upper body. It also protects the organs that are located in the pelvis. And the strong muscles lining the pelvis close the bottom of the bowl to keep your insides from falling through.

## The Thigh

The thigh is the section of the lower limb from the hip to the knee.

The lower end of the femur makes up part of the knee joint.

## The Leg

The leg is the part of the lower limb from the knee to the ankle. The leg is composed of two bones, the tibia and the fibula. These two bones help form the knee joint at one end and the ankle at the other. The tibia, or shinbone, is the larger of the two bones and bears almost all of the weight. The fibula does not bear any significant amount of the body's weight. That does not mean it is unimportant. The fibula provides a place of attachment for several muscles, and it forms the outer part of the ankle joint.

## The Ankle and Foot

Just as with the wrist and hand, the ankle and foot are composed of many bones. On each side in the ankle and foot are 26 bones. So our ankles and feet altogether have 52 bones! Again, this indicates the complex nature of our feet and ankles. Without them we could not walk or move as effortlessly as we do. Just like our hands, our ankles and feet are engineering marvels.

The ankle is made up of the distal ends of the tibia and fibula, along with one of the tarsal bones in the foot, called the *talus*. You can feel a rounded part of the fibula on the outside of your ankle and a rounded part of the tibia on the inside of your ankle. Although not quite as flexible as the wrist, the ankle is quite mobile in its own right.

The posterior portion of the foot is composed of seven bones called tarsals.

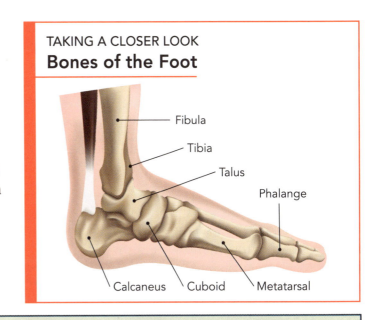

### TAKING A CLOSER LOOK
### Bones of the Foot

## What Is a Sprained Ankle?

Lots of folks have sprained their ankles. It's a pretty common injury.

So what is a sprain exactly?

An ankle sprain is a situation where a ligament in the ankle is torn. A severe twisting or turning of the ankle causes a sprain. Perhaps the most common cause of a sprain is an inward twisting (or inversion) of the ankle. This results in a tear in one of the ligaments in the ankle (most often the *anterior talofibular ligament*).

This injury results in pain and swelling of the ankle. There is often bruising or discoloration in the area of the injury. If you have ever sprained your ankle, you know how uncomfortable it can be.

Treatment of an ankle sprain usually consists of rest, intermittent ice packs, and some sort of support. The support can be a walking boot, an elastic wrap, or even a small air cast.

Most ankle sprains heal in two to four weeks. However, depending on the severity of the sprain, some injuries can take up to a year to heal.

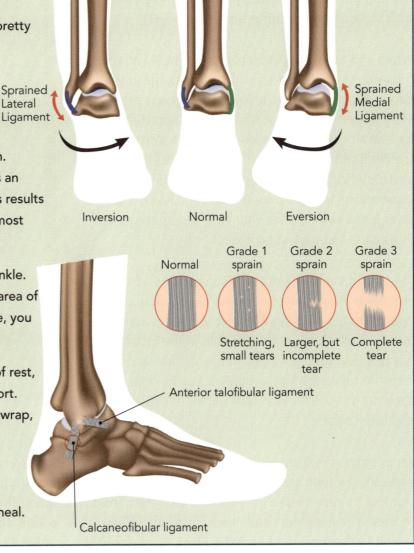

One of these bones helps form the ankle joint as just mentioned. Another of the tarsal bones is called the *calcaneus*, the heel bone.

The mid-portion of the foot (again like the hand) is composed of five bones; these are called *metatarsals*. And again, as in the hand, the toes are made of three phalanges each, save for the big toe, which has only two.

The tarsals and metatarsals are each shaped and arranged to work together. They are strapped together by many ligaments, tendons, and sheets of tough connective tissue, but they are strapped together just loosely enough to allow them to shift a little. Does this seem strange? Well, if the bones in the foot didn't give a little, you would have a hard time walking and running smoothly.

When you take a step, you plant your heel and rock forward. As the heel strikes the ground, the joints between the bones in the foot loosen just a little. They loosen *just enough* to allow the foot to change its overall shape as the weight above it shifts during the beginning of each step. Then the bones shift again, sliding into place against each other and interlocking to form a somewhat rigid lever. That lever helps you transfer your body's weight forward during the last part of your step.

These slight movements between the foot's bones also allow the foot to constantly change shape enough to remain stable as we walk or run across uneven surfaces. Tough ligaments in the foot help our feet bear all the weight from the thousands of steps we take each day, keeping its many bones in their proper positions as they do all that shifting and interlocking.

An important feature of the human foot is the arch. Actually, the foot has several arches. But most people think of its easy-to-see longitudinal arch from front to back. The ligaments and other connective tissue maintain the foot's arches. Take a look at your foot

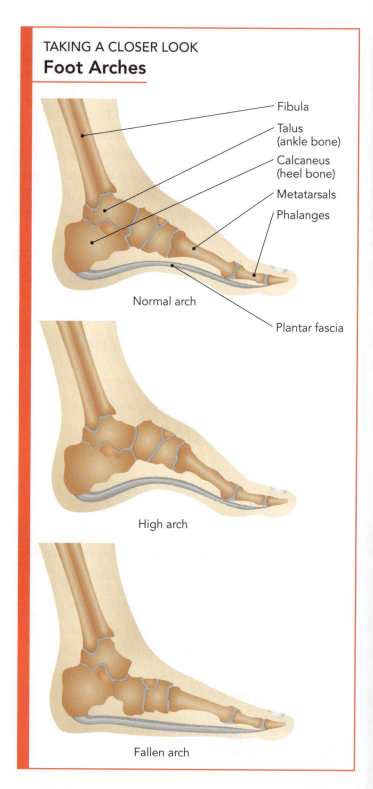

**TAKING A CLOSER LOOK**
## Foot Arches

Normal arch — Fibula, Talus (ankle bone), Calcaneus (heel bone), Metatarsals, Phalanges, Plantar fascia

High arch

Fallen arch

starting at the base of the big toe and moving toward your heel. See the curve in your foot? That is the longitudinal arch. This curve keeps part of the foot up off the ground. This is a better arrangement for people than a flatter foot would be because we walk upright on two legs. God designed us to walk that

# THE SKELETON

## Bipedal People

God designed human beings to walk upright on two legs. This is called a bipedal gait. The most efficient way for humans to walk is upright. Apes are designed to move through trees using their hands and feet, to climb while gripping branches with their feet, to walk along branches, and also to move along the ground. An ape can walk on two legs for a time, but a two-legged walk is not an efficient way for an ape to walk.

Some people claim that humans evolved from ape-like ancestors. They say they learned to walk upright and, having hands free to use tools and explore the world, eventually developed larger, smarter brains and became human. Of course, walking on two legs could never make it possible for an ape-like animal to become a more human-like one. There have never been any ape-men, only apes and humans. God made the first man and woman in His own image on the same day He made apes and monkeys. He did not use evolution.

When evolutionists find fossilized bones belonging to extinct apes (like the australopithecine ape they call "Lucy"), they often claim that the bones show the extinct apes were bipedal evolutionary ancestors of humans. But fossils do not walk. Even when paleontologists find lots of bones, despite the claims of evolutionists, those shapes do not show that the ape was truly bipedal. Sometimes paleontologists also find fossilized footprints. You may have heard of the fossilized Laetoli footprints found in Africa. Evolutionists claim the footprints were made by a family of our ape-like ancestors. But we know that the footprints were made by human beings, because they show the feet had arches! Only human feet have arches, because only human feet were designed for a lifetime of walking upright on two feet!

way, with the muscles and bones in our hips, legs, and feet all working together.

The foot's arches are important because when you walk, you push your weight off of your toes and you land with your weight on your heel. This is repeated with each step. The longitudinal arch provides both support and flexibility to your feet. When you step, the arch stretches a little, making walking much more efficient. In a way, you might say the arches in your feet give you a "spring" in your step!

**Walking from Head to Toe**

Many features of the human skeleton make walking upright on two legs just right for humans. The way the skull attaches to the neck makes looking forward comfortable for us. The way the muscles and bones of the hip are arranged allows us to step forward without swinging our thighs out to the side. The curves in the back, which develop once a child begins walking but are made possible by the shapes of the vertebrae, help us keep our balance. The way our arms attach to our shoulders allows them to swing just right, helping us shift our weight smoothly from one leg to the other. The way the bones in the foot constantly shift just enough to allow the foot to change shape as we begin each step helps us adjust to bear our constantly shifting body weight. The way those same bones then slide into a locking position transforms the foot into a rigid lever to propel our upright body forward.

It is amazing how the bones, ligaments, and muscles in the hips, legs, and feet are all arranged at just the right angle to make the bones in the foot slide into the optimal positions to become semi-rigid during the last part of every step. The arches of the foot make it springy enough to bear our weight through hours of standing and walking but allow the foot to become that rigid lever needed to walk upright efficiently. If the foot had no arch, it would bend like an ape's foot with every step, making it difficult to propel the body forward.

# SECTION 3
# THE MUSCULAR SYSTEM

Now that we've seen what a busy place a bone is, you'll never again think a skeleton is just a frame to give your body shape. A lot is going on inside your bones. But as cool as our skeletons are, they really aren't going anywhere on their own. They are going to need some serious help. We need to learn about muscles. Without muscles, all this stuff about bones and joints would not mean a thing. Muscles cover your skeleton and help give you the movement you need!

Muscle cars. Muscle builders. Strength is often associated with muscles. The stronger our muscles the more we can often endure, the more we can lift, and the more weight we can move. So let's see what the muscle is made of, the characteristics of muscle, the different types of muscle, what it is they do to move you, and more! With over 600 muscles in the body, you're bound to discover a few amazing things about how God designed you and formed you for His beautiful purpose for your life.

*"Have I not commanded you? Be strong and of good courage; do not be afraid, nor be dismayed, for the Lord your God is with you wherever you go" (Joshua 1:9).*

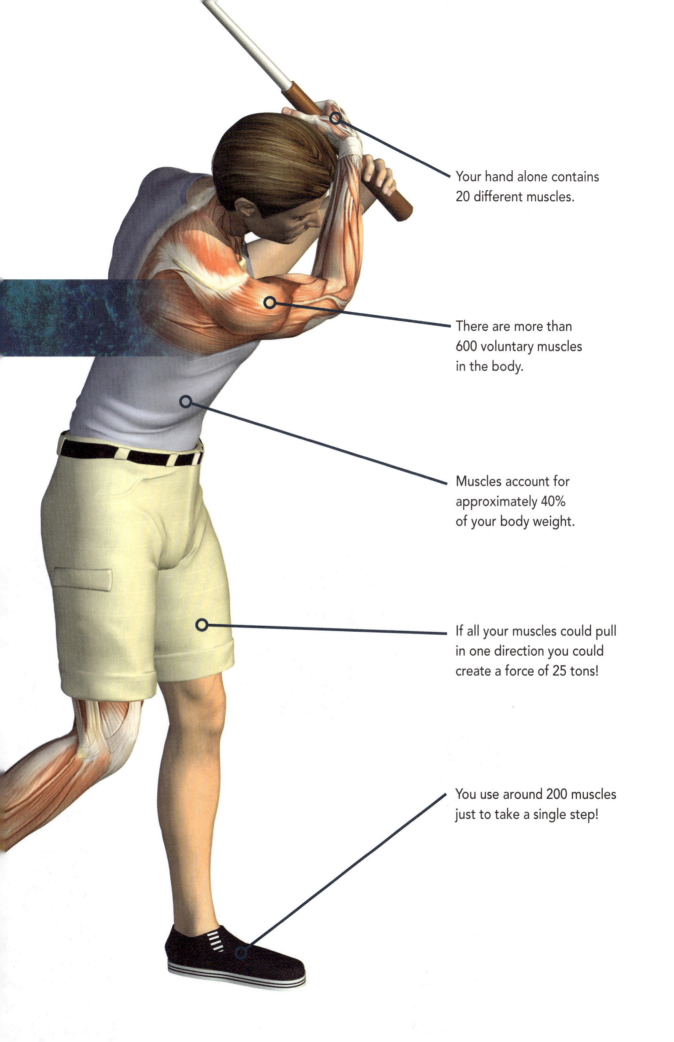

Your hand alone contains 20 different muscles.

There are more than 600 voluntary muscles in the body.

Muscles account for approximately 40% of your body weight.

If all your muscles could pull in one direction you could create a force of 25 tons!

You use around 200 muscles just to take a single step!

# MUSCLE BASICS

Muscles do one thing: they contract. *Contract* means get smaller or shorter. Muscles pull on whatever is attached to them when they contract. And even though *you* are able to pull and push on things, *your muscles can only pull. A muscle cannot push.* Does that surprise you? We'll see later that many muscles are paired to pull in opposite directions, but each muscle itself can only pull. By contracting, muscles perform an amazing variety of jobs.

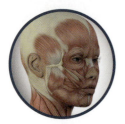

Skull

Hand

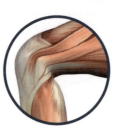

Knee

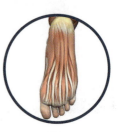

Foot

Because some muscles are attached on each end to bones, their contraction makes those bones move. The kind of movement that happens depends on where the muscles are attached. Throwing, running, swimming, playing the piano — all are movements produced when particular combinations of muscles contract. Without muscles, we wouldn't be going anywhere.

When the muscle tissue that makes up your heart contracts, your heart squeezes and pumps blood. Muscles in the wall of your digestive tract propel your food along its way as it is digested. Muscle tissue in the ureters help push urine into the bladder.

Second, muscles help to maintain your upright posture. Without the continuing activity of muscles along your spine, in your abdomen, in the neck, and attached to your pelvis, you could not stay upright against the constant pull of gravity. You would just fall down in a heap!

Next, muscle help strengthen our joints. The tension generated by muscles helps keep our joints strong and stable.

Muscle contraction also generates heat. One of the ways we can stay warm in cold weather is from the heat generated by our muscles. If you have ever stood outside on a very cold day, you might have begun to shiver. This shivering is due to muscle contractions. This is a way the body generates heat to keep you warm.

## Characteristics of Muscle

The most obvious characteristic of muscle tissue is its *contractility*. This means it can contract, or shorten, with great force. No other tissue does this.

The next characteristic is its *elasticity*. This means that when a muscle is stretched, it has the ability to return to its resting length. In other words, it can recoil.

Another characteristic is *excitability*. This means that muscle can respond to a stimulus or a trigger. For example, a muscle cell can be stimulated to contract by a signal from a nerve cell, or from a chemical messenger.

The last characteristic is *extensibility*. This means that muscle can be stretched. When relaxed, muscle can even be stretched beyond its resting length when circumstances dictate.

Muscle is quite amazing, isn't it!

---

### Movement

*So Simple Yet Designed by the Master So Complex*

Have you ever stopped to consider just how lucky we are? Our bodies have been designed to allow us to move around in our environment. How dull would it be if we were like trees or flowers and had to stay in the same place our whole life?

Our wonderful Creator has given us the privilege of moving freely in the world He made for us. We can walk, run, climb, or swim.

Not only have we been designed to move about in the world, we have the ability to interact with the world and the people around us. We can hold a pen to write a letter. We can move our lips to speak. We can use a hammer to build a new doghouse for our favorite pet. It goes on and on.

Our muscles make all these things possible. It is just one more reason that we should always stop and give praise to our Creator, the One who made our incredible bodies!

# Types of Muscle

In our discussion of tissue types, we saw that there are three types of muscle: skeletal muscle, cardiac muscle, and smooth muscle. Let's review these again.

Skeletal muscle is the kind of muscle that moves the body. These muscles attach to the skeleton, and their contraction causes the bones to move and propel the body. Skeletal muscles also make good posture possible. This type of muscle accounts for about 40 percent of our body mass.

Skeletal muscle is also known as *voluntary muscle*. This means that skeletal muscle contracts on your command. You can consciously control it. If you want to open a door, your brain sends a signal down nerve fibers to the muscles telling your upper limb to reach forward. Then your brain sends nerve signals to the muscles in your hand to grab the doorknob and twist it.

Also, skeletal muscle is *striated*, which means "striped." A distinctive pattern of stripes is visible under the microscope. We will see why later.

Smooth muscle is found in the walls of most hollow organs in the body. It is in the walls of the stomach, the urinary bladder, the blood vessels, and even respiratory passages. Smooth muscle is involuntary, so it contracts without your having to consciously think about it. We will not deal with smooth muscle here but will discuss smooth muscle in more detail when we explore other organ systems.

Cardiac muscle is the third type of muscle. As we have mentioned, this type of muscle is found only in the heart. It is also involuntary muscle, although cardiac muscle is striated and has some similarities to skeletal muscle. However, there are also distinct differences. We will examine these differences when we study the cardiovascular system.

*Photomicrograph of skeletal muscle*

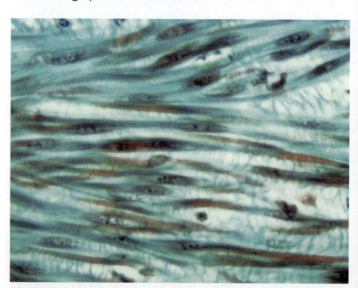
*Photomicrograph of smooth muscle*

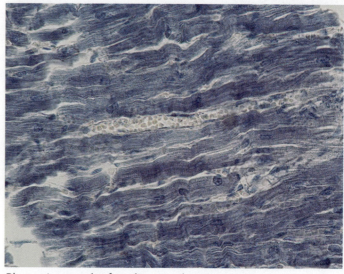
*Photomicrograph of cardiac muscle*

# Skeletal Muscle Structure

To better understand how skeletal muscles work, we need to see how muscle tissue is organized. Let's start with a cross section of a typical skeletal muscle so that we can see how muscle cells are bundled together and how those bundles are bundled together. Because muscle tissue is organized in bundles and bundles of bundles, its parts can work together very efficiently.

A muscle cell is long and thin, like a string, so another name for a muscle cell is a *muscle fiber*. A bundle of muscle fibers wrapped in a connective tissue sheath is called a fascicle. The fibers in a fascicle all contract at the same time. The muscle is composed of many fascicles.

Remember, a *muscle fiber* is the same thing as a *muscle cell*. A muscle fiber can be very long, often extending the entire length of the muscle. The longest muscle fibers are in the thigh muscles. We think of cells as very small things, but the muscle cells, or muscle fibers, in the thigh can be 12 inches long. Each muscle cell is surrounded by its own connective tissue sheath, and then each fascicle is bundled together by another connective tissue sheath.

All the connective tissue sheaths in a muscle are continuous with one another. In addition, they extend on each end of the muscle to the *tendons* that attach the muscle to the bones. (Remember, ligaments attach bones to bones, and tendons attach muscles to bones.) When the muscle fibers contract, this force is ultimately transmitted through the network of connected connective tissue sheaths to the tendons, and the bones are thus moved.

Muscles need a lot of fuel and oxygen to do their work. Therefore, muscles must have a very rich blood supply. Blood brings oxygen and sugar and other important materials to the muscle and removes the waste products generated by all the muscle activity. Without a good blood supply, muscle fibers wouldn't have the fuel and oxygen to efficiently produce the energy to move.

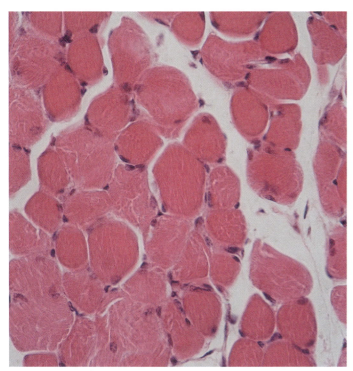

*Cross section of skeletal muscle*

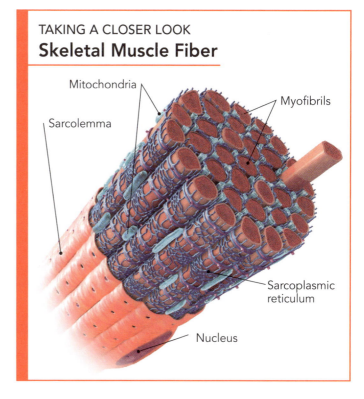

**TAKING A CLOSER LOOK**
### Skeletal Muscle Fiber

- Mitochondria
- Sarcolemma
- Myofibrils
- Sarcoplasmic reticulum
- Nucleus

# THE MUSCULOSKETETAL SYSTEM

## Organized to Move

As we noted, a skeletal muscle fiber is the same thing as a skeletal muscle cell. Each long cell has more than one nucleus and many mitochondria. Each also contains many special protein molecules called *actin* and *myosin*. Muscles cells are only able to contact because of the way actin and myosin molecules tug on each other. Let's zoom out to see how each muscle cell is organized to take advantage of the unusual properties of actin and myosin molecules.

Many of the words describing muscles and their parts contain the prefix "myo." Remember that we said words that begin with the prefix "osteo" refer to bone. Well, words that begin with the prefix "myo" refer to muscle. Thus you see that one of the important molecules that makes a muscle contract — myosin — begins with "myo."

*Myofilament* is another of those important "myo" words. *Myofilaments* can be made of myosin molecules or actin molecules, and there are a lot of myofilaments in every muscle fiber. *Thick myofilaments* are made of myosin. *Thin myofilaments* are made of actin. They are arranged like cordwood next to each other. When a muscle fiber is stimulated to contract, the thick and thin myofilaments slide past each other because of the way actin and myosin molecules interact.

Of course, just a few bundles of molecules tugging past each other wouldn't move your leg. It takes a lot of them and they must be organized just so in order to work together. The simplest *contractile unit* of a muscle is called the *sarcomere*. That means that each sarcomere contracts. Each sarcomere is made of thick and thin myofilaments. Sarcomeres attached end-to-end are called *myofibrils*. There's another "myo" word! Because each sarcomere contracts, and the sarcomeres are attached end-to-end, the myofibril contracts. Hundreds of myofibrils are packed into each muscle fiber, or cell. And of course each muscle contains many such cells.

Thus, we see that each muscle fiber is made of myofibrils. Myofibrils are made of sarcomeres hooked end to end. And each myofibril is made of thick and thin myofilaments. Those myofilaments are made of

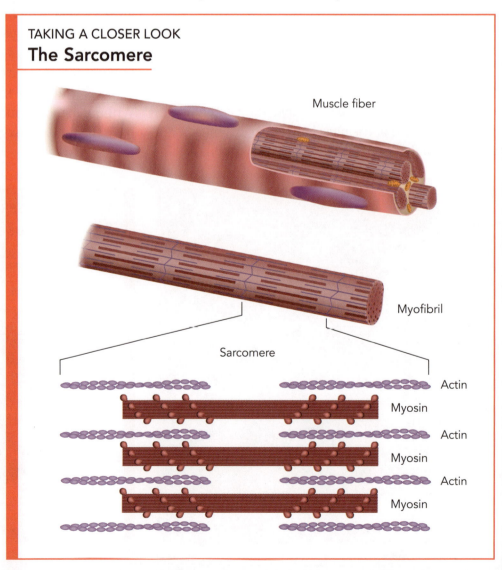

### TAKING A CLOSER LOOK
### The Sarcomere

actin and myosin. (We'll learn more about how those molecules work soon.) In the illustration, you can see how these parts are arranged.

Remember we said that muscle is striped, or *striated*? Skeletal muscle looks striped under the microscope because of the alternating pattern of thin/thick/thin filaments.

## Stimulated to Move

This organization of the myofibrils is vitally important to the function of muscle tissue. It is the interaction of the actin and myosin that produces the contraction of muscle. You see in the illustration that the thick myofilament, made of myosin, has small "heads" along its length. These heads can actually form links with the thin actin myofibrils and pull them in a fashion that shortens the sarcomere. When each sarcomere in a fiber contracts, the muscle fiber itself obviously contracts! And because each fiber and each bundle of fibers is wrapped in connective tissue that extends into the tendon attaching the muscle to the structure (like a bone) it is supposed to move, the contraction of those strings of sarcomeres is translated into a movement of the entire muscle. Simply, this is how a muscle works.

Well, as it turns out, it's not really that simple.

You remember that skeletal muscle is voluntary muscle. That is, it moves when you want it to.

So, let's back up to the beginning. You know, to the "I want to move my hand!" part, and take it step by step.

First of all, you decide, "I want to move my hand!" The part of your brain that controls muscle activity sends a signal, through the spinal cord, to the nerves going to the hand. (We will examine how this nerve impulse travels when we study the nervous system. For now, think of the message getting sent along a nerve like a line of dominoes falling down.)

When the nerve signal reaches a muscle fiber (a muscle cell), another signal is generated in the muscle fiber itself. This signal causes calcium to pour out of an organelle called the *sarcoplasmic reticulum*. Remember when we studied the parts of a cell? You may recall a structure called *smooth endoplasmic reticulum*. The sarcoplasmic reticulum is merely the specialized form of smooth endoplasmic reticulum found in muscle cells. It is designed to store large amounts of calcium.

### TAKING A CLOSER LOOK
### Actin and Myosin Interaction

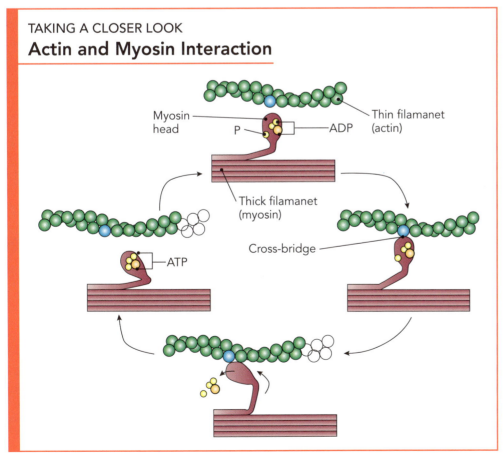

You see, in this sequence, calcium is the key. When calcium is released into the muscle fiber, it binds to specific sites on its actin myofilaments. The binding of calcium triggers the myofilament surface to change shape. When the thin myofilaments change shape, myosin molecules in the thick myofilaments can connect to the actin myofilament, forming what is called a *cross bridge*. When this attachment is made, the heads of the myosin then flex and pull the actin, causing contraction. This is called the *sliding filament model* of muscle contraction. (Well, I guess that wasn't so terribly complicated, was it?)

When a muscle fiber is stimulated to contract, the entire fiber contracts. That is, all the sarcomeres in a muscle fiber contract or none of them do. This is called the *all-or-none law*. There is no way for a single muscle fiber to be partially contracted.

One more thing — this entire process takes energy. Lots of energy. That is why muscle cells have so many mitochondria. Mitochondria are the powerhouses of cells. They convert fuel, like sugars, into usable energy, constantly recharging each cell's "batteries" so the fiber has the energy to contract. Lots of mitochondria are packed into each fiber to provide plenty of energy for active muscle.

As long as nerve stimulation is applied, calcium is released into the muscle fiber causing contraction. But muscles need to relax also, don't they? Obviously, they do. So how does this happen? Relaxation of the muscle occurs when the nerve stimulation to the muscle stops.

When a muscle fiber is no longer receiving nerve signals, there is no further stimulation of the sarcoplasmic reticulum to release calcium. In fact, without nerve stimulation, the sarcoplasmic reticulum actually pumps the calcium back inside itself. Then calcium is not available to trigger muscle contraction. So the muscle fibers in the muscle relax, and the muscle relaxes.

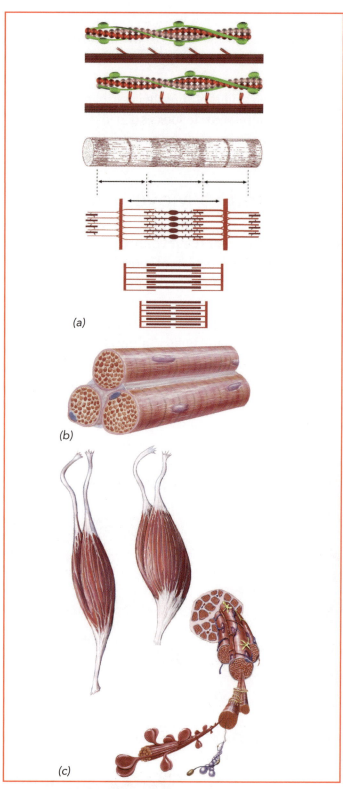

The muscle structure showing myofibrils and how they are bound together to form muscle fibers, then muscles. (a) Muscle cells contain many myofibrils, organelles that are composed of actin and myosin bound together end to end. (b) The myofibrils are secured at either end to the cell membrane. A group of cells are then bound together to form a muscle fiber. (c) Many fibers are in turn bound to form muscle tissue.

# Energy to Move

To move our body around, we need lots of energy. Just carrying the weight of your body around from day to day takes an enormous amount of energy. Even holding your body upright in a sitting or standing position instead of flopping over like a rag doll takes energy. How much more energy is needed to play sports or clean your room or ride your bike! We need to take a closer look at how muscles get the energy they require.

The first process by which muscles generate energy is called *aerobic respiration*. Aerobic respiration takes place through a series of chemical reactions that *require oxygen*. That is why it is called *aerobic*. (Do you see that "aer" is sort of like the word *air*, which is where we get our oxygen?) In aerobic respiration the energy stored in a sugar, called *glucose*, is used to generate a special energy storage molecule called adenosine triphosphate (ATP). This process of producing ATP takes place in the mitochondria of the cell. Since muscle cells require so much energy, is it any wonder that they each contain many mitochondria?

Glucose (in the presence of oxygen) is broken down into water and carbon dioxide. This process generates energy. When you build a campfire, you burn wood as a fuel, in the presence of oxygen, to get heat energy. But inside each cell, the energy produced from the fuel (glucose) must be captured and stored so it can be used. The energy produced by "burning" glucose would be useless without a way to store it. That is where ATP comes into the picture.

In aerobic respiration, the energy released from the breakdown of glucose is used to turn a molecule called adenosine diphosphate (ADP) into ATP. This is done by attaching an additional phosphate group (P) to ADP. Adding a phosphate group (P) converts adenosine diphosphate (ADP) to adenosine *triphosphate* (ATP). The key here is the additional phosphate. The chemical bone that attaches the phosphate contains a lot of stored energy.

Now when a muscle cell needs energy, it breaks down ATP into ADP. With the release of the phosphate group (P), energy stored in that chemical bond is released for the cell to use.

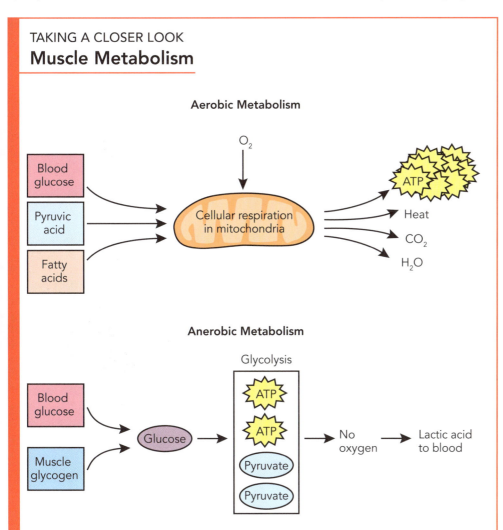

TAKING A CLOSER LOOK
## Muscle Metabolism

One way to understand this is to think of ATP as a charged battery in the cell. When the cell uses its stored energy, ATP becomes ADP, which is like an uncharged battery. When this uncharged battery (ADP) is recharged, it becomes ATP.

The problem is that this process of aerobic respiration does not happen instantly. It takes a little time. It does generate lots of ATP, but it takes time. So what if you need lots of ATP quickly? You then need *anaerobic respiration*.

Anaerobic respiration is the process where glucose is broken down *without oxygen being present*. This can take place very quickly, but it does not produce as much ATP as aerobic respiration does. Anaerobic provides a quick burst of energy to get movement started, providing the time needed to generate the large amounts of ATP needed for sustained activity.

In the process of breaking down glucose to generate ATP, anaerobic respiration does not produce water and carbon dioxide like aerobic respiration does. Instead, it produces something called *lactic acid*. Lactic acid diffuses out of the muscle cells and into the blood to be disposed of. However, if anaerobic respiration goes on too long, then a great deal of lactic acid is generated. In that case, the lactic acid can build up in the muscle cells, causing the muscle to become fatigued.

When you exercise regularly, your muscles can actually develop an improved network of capillaries to deliver oxygen-rich blood more efficiently. Then the oxygen you need for quick and prolonged muscle activity can reach the muscle fibers quickly and in large quantities. That way the muscle fibers can use aerobic respiration to generate a lot more energy from glucose. These capillaries also carry away the lactic acid quickly, preventing fatigue. Thus, when you are "in shape," your muscles produce more energy and don't tire as quickly.

## What Is a Muscle Cramp?

Have you ever gotten a muscle cramp while you were exercising? It really hurts, doesn't it? Exercise-related muscle cramps can happen when muscle fatigue causes the nervous system to *overstimulate* the muscle. Muscles are more likely to become fatigued if you exercise when you are dehydrated or your body is salt-depleted. Muscle fatigue can also result from a temporary build-up of *lactic acid* in muscles during fast sprinting or heavy exercise that requires the muscles to rely on *anaerobic metabolism*. Lack of regular exercise can also cause muscles to fatigue more easily and cramp. The best thing to do if you get a cramp is to *very gently* stretch the muscles. About half a minute of gentle stretching can reduce the excessive stimulation of the muscle and stop the cramp. Be gentle and don't stretch a cramping muscle too far or forcefully, however, or you may damage the muscle.

# Muscle Growth and Performance

Your body is designed to work and exercise. In fact, exercising and using your muscles is the best thing you can do to stay healthy. Lots of activity helps bones and muscles stay strong. It makes your heart and lungs work more efficiently. Most people even say that they think more clearly when they exercise regularly! Now I won't say that regular exercise will make you a straight A student, but it couldn't hurt, right?

Since in this section we are mainly concerned with muscle tissue, what happens to muscle when you exercise? How does muscle grow?

Well, when you exercise regularly, you put your muscles under stress. The muscles respond to this stress by becoming bigger and stronger or by increasing in endurance. How this occurs is interesting.

If you want to increase your endurance, you might begin running, biking, or swimming. Here, muscle is put under less stress, but for longer periods of time. For the most part, these types of exercises do not cause muscle to increase in size. They help increase muscle endurance by increasing the number of mitochondria in the muscle tissue. In addition, the number of capillaries increases. Capillaries bring in lots of oxygen and glucose and carry away lactic acid. In this way, the muscle is able to perform for longer before tiring.

## Muscle and Steroids

If you want to build stronger, bulkier, more powerful muscles, you should eat a balanced diet containing sufficient protein — because muscle is mostly made of protein — and you should exercise sensibly and regularly. You can adjust the amount of exercise and the kind of exercise to achieve the results you desire. But some athletes are so determined to build their muscles that they resort to using "anabolic steroids" to promote muscle growth and improve their performance. Use of these medications is *very dangerous* and has resulted in athletes being disqualified from the sports in which they worked so hard to excel.

*Anabolic steroids* are synthetic male hormones taken to enhance muscle growth and performance. To see why that would work, you need to know that hormones help regulate muscle growth. Have you ever wondered why men are able to build bulkier muscles than women? It's because men have more of the natural hormones (like testosterone) that promote muscle bulk.

Well, anabolic steroids are drugs that have effects similar to the natural male hormones. They cause the

muscle to make more of the protein molecules that fill muscle fibers. However, enhanced muscle mass and strength come at a terrible price. Anabolic steroids damage the liver, cause tumors, increase the risk of heart disease and high blood pressure, decrease the ability of the body to fight infection, trigger long bones to stop growing prematurely, cause psychological disturbances, and upset the proper balance of cholesterol. You should never use anabolic steroids. Build your muscles the safe and sensible way with a well-designed exercise program and good nutrition. Anabolic steroids are a shortcut and a cheat that can ruin your health, cut you out of the sports you wish to play, and even cut your life short.

If you were more interested in increasing the size and strength of muscle tissue, then you would want to begin resistance exercise, such as weight lifting. This type of exercise puts the muscle under more stress, but for shorter periods of time. The goal here is increasing strength rather than endurance.

Resistance exercise helps increase the size of muscle, but not by increasing the number of muscle fibers. Rather, the individual muscle fibers increase in size. The usual pattern here is short periods of intense exercise, focusing on individual muscles or sets of muscles. Then a break is taken, either by taking a day off or by engaging in a less strenuous workout. This allows the muscle to recover, providing time for the fibers to produce more of the proteins they need to contract.

## Muscle Tone

Have you ever heard the term *muscle tone*? This refers to the fact that there is some tension in a muscle even when it is not being actively contracted.

Well, if there is tension in a muscle, doesn't there have to be some contraction of muscle fibers? The answer is yes, of course. But wait a minute, it was noted earlier that a muscle fiber was either relaxed or contracted (remember the all-or-none law). So how could muscle tension be there without muscle movement? The answer is simple.

Muscle tone is generated by stimulation of some select muscle fibers, not the entire muscle. There are multiple nerve fibers connecting to muscle. Alternatively, different nerve fibers stimulate different muscle fibers to contract, but in this case, not the entire muscle itself. Thus, the all-or-none law is not violated!

This mild, intermittent stimulation is helpful in keeping muscle ready to respond promptly to move intense nerve impulses.

## Ever Pulled a Muscle?

Have you ever pulled a muscle? A "pulled muscle" is also called a "muscle strain." Muscle pulls often happen with sudden exertion or during athletic activity. A pulled muscle has some tears in the muscle or in the tendon that attaches it to the bone. The tears are usually small but are very uncomfortable due to the irritation of the nerve endings in the region and damage to small blood vessels with bruising inside the muscle. Swelling, bruising, weakness in the muscle, pain in the muscle and in the joint it moves during movement, and pain at rest are typical of a pulled muscle. An ice pack can help decrease the symptoms. (Be sure to avoid putting an ice pack directly on your skin. Twenty minutes of ice every hour is a good rule of thumb.) Elevating the muscle can also decrease swelling. So can an elastic bandage, but don't wrap it tightly. Rest the pulled muscle in its gently stretched position and give it time to heal before resuming the activity that injured it.

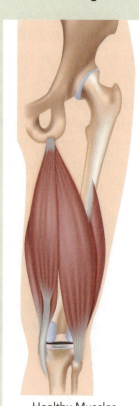
Healthy Muscles

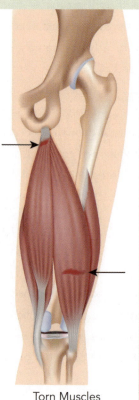
Torn Muscles

# Muscles Get Weak Too

It has been stressed over and over that to keep muscle healthy, it must be used and exercised. Unfortunately, in life there are many circumstances where muscles can go unused.

Have you ever known someone who broke an arm? The arm probably had to stay in a cast for several weeks. Do you remember what their arm looked like after the cast was taken off? The muscles were likely much smaller in the arm that was broken. Why? It's because those muscles had not been used regularly for those weeks. This is called *muscle atrophy*. The good thing is that the injured arm will soon return to normal as the arm begins to be used once again. Again, remember the phrase "use it or lose it!"

Another example is a person with a long illness that keeps them in bed for several weeks. They get progressively weaker because their muscles are not being used. Often, after an extended illness, patients will need weeks of special exercise therapy with good nutrition to help them get their strength back — to literally rebuild the protein molecules inside their muscle fibers. Remember, muscles like to be used!

## Rigor Mortis

Have you ever heard that the body becomes stiff after death? This is called *rigor mortis*. Rigor mortis occurs because the muscles all over the body contract and are unable to relax. Does it seem strange that muscles would contract after death?

Ordinarily, muscles contract because calcium flows out of the sarcoplasmic reticulum and triggers actin and myosin myofilaments to form cross-bridges. Do you remember why a muscle relaxes? Muscles relax when energy from ATP is used to pump calcium away from the myofilaments and back into the sarcoplasmic reticulum. Then the cross-bridges disengage.

After death, cells cannot make ATP and soon run out of energy. Calcium ions leak out of the sarcoplasmic reticulum and trigger cross-bridges to form in muscle cells all over the body. This begins about 3 hours after death. With no more ATP being made, the calcium never gets pumped away, and the contractions stiffen the entire body. The stiffness reaches a maximum around 12 hours after death. About 36–72 hours after death, the body relaxes again because actin and myosin molecules begin to disintegrate.

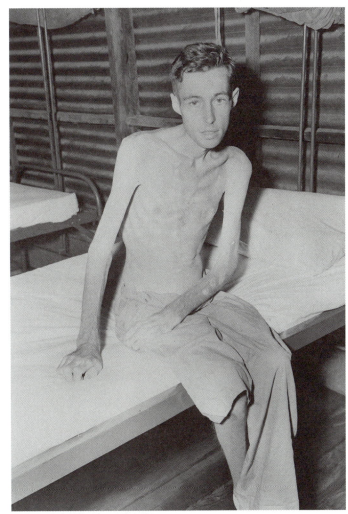

*This ex-prisoner of war, Corporal Noel Havenborg (9/21/1945 in Luzon, Philippines), has suffered severe muscle loss as a result of starvation. After depleting its stored fats, the body begins to use the proteins in its muscles for fuel. Muscles may atrophy as a result of malnutrition, physical inactivity, aging, or disease.*

# MUSCLES . . .
# HOW THEY MOVE ME!

It is time to take a look at a few of the major muscle groups in the body. Understandably, we will not be looking at every single muscle. After all, there are 640 skeletal muscles in your body, so trying to deal with all of them would be a bigger task than we can tackle today. We are just going to hit the high points.

So relax and enjoy our study of muscle groups.

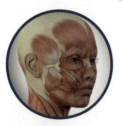
Skull

Hand

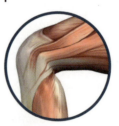
Knee

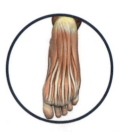
Foot

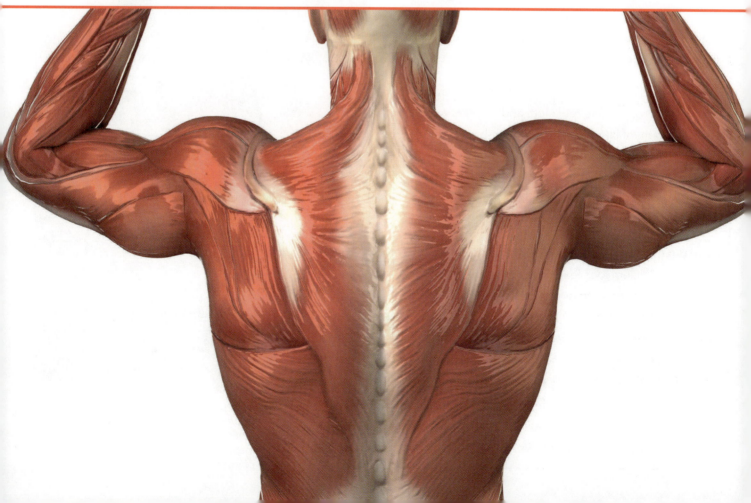

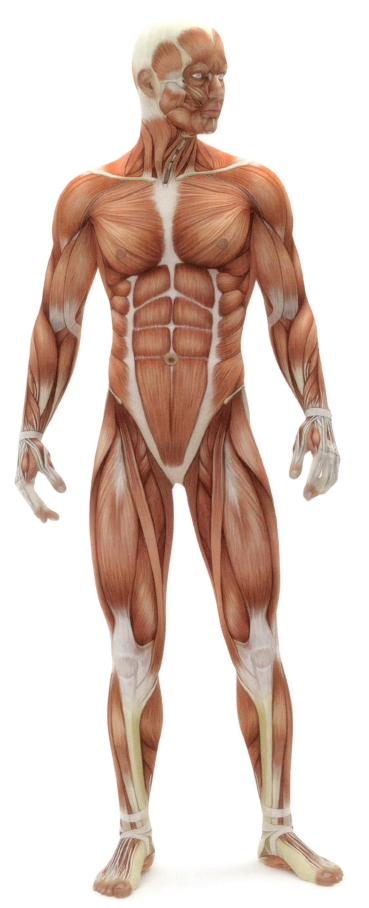

## Describing Movement

Muscles move things. It is pretty much that simple. In the typical circumstance, a muscle is attached to two different bones across a joint. The muscle is connected to the bone on each end by a tendon. When the muscle contracts, the bones are pulled toward one another.

Now what do you do if you want these two bones to be farther apart again? The muscle you just used can't do it, because *muscles only pull, they don't push.* To move these bones apart, you need another muscle to do the job. You need a muscle positioned to pull in the opposite direction. Two or more muscles "oppose" one another to move bones back and forth. Muscles that work opposite each other like this are often said to be *antagonistic*. This just means that the muscles in question have opposite purposes, not that they are hostile toward one another!

It is a simple process. One muscle contracts to pull a bone in one direction while another muscle relaxes to allow the movement. To move the bone back, the first muscle relaxes while the muscle on the opposite side contracts to pull the bone in the opposite direction.

## Whole Body View

Take a quick look at an overview of the major muscles of the body. I'm sure you will see many muscle names you already recognize. Perhaps you have heard of the pectoralis muscle or the gluteus maximus. Spend some time looking over the major muscle groups. Then we will explore some of these in more detail!

## Naming Muscles

*SO SIMPLE YET SO COMPLEX — Designed by the Master*

As we mentioned, there are over 600 muscles in the human body. At first glance, it may seem an impossible task to remember their names. But it is easier than you might think. Muscle names often *describe* the muscle in some way! Some muscles are named by their location. Some muscles are named for their shape. Some muscles are named to reflect their size. For example, the *intercostal* muscles are located between the ribs. (*Costal* means "rib," and *inter* means "between."). The *orbicularis oculi* muscle surrounds the eye. (*Oculi* means "eye," and *orbicularis* sounds a little like the word "orbit," doesn't it?) The *gluteus maximus* and the *gluteus minimus* are two of the muscles in the buttock. Can you guess which one is larger? See how easy this can be! The *trapezius* muscle in the upper back is shaped like a trapezoid. Clever, huh? Pay close attention and see if you can find other examples as we examine the major muscle groups. How muscles are named can tell you a lot about them!

## The Upper Limb

If you examine the shoulder, you will see most prominently a round muscle over the cap of the shoulder. This is called the deltoid muscle. The deltoid helps rotate the arm. It also *abducts* the arm. That is, it raises the arm out to our side, *away* from the body. (Remember, *abduction* is movement away from the center of the body, and *adduction* is movement *toward* the middle of the body. If you've ever studied Latin, you'll probably recognize that *ab* means "away from" and *ad* means "toward.")

There are four muscles that hold the humerus in the shoulder joint, keeping your arm in its socket. Their names are the *supraspinatus*, the *infraspinatus*, the *teres minor*, and the *subscapularis*. These muscles allow the shoulder to have an incredible range of motion. Together, these four muscles are called the *rotator cuff*.

Remember the scapula? The large oddly shaped bone in your shoulder? Notice that the scapula has a "spine" protruding from its back side. The *supra*-spinatus muscle attaches *above* this spine, and it is named accordingly. "Supra" means "above." Likewise, the *infra*-spinatus attaches "below" the spine. Because of the way they attach to the humerus, they pull in slightly different directions. The *supraspinatus* moves the humerus up and away from the body. The *infraspinatus* and the *teres minor* externally rotate the humerus. The *subscapularis* muscle, which is attached to a depression on the lower part of the scapula ("sub" means "under"),

### TAKING A CLOSER LOOK
### The Chest and Shoulders

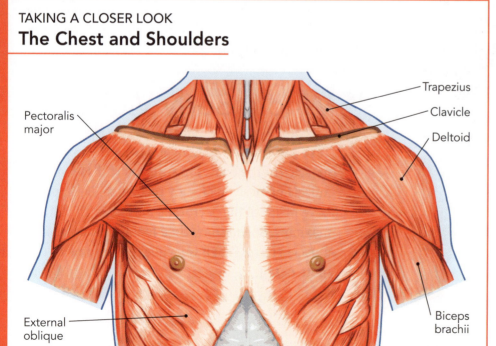

- Pectoralis major
- External oblique
- Trapezius
- Clavicle
- Deltoid
- Biceps brachii

## TAKING A CLOSER LOOK
### Rotator Cuff Muscles

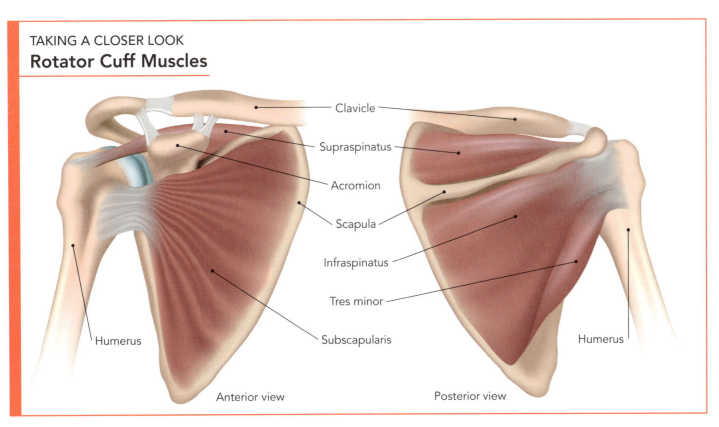

Anterior view — Clavicle, Supraspinatus, Acromion, Scapula, Infraspinatus, Tres minor, Subscapularis, Humerus

Posterior view — Humerus

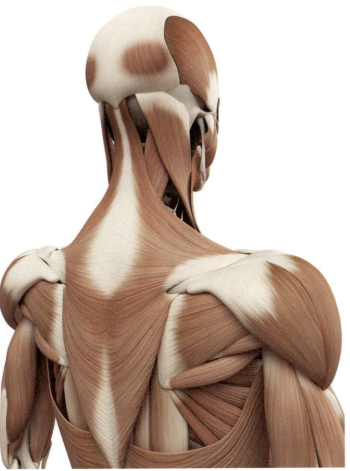

internally rotates the humerus. Thus, each muscle, because it is attached to a particular place on the scapula and on the humerus, tugs the arm from a different angle. When they work together, an infinite number of motions are possible.

The last muscle that we will mention here is called the *trapezius*. It is the triangular-shaped muscle in the upper back. It helps connect the pectoral girdle to the thorax. It helps move the scapula (shoulder blade). When you shrug your shoulders, you are using the trapezius.

Two of the best-known muscles in the body are found in the upper limb. These are the biceps, or as it is officially known, the *biceps brachii* and the *triceps brachii* (or triceps, for short). These are the muscles that flex and extend the forearm. Let's take a closer look at them.

We have already seen that muscles only pull. They do not push. So in order for full movement of the body to occur, there are muscles that work opposite one another. We called these muscles *antagonistic*. The biceps and triceps are the classic example of this arrangement.

To flex the forearm, the biceps contract (they pull the forearm, bending the elbow). However, for this movement to occur when the biceps contract, the triceps must relax. Otherwise, the forearm would not move.

To extend the forearm, the triceps contract and the biceps relax. You see, they work opposite one another.

You will see other examples of this. Can you think of others? How about your knee? One group of muscles bends the knee; another extends it so that you can kick forward. Look at your fingers. They can bend and straighten. Once again, antagonistic pairs of muscles make this happen.

The muscles of the forearm can be divided primarily into two groups, the anterior flexors and the posterior extensors. The anterior flexors flex the wrist and fingers. The posterior extensors extend the wrist and fingers.

Wait a minute! Did you say fingers? Yes, fingers. You see, even though these muscles are located in the forearm, they have long tendons to control the movement of the fingers. In this way our wonderful Creator allowed us to have very fine control of our hands while at the same time having great strength in the hands. This would not be possible if the major muscles were located in the hands themselves. There are small muscles located in the hand itself. However, the big muscles needed to supply great strength would take up so much room that our hands could not function properly.

### TAKING A CLOSER LOOK
## Bicep and Tricep Action

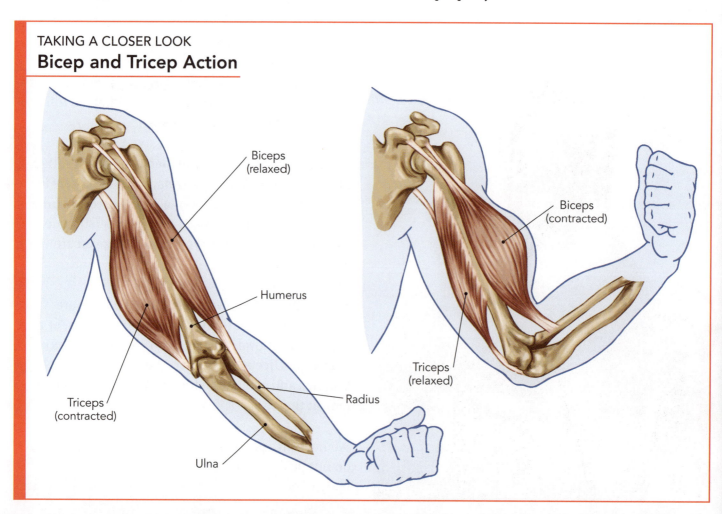

# MUSCLES...HOW THEY MOVE ME!

## The Strongest Muscle

*So Simple Yet So Complex — Designed by the Master*

What is the strongest muscle in the body? The right answer is: it depends. It depends on what you mean by "strong." If by "strongest" you mean "biggest," then the *gluteus maximus* muscle is the strongest. It is the largest muscle in the body. If by "strongest" you mean "works hardest," then the heart is the strongest. The heart does the most work throughout our lives. If by "strongest" you mean which muscle generates the most force compared to its size, then the masseter—the powerful chewing muscle that closes your jaw—is generally considered the strongest.

## The Chest and Abdomen

Let's learn about a few important muscles in the chest and abdomen. You may have even done exercises like sit-ups or weight lifting to make some of them stronger.

First is the *pectoralis major* muscle. Most likely you already know about this muscle. It is the primary muscle of the chest. This muscle helps flex the arm. People who work out with weights usually have very prominent pectoralis muscles!

The other important muscles in the chest are the *intercostals*. These are the muscles between the ribs. The intercostals help us breathe. We will learn more about the intercostals when we study the respiratory system.

In the abdomen, take note of the *rectus abdominis*. This long muscle on the front wall of the abdomen helps keep the pelvis stable as we walk. It also helps flex the trunk. People do sit-ups or crunches to make it stronger.

To each side of the rectus abdominis are three muscles: the *external oblique*, the *internal oblique*, and the *tranversus abdominis*. These muscle sheets are oriented in different directions. Together they support and protect the abdominal organs below them.

### TAKING A CLOSER LOOK
### The Chest and Abdomen

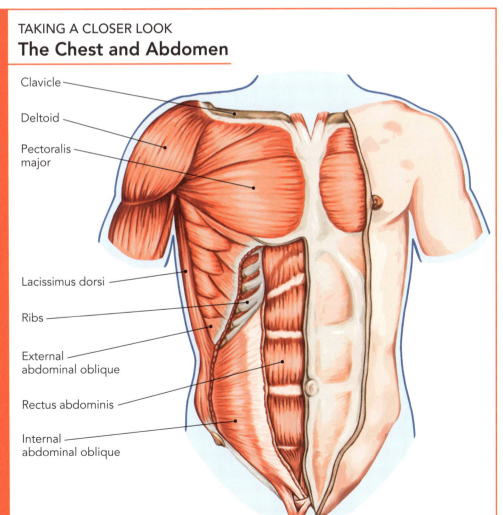

- Clavicle
- Deltoid
- Pectoralis major
- Lacissimus dorsi
- Ribs
- External abdominal oblique
- Rectus abdominis
- Internal abdominal oblique

# The Lower Limb

The muscles in our legs are strong and powerful. They have to be in order to carry us around all day long.

One of the primary muscles for you to know is *rectus femoris*. It is one of four muscles in a group known as the *quadriceps*. Your "quads" extend the knee and flex the hip.

On the rear of the thigh are the *hamstring* muscles. This is a group of three muscles, the most prominent of which is called the *biceps femoris*. The hamstrings flex the knee and extend the hip. The hamstrings are the antagonists of the quadriceps.

One other muscle to mention here is the *gluteus maximus*. It forms a large part of the buttock. But your gluteus maximus does much more than give you a place to sit. This muscle extends the thigh, moving it backward.

Below the knee we find the muscles of the leg. (Remember, the upper part of the lower limb is called the thigh, and the "leg" officially refers to the lower part of the lower limb!) As with the forearm, it is easiest to consider the muscles in groups. There is the *anterior compartment* and the *posterior compartment*. Since *anterior* means "in the front" and *posterior* means "in the back," you can probably guess where these muscle groups are located. That gives you a great clue to what they do.

The muscles in the *anterior compartment* pull on the toes and *dorsiflex* the foot. That is, they flex the ankle to move the foot and toes upward.

The muscles in the *posterior compartment* flex the toes and *plantar flex* the ankle. In other words, they

### TAKING A CLOSER LOOK
### The Lower Limb Muscles

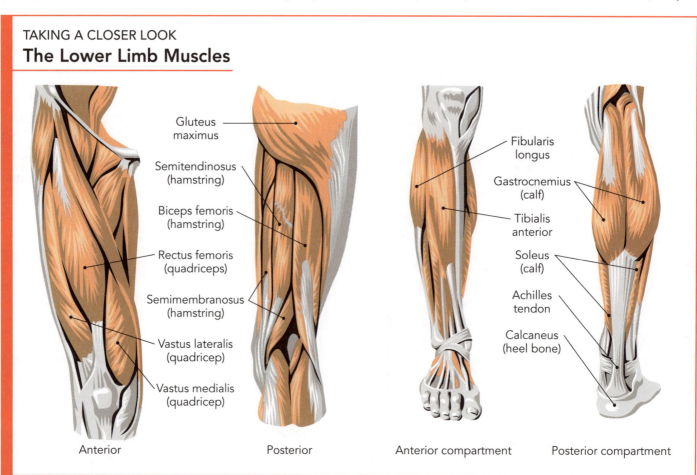

pull your foot downward, as if you were pressing on the gas pedal of a car, and they curl your toes. In the posterior compartment you will find the *gastrocnemius* muscle and the *soleus* muscle. These are called "calf" muscles and they attach to the *calcaneus* (heel bone) via the *Achilles tendon*. The Achilles tendon is the strongest tendon in the body, and it is named after a character in a Greek myth who supposedly was defeated in battle when that tendon was cut.

As with the hand, the powerful movements of the foot and toes are controlled primarily by muscles in the leg.

# The Head and Face

There is a wide array of muscles in the face. Many of these muscles anchor not to bone but to skin. Movement of these facial muscles is responsible for the amazing variety of facial expressions that human beings can make. Of course, muscles in the head and face also make chewing our food possible!

The *orbicularis oculi* muscle encircles each eye. This ring of muscle is an example of a *sphincter*. With this sphincter you can close your eyes. Surrounding the eyes and being embedded in the eyelids, it helps us blink, wink, and squint.

The main muscle across our forehead is the *occipitofrontalis*. This muscle elevates the eyebrows. If you've ever seen someone raising their eyebrows with a surprised look on their face, the occipitofrontalis was involved.

There are several muscles that move the lips and cheeks. These muscles help us eat, smile, and speak.

The *orbicularis oris* is the muscle that surrounds the mouth. This is another ring-shaped sphincter muscle. (Did you notice that the sphincters around the eyes and the mouth both have *orbicularis* in their names? Doesn't that sound like "orbit," or circling

## Monkeys and Muscles

There are actually three gluteal muscles on each side of our body. (They are the gluteus maximus — the biggest one — and the gluteus medius and the gluteus minimus.) Together, they extend your thigh in a way that makes it possible for you to walk efficiently.

Apes have gluteal muscles too. If a gorilla or a chimpanzee tries to walk upright on two legs, it can manage it for a while, but must swing its thigh far out to the side. This is not an efficient way for apes to walk. God designed apes to move through the trees, but He designed people to walk. Therefore, the shape of the pelvic bones (the iliac crest, specifically), in a human, curves to allow the gluteal muscles to pull the thigh in just the right direction for us to walk.

The gluteal medius muscle tugs our thigh sideways as it extends it, allowing us to keep our balance as we walk. In an ape, the iliac crest doesn't curve but sticks out to the side; this prevents the gluteal medius muscle from pulling the thigh sideways. Therefore, an ape must sway and swing its leg to the side to keep from falling over when it tries to walk like a man.

# THE MUSCULOSKETETAL SYSTEM

something?) Just as the orbicularis oculi closes your eyes, so the orbicularis oris helps you close your mouth. It helps control movement of the lips. This is the muscle you need to pucker up and kiss!

The *mentalis* muscle on the front of the lower jaw helps wrinkle the chin.

The *buccinator* compresses the cheek and helps keep food in place as we chew. It is involved with sucking movements, so nursing infants use the buccinators quite a bit!

Farther below you find a large, flat muscle that extends from the neck to the chin. This is the *platysma*. This muscle tenses the skin of the neck and pulls the corners of the mouth down. When you frown or grimace, you are using your platysma muscle.

## SO SIMPLE YET SO COMPLEX — Designed by the Master

### The Muscle with the Longest Name

You have over 40 facial muscles. Some help you chew, close your eyes, and do other critically important things. But some facial muscles mainly create facial expressions: smiles and frowns, wrinkled noses and pursed lips, wrinkled forehead and raised eyebrows. You even have a muscle to flare your nostril while lifting the corner of your upper lip! It is nicknamed the "Elvis Muscle" because the famous singer was often photographed with that expression. That muscle has the longest name in the medical dictionary: the *levator labii superioris alaeque nasi*, which is Latin for "lifter of the upper lip and the wing of the nose."

## TAKING A CLOSER LOOK
### Head and Facial Muscles

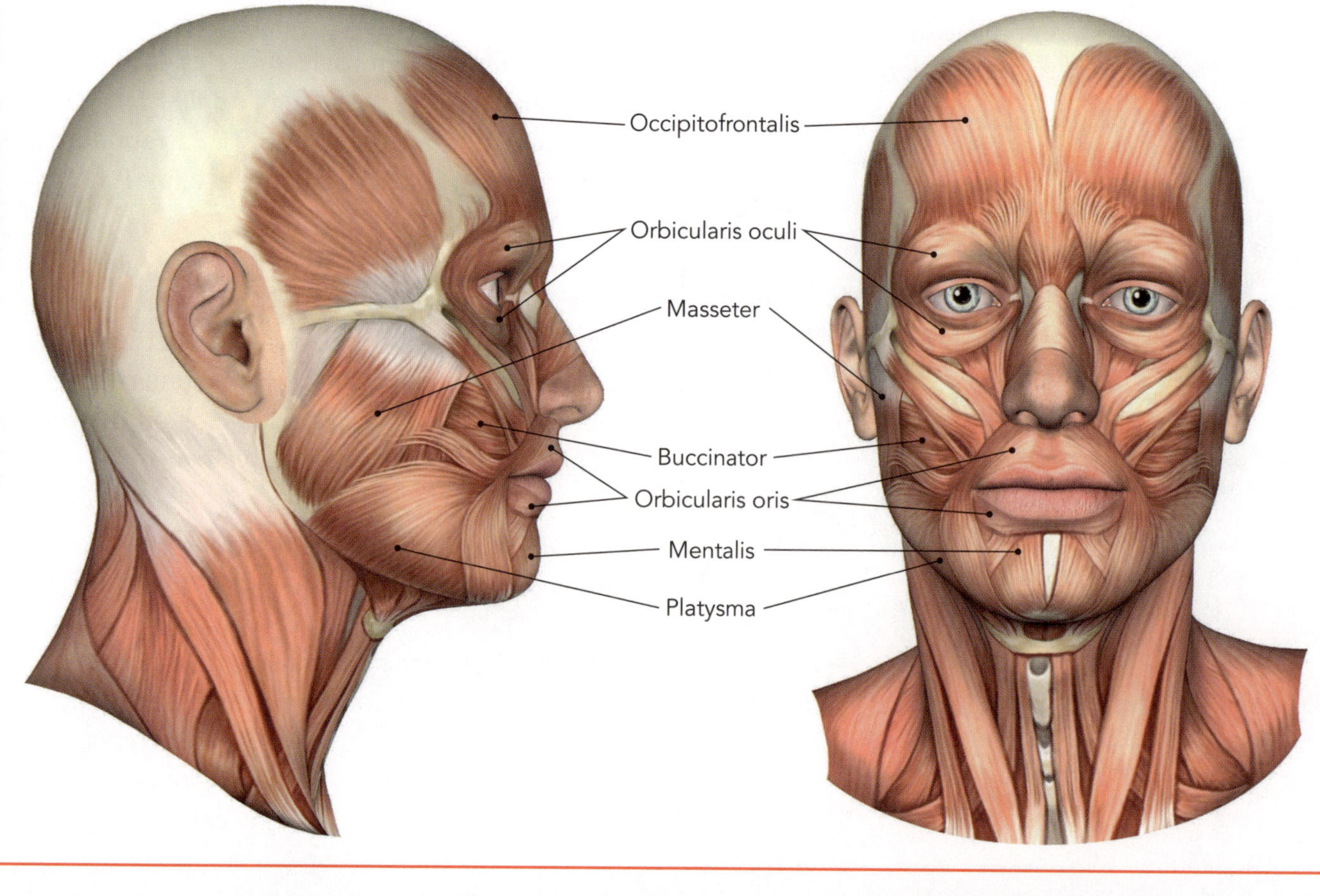

- Occipitofrontalis
- Orbicularis oculi
- Masseter
- Buccinator
- Orbicularis oris
- Mentalis
- Platysma

# 103
MUSCLES...HOW THEY MOVE ME!

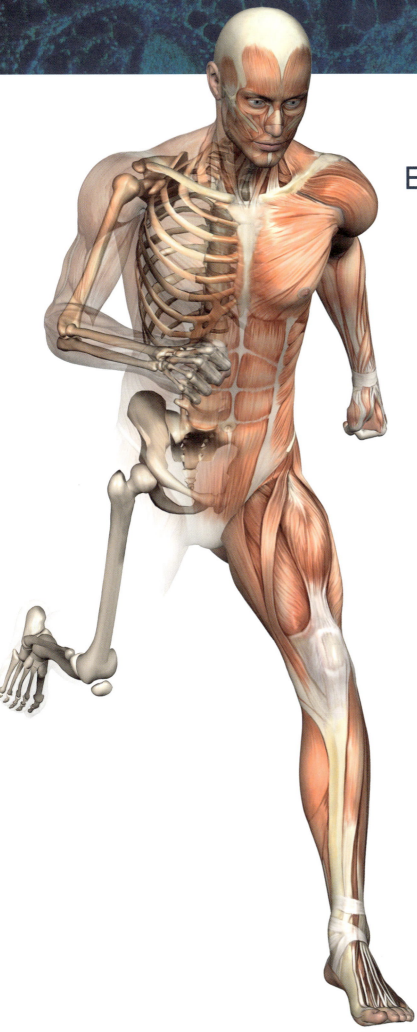

# End of Unit 1

# UNIT 2
## CARDIOVASCULAR & RESPIRATORY SYSTEMS

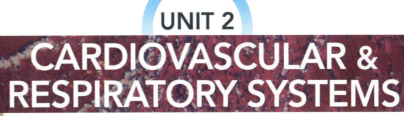

wonders of the
HUMAN BODY

# SECTION 1
# THE CARDIOVASCULAR SYSTEM

The heart must ceaselessly move blood around your body to keep you alive. It pushes that blood through a system of blood vessels. Those vessels branch out to carry blood all over your body, making oxygen, nutrients, water, and dissolved electrolytes available to every cell in your body. They also carry away waste materials for disposal or recycling. The heart, with all its associated vessels, is called the *cardiovascular system*. This name — cardiovascular — is one of those anatomy word puzzles: cardio- means "heart" and vascular means "vessels."

> For You formed my inward parts;
> You covered me in my mother's womb.
> I will praise You, for I am fearfully and wonderfully made;
> Marvelous are Your works,
> And that my soul knows very well.
> My frame was not hidden from You,
> When I was made in secret,
> And skillfully wrought in the lowest parts of the earth.
> Your eyes saw my substance, being yet unformed.
> And in Your book they all were written,
> The days fashioned for me,
> When as yet there were none of them.
>     (Psalm 139:13-16)

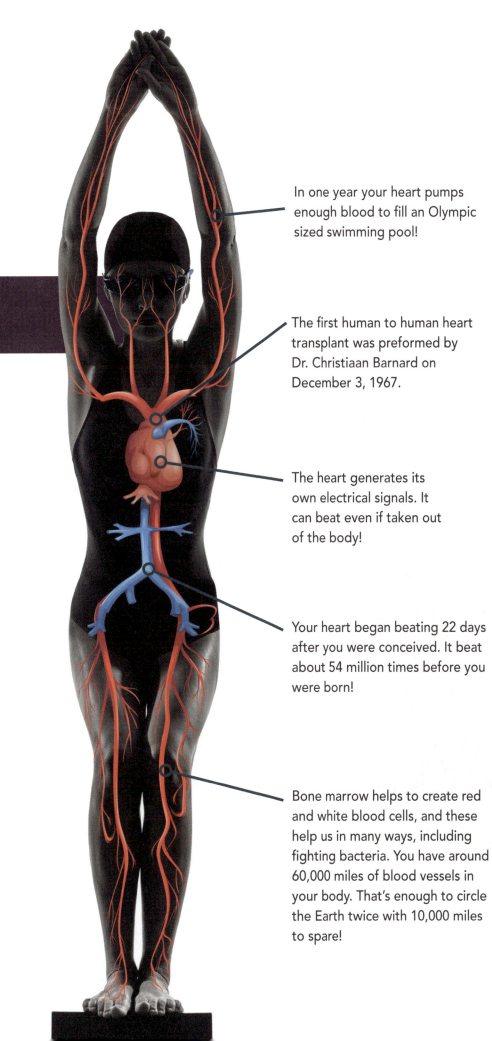

In one year your heart pumps enough blood to fill an Olympic sized swimming pool!

The first human to human heart transplant was preformed by Dr. Christiaan Barnard on December 3, 1967.

The heart generates its own electrical signals. It can beat even if taken out of the body!

Your heart began beating 22 days after you were conceived. It beat about 54 million times before you were born!

Bone marrow helps to create red and white blood cells, and these help us in many ways, including fighting bacteria. You have around 60,000 miles of blood vessels in your body. That's enough to circle the Earth twice with 10,000 miles to spare!

# INTRODUCTION

Have you heard your heart beat or felt your pulse? Have you ever blown up a balloon or had your milk "go down the wrong way"? Do you have any idea why people sneeze or cough?

Do you have a friend with asthma? Do you know what a heart attack is? Has someone in your family had heart surgery? Have you wondered how CPR works? Wouldn't it be great to know these things?

The purpose of this book is to explain how God's amazing designs enable your heart and lungs to move blood and oxygen around in your body for a lifetime. Once you understand how these systems work, you'll be able to understand many of the things that go wrong with them and the things you can do to keep yourself as healthy as possible.

 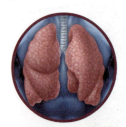 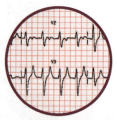 

The human body is a collection of organ systems which all work together to keep you going. Your heart, lungs, kidneys, stomach, and liver are examples of organs. An *organ system* is a group of organs working together to do an important job. Your *circulatory system* consists of all the parts of your body that move blood around. The heart and many blood vessels, large and small, make up your circulatory system.

Another system, the *respiratory system*, gets oxygen from the air; you need oxygen to live. The respiratory system also gets rid of the carbon dioxide your body makes. The respiratory system consists of the lungs and all the tubes (called airways) that air must travel through.

The circulatory (or cardiovascular) system and the respiratory system work together. The oxygen your lungs obtain from the air must be carried to all parts of your body, even into the tiniest places far from your lungs. How these systems work together so precisely is a testimony to our marvelous Creator, the One who designed our bodies with great care.

## How We'll Proceed

The body has many organ systems, but since all parts of the body work together, we'll mention other organ systems a lot. For example, your brain and nervous system help control your respiratory system. We'll talk a little about those systems whenever we need to right here in this volume.

When we learn about an organ system, we first will show you its parts and learn their correct names. Learning the names for things in science is like a puzzle: a lot of the names are built of little words and syllables which help you guess and remember the names of other things in science. We'll use lots of pictures and illustrations to show you anatomy — the way your parts are put together.

*Organs* are made of tissues, and tissues are made of cells. Sometimes we will show you pictures of what those tissues and cells look like under a microscope, amazing details too small to see with the naked eye. Those "photomicrographs" not only show you the anatomy but also help us to understand how the organs work.

Once you see the anatomy of an organ system and know its parts, you'll be able to understand how the system works. How the systems work is called physiology. When you finish this volume, you'll know where the organs are (anatomy), how they work (*physiology*), and what you can do to keep them healthy.

Often, learning about what happens when things don't work right helps us understand how organ systems work in the first place, so we'll discuss some diseases and how they affect the heart and lungs.

### In the Beginning ...

*SO SIMPLE YET SO COMPLEX — DESIGNED by the Master*

You may have heard that the incredible systems in your body evolved little by little over millions of years, but in fact, God created them perfect and complete in the first man and woman, Adam and Eve, about 6,000 years ago. Their hearts and lungs would have worked perfectly forever if they had not sinned, but disobeying God caused disease and death to enter a perfect world. When we learn about diseases, we are learning about the many things that have gone wrong in the world since Adam and Eve first sinned. In this book we'll talk a lot about the heart that moves your blood around, but in the Bible you can learn about another kind of heart — not the physical heart that beats in your chest, but the invisible heart that can believe in Jesus Christ. Look in the Book of Romans, chapter 10, verse 9. God wants you to pay attention to both kinds of heart.

# THE HEART

A normal heart is about the size of a person's fist. It is mostly made of **cardiac muscle**. There are two other kinds of muscle — skeletal muscle and smooth muscle. Muscles that enable you to walk or use your hands are examples of skeletal muscles. So is your diaphragm. Muscles that move your food through your digestive tract and the muscles that surround your arteries in order to allow them to influence your blood pressure are examples of smooth muscles. Cardiac muscle cells are designed to communicate efficiently with each other to pass along the electrical impulses that cause the heart to contract. Cardiac muscle cells are packed with **mitochondria**, tiny power-generators that keep the heart muscle continually supplied with energy. Incredibly, the heart only rests for about a fourth of a second during each "heartbeat." After all, the heart cannot afford to take a break!

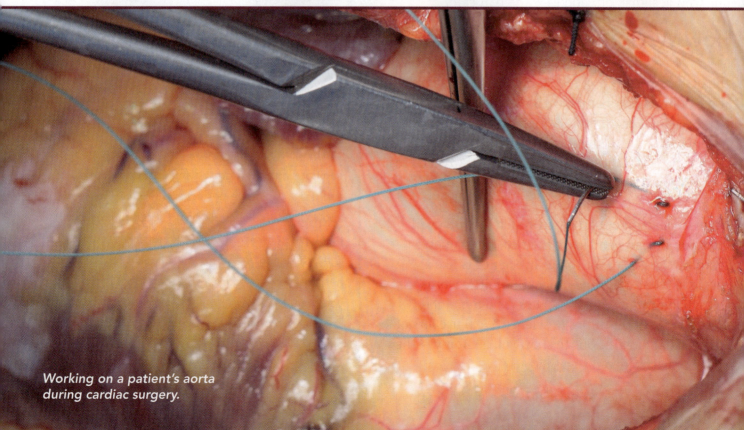

*Working on a patient's aorta during cardiac surgery.*

# THE HEART

The heart in an average adult pumps around 5 liters of blood every minute when resting. In a trained athlete, the heart can pump up to 33 liters per minute during vigorous exercise. On average, the heart moves 7,200 liters of blood per day. You've only got about 5 liters of blood altogether, so you can imagine that the blood circulates throughout the entire cardiovascular system many, many times in a day.

The heart "beats" on average around 72 times a minute when at rest. A young, healthy person's heart may beat up to 200 times a minute while exercising vigorously.

To keep up this steady pace, the many mitochondria in the muscle cells constantly use oxygen to convert glucose (a form of sugar) to energy. Therefore, those cells must be constantly supplied with oxygen. Without oxygen they cannot contract or even survive. If cardiac muscle cells are damaged by lack of oxygen, they have very little capacity to regenerate or replace themselves. Dead cardiac cells are replaced with scar tissue, but scar tissue cannot help pump. When people eat "heart healthy" foods and do "aerobic exercise," they are trying to keep their heart tissues in good shape to work well for a lifetime.

## The Heart, a Workhorse

To really understand how much work the heart does, let's do some calculations.

We will base our calculations on a person with an average heart rate of 72 beats per minute. At rest, the heart pumps roughly 70 mL (2.4 ounces) per beat. So . . . if the heart beats 72 times a minute, that means it beats 4,320 times in an hour, 103,700 times in a day, 37,843,000 times in a year. So, in a person who is 70 years old, for instance, the heart has already beat roughly 2,649,000,000 times. That is almost 3 billion heartbeats (yeah, that's billion, not million)!

*The average heart pumps 5 liters of blood a minute.*

Looking further, if the heart pumps 70 mL per beat, that means it pumps 5 liters a minute, 302 liters per hour, 7,257 liters (1917 gallons) per day, 2,649,000 liters (699,798 gallons) per year. So the heart of our 70-year-old would have pumped 185,431,680 liters (48,985,000 gallons)!

And your heart does all this without taking any time off. It works 24 hours a day, seven days a week. So you would think it wise to keep your heart healthy, right?

## Location of the Heart

Your heart is in the center of your chest, under your *sternum*, or breastbone. The heart is shaped sort of like an upside-down pyramid. It is pointed so that its apex is below the middle of your left collarbone. That is why when you put your hand over your heart to say a pledge, you place your hand a little to the left of the sternum, because this is where the "beats" of the heart can be easily felt.

Your thoracic cavity, or chest cavity, has three main compartments. The left and right are occupied by your lungs. Your heart is in the middle one — the *mediastinum*. (The word comes from the Latin word

# CARDIOVASCULAR & RESPIRATORY SYSTEMS

for "middle.") The heart isn't alone in this space. Also in the mediastinum are some important nerves, the large blood vessels (and lymphatic vessels) that enter and leave the heart, and the esophagus and trachea. The esophagus carries the food you swallow to your stomach. The trachea carries the air you breathe to your lungs. There is a lot of traffic in the mediastinum, and with the ever-beating heart the mediastinum is a busy place!

If we look at the mediastinum from front to back at the level of the heart, we'd see the sternum in front, then the heart. Behind the heart is the esophagus, but not the trachea. The trachea splits into the right and left bronchi before it reaches as low as the heart. Behind the esophagus is the descending aorta, and then the spine.

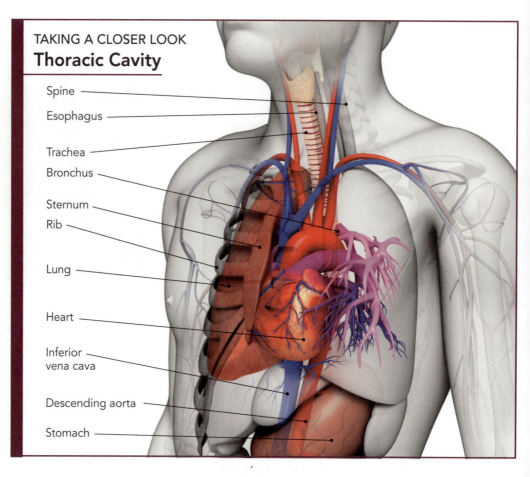

**TAKING A CLOSER LOOK**
**Thoracic Cavity**

- Spine
- Esophagus
- Trachea
- Bronchus
- Sternum
- Rib
- Lung
- Heart
- Inferior vena cava
- Descending aorta
- Stomach

Then, below the mediastinum is the diaphragm. The diaphragm is a large sheet of skeletal muscle that separates the chest cavity from the abdominal cavity.

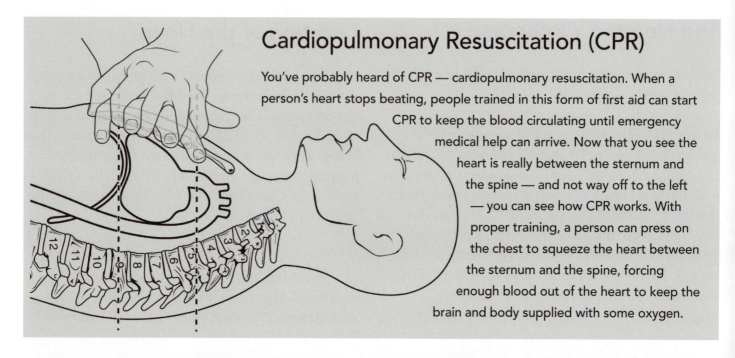

## Cardiopulmonary Resuscitation (CPR)

You've probably heard of CPR — cardiopulmonary resuscitation. When a person's heart stops beating, people trained in this form of first aid can start CPR to keep the blood circulating until emergency medical help can arrive. Now that you see the heart is really between the sternum and the spine — and not way off to the left — you can see how CPR works. With proper training, a person can press on the chest to squeeze the heart between the sternum and the spine, forcing enough blood out of the heart to keep the brain and body supplied with some oxygen.

# The Pericardium

As the heart pumps, it constantly rubs against the other structures in the mediastinum. You might think that would create a lot of friction. Friction would generate heat and lots of wear and tear on the outer surface of the heart. To prevent this, God designed the heart with its own lubrication system. (After all, blisters from friction like you get on your feet wouldn't do your heart any good!)

Like many other organs that we'll learn about, the heart grows inside a pushed in, double-layered, balloon-like sac during embryonic development. Imagine a slightly inflated balloon containing a tiny bit of lubricating fluid. Now imagine pushing your fist into the balloon so that two layers of rubber are against your fist. Try it yourself with a few drops of cooking oil inside a slightly inflated balloon. Is your hand inside the balloon? Not exactly. But when you wiggle your fist, the oiled rubber surfaces should slide smoothly against each other. The oil prevents friction.

Your heart is inside just such a sac, the *pericardium*. *Peri* means "around." This sac goes around the heart. The *pericardial sac* has an outer layer called the *fibrous pericardium* and an inner layer called the *serous pericardium*.

The fibrous pericardium is composed of tough, inelastic connective tissue. It serves to protect the heart, and to hold the heart in position in the chest.

The serous pericardium itself is made of two layers. The inner layer of the serous pericardium is called

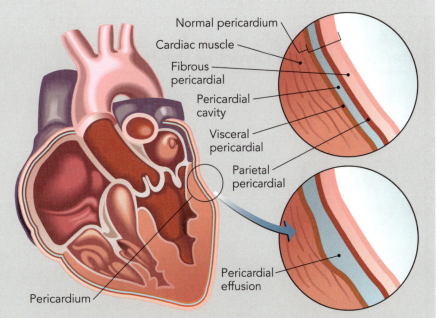

## Pericarditis

Occasionally, the pericardium can become inflamed. This condition is known as pericarditis.

It can occur suddenly, and it causes chest pain that is quite often severe. This pain sometimes radiates to the left shoulder and can be mistaken for a heart attack. The inflammation can be the result of a viral, bacterial, or fungal infection. Other causes include malignancy (cancer), heart attack, and trauma.

Some cases of pericarditis are quite mild and are treated with medication that controls inflammation. Other cases can be more aggressive and cause thickening of the pericardial sac, which can limit the movement of the heart. At times, the inflammation is severe enough that fluid begins to collect inside the pericardial sac. (This is called a **pericardial effusion**). Small amounts of fluid are easily tolerated and often resolve with treatment. However, in certain cases the amount of fluid that accumulates in the pericardial sac is enough to compress the heart and alter its ability to pump blood. This dangerous condition is a medical emergency known as **cardiac tamponade**. It is most often treated by inserting a needle into the pericardial sac and draining the fluid.

the visceral pericardium. The *visceral pericardium* is a thin layer stuck to the outer surface of the heart, just like the inner layer of balloon rubber was against your fist. The outer layer of the serous pericardium is called the *parietal pericardium*. The parietal pericardium is fused to the fibrous pericardium.

The visceral pericardium secretes a small amount of fluid, known as *pericardial fluid*, that provides lubrication between the visceral pericardium and the parietal pericardium. This fluid minimizes friction as the heart beats. You see, our Master Designer thought of everything!

If we peeled back the pericardium, we'd see the great vessels emerging from the upper part of the heart. The upper end of the heart is called the *base*, even though it is on the top, because it forms the broader part of the pyramid-like heart's shape. (The *apex* is the pointy bottom end.) Peeling back the pericardium would also reveal the coronary arteries and the cardiac veins running across the surface of the heart and sending their smaller branches down into the muscle of the heart.

## The Layers of the Heart

The wall of the heart consists of three layers: the *epicardium*, the *myocardium*, and the *endocardium*. Now you can see how thinking of anatomical names as word puzzles can help you! *Peri*, as in "pericardium," means "around," and the pericardium surrounds the heart. *Epi* means "outer," *myo* means "muscle," and *endo* means "inner." And of course *cardium* means "heart"! Therefore, these words are names for the layers of the heart itself.

Remember, we said that the pericardium consists of the outer parietal pericardium and the inner visceral pericardium, which is plastered to the surface of the heart. The outermost layer of the heart is actually the visceral layer of the pericardium. Where this membrane contacts the heart it is called the *epicardium*. It is made mostly of connective tissue and provides a protective covering for the surface of the heart.

The middle layer forms the bulk of the heart and is called the myocardium. As you might expect, knowing that *myo* means "muscle," this layer

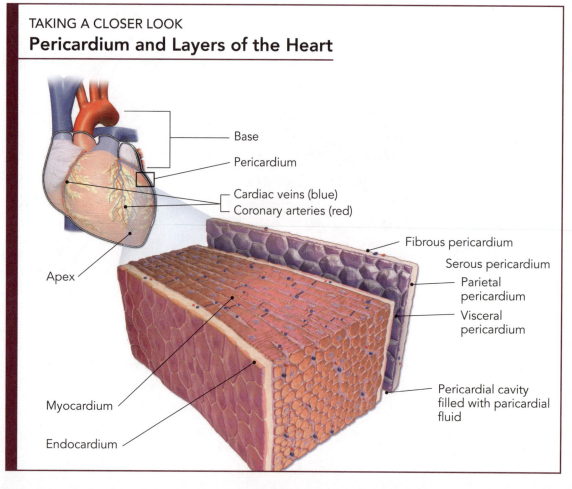

TAKING A CLOSER LOOK
**Pericardium and Layers of the Heart**

is primarily cardiac muscle. The myocardium makes up about 95 percent of the mass of the heart. This is the layer that is responsible for the contraction of the heart. There is also some connective tissue in the myocardium. This connective tissue helps hold the cardiac muscle fibers in proper orientation so they can work together to make the heart contract properly.

The innermost layer of the heart wall is a smooth, thin lining called the endocardium. The *endocardium* lines the heart chambers and covers the valves of the heart. It also extends into the blood vessels attached to the heart. Because it is very smooth, the endocardium minimizes friction as blood passes through the heart. Healthy endocardium keeps blood from clotting as it moves through the heart.

## Cardiac Muscle

Let's take some time to examine the myocardium in more detail.

You have learned that there are three types of muscle: skeletal muscle, smooth muscle, and cardiac muscle. The myocardium is mainly composed of cardiac muscle. As we will see, cardiac muscle is both similar to and different from skeletal muscle.

Like skeletal muscle, cardiac muscle is striated. However, the striations are not as easily seen in cardiac muscle. Cardiac muscle cells are shorter and fatter than skeletal muscle cells. Also, cardiac muscle cells branch and connect with one another in a somewhat irregular pattern. Like all cells, cardiac muscle cells are surrounded by a plasma membrane (also called a cell membrane). At the end of cardiac

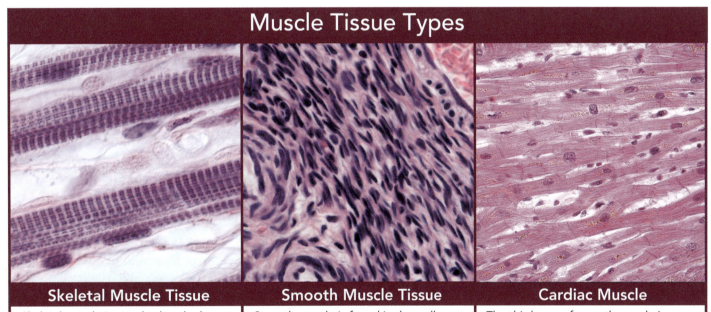

### Muscle Tissue Types

**Skeletal Muscle Tissue**
Skeletal muscle is attached to the bones of the skeleton. When it contracts, it allows us to move our arms and legs, or grasp something with our hands, or smile when we're happy. It has a structure that is distinct from other types of muscle.

**Smooth Muscle Tissue**
Smooth muscle is found in the walls of most of the hollow organs of the body. For example, it is found in the walls of our digestive tract where it helps push our food as it is digested. Smooth muscle is found in blood vessels, the urinary tract, the respiratory tract, the prostate, among other places. Smooth muscle is not under our direct control, and is sometimes referred to as involuntary muscle.

**Cardiac Muscle**
The third type of smooth muscle is cardiac muscle. It is found only in the walls of the heart. This type of muscle is also an involuntary muscle.

muscle cells are thick areas of the surrounding plasma membrane called *intercalated discs*. These intercalated discs form a special interlocking connection between the cells. Each intercalated disc contains two special structures that are very important to the proper function of cardiac muscle. One of these is called a *desmosome*, which helps hold the muscle fibers together as they contract. Also found in the intercalated disc are *gap junctions*. The junctions provide a route for electrical signals to be transmitted from muscle cell to muscle cell. These gap junctions ensure efficient transmission of electrical signals, which allows the cardiac muscle to contract in a coordinated fashion.

Cardiac muscle also differs from skeletal muscle in the number of mitochondria it contains. Mitochondria generate energy for the cell, and even though skeletal muscles need energy, they don't need nearly as much as the heart's muscle. Mitochondria make up about 25 percent of the volume of a cardiac muscle cell. In contrast, mitochondria account for only about 2 percent of the volume of a typical skeletal muscle cell. This, of course, makes perfect sense when you think about it, right? A large part of the time a skeletal muscle is at rest so its energy needs would be low. On the other hand, cardiac muscle is constantly active, constantly beating. The much greater number of mitochondria would give the cardiac muscle the energy production necessary to support this high level of activity.

Skeletal muscle responds to the voluntary control of your nervous system. Your conscious command can make skeletal muscle contract. On the other hand, cardiac muscle is involuntary. It does not require conscious command to contract. It is not under your conscious control. This is really the only way the heart could work. None of us would live very long if we had to think about every heartbeat!

## Two Pumps in One

We said the heart is a pump, but really, it is two pumps. The heart is two pumps operating side by side, simultaneously. The right side of the heart pumps blood to the lungs. The left side of the heart pumps blood to the brain and the body. One heart, two pumps.

> **TAKING A CLOSER LOOK**
> **Cardiac Muscle**

The heart's two pumps must be perfectly synchronized. Deoxygenated blood has given up most of its oxygen supply to the body's tissues. This deoxygenated blood returns to the right side of the heart and gets pumped out to the lungs. There it will be resupplied with oxygen. At exactly the same time, oxygenated (oxygen-rich) blood returns to the left side of the heart from the lungs and gets pumped out to the brain and body. If there is even the slightest mismatch between the two sides, problems can develop quickly. A healthy heart is perfectly balanced and keeps blood moving in a coordinated fashion, shuttling it first through the right-side pump, then to the lungs, and then through the left-side pump.

Since the pump on the right circulates blood to the lungs, the right-sided circulation is called the *pulmonary circulation*. *Pulmonary* means "lung." The pump on the left sends blood to all the body's other *systems*, so the left-sided circulation is called the *systemic circulation*.

We will learn the names for the large blood vessels entering and leaving the heart, but we'll first need to learn the difference between an artery and a vein. An *artery* is the name given to a blood vessel in which blood moves *away* from the heart. When blood leaves the heart to go to the lungs, it travels in arteries. And when blood leaves the heart to go to the body and brain, it also travels in arteries. Of course, the blood going to the lungs is deoxygenated, and the blood going to the body is oxygenated. So the blood in arteries can be carrying lots of oxygen or very little.

Vessels carrying blood *toward* the heart are called *veins*. Now you know that both oxygenated and deoxygenated blood can be carried in arteries. What about veins? The same is true. Some large veins (called *vena cavae* — a word that means big "cavernous" veins) carry deoxygenated blood back to the right side of the heart. And some other large veins (*pulmonary veins*) carry freshly oxygenated blood from the lungs to the left side of the heart. So, as with the arteries, veins can be carrying blood rich in oxygen or blood with very little.

Confusing, right? Well, we will try and give you a hand.

You may have seen drawings of the circulatory system and noticed that some of the blood vessels are colored red and some blue. Artists often draw the blood vessels this way to show you which vessels carry oxygenated blood and which vessels carry deoxygenated blood. Oxygenated blood has recently passed through the lungs to pick up a full load of oxygen using the hemoglobin in its red blood cells. Deoxygenated blood has already dropped off most of its oxygen supply in the tissues and is ready to be sent back to the lungs to pick up some more. All blood is red, but oxygenated blood is a brighter red and deoxygenated blood has a more purplish-red color. Even though deoxygenated blood is not really

### TAKING A CLOSER LOOK
### Pulmonary vs Systemic Circulation

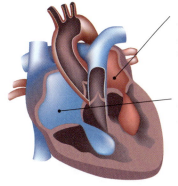

Systemic circulation - The left side pump fills with oxygen-filled blood from the lungs.

Pulmonary circulation - The right side pump fills with oxygen-depleted blood from the body.

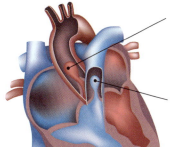

Systemic circulation - The left side pump pushes the oxygen-filled blood to the body.

Pulmonary circulation - The right side pump pushes the oxygen-depleted blood to the lungs.

blue, the blood vessels carrying it are most often illustrated as blue to help people see the difference more clearly.

## Chambers of the Heart

The human heart has four chambers.

Two chambers belong to the pump on the right — the right atrium and the right ventricle. These chambers are responsible for circulating blood to the lungs. Again, this is known as the pulmonary circulation.

The other two chambers belong to the pump on the left — the left atrium and the left ventricle. These chambers work to push blood out to the body tissues to supply them with oxygen and nutrients. This is the systemic circulation.

The word *atrium* means "entry room" or "receiving room." The *atria* (plural of atrium) collect blood as it returns to the heart. Blood that has already dropped off most of its oxygen supply enters the right atrium. (This is *deoxygenated* blood.) The left atrium collects oxygen-rich blood returning from the lungs.

Do arteries or veins bring this blood to the heart's atria? Hopefully, you said, "veins." Remember, *veins* bring blood *to* the heart. The veins that bring blood from the lungs to the left atrium are called *pulmonary veins* because they *come from the lungs*. The veins that bring blood back from the brain and the body are called *vena cavae*. The big vein from the upper body and brain is called the *superior vena cava*, and the big vein from the lower body is called the *inferior vena cava*. The name *vena cava* means "hollow vein," and *cavae* is the plural of *cava*. The words *superior* and *inferior* mean "upper" and "lower," respectively.

What kind of blood would you find in the superior and inferior vena cavae?[1] How about the pulmonary veins?[2] See, it's not really all that hard, is it?

The right and left atria collect blood and then send it on to the ventricles. As the atria fill, the pressure within the atria rises as a result of the increasing amount of blood. Then, when the ventricles relax, this pressure starts pushing blood from the atria

---

[1] Deoxygenated blood returns to the heart via the superior and inferior vena cavae.
[2] Oxygenated blood returns to the heart from the lungs through the right and left pulmonary veins.

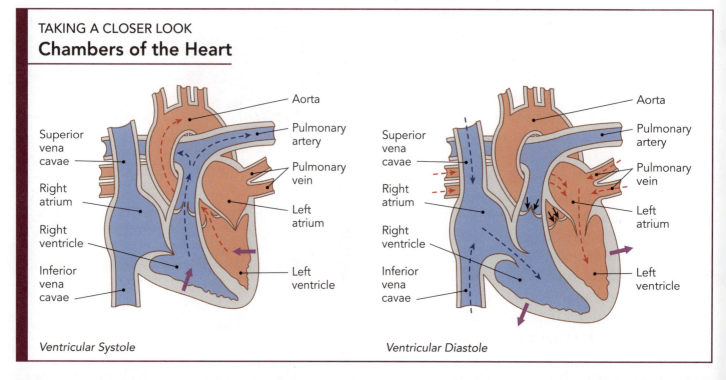

### TAKING A CLOSER LOOK
### Chambers of the Heart

Ventricular Systole · Ventricular Diastole

into the ventricles through the valves connecting them even before the atria contract. Just before the ventricles pump, the atria squeeze to push an extra bit of blood into the ventricles. After the atria empty, it's time for the ventricles to squeeze hard and push blood out to the lungs and body.

The right ventricle is part of the pump on the right, and it pushes oxygen-poor (deoxygenated) blood out through the pulmonary artery to the lungs. The left ventricle is part of the pump on the left, and it pushes blood out through a large artery called the *aorta*. This oxygen-rich (oxygenated) blood is sent through the aorta's branches to the brain and to the entire body.

The walls of the ventricles are made of thicker muscle than the atrial walls, but the ventricles are not the same. Remember, the right and left sides must always have the volume of blood they pump in and out perfectly matched. Even though this balance must be maintained, the two ventricles are different from one another. You see, the right ventricle only has to pump blood to the lungs, a short distance away. And it doesn't take much pressure to push blood through the pulmonary circulation. In contrast, the left ventricle pumps blood out to the entire body. It must push blood through the miles and miles of blood vessels that make up the systemic circulation. The pressure in the systemic circulation is much higher than in the pulmonary circulation. Therefore, the muscle of the left ventricle is much thicker than that of the right ventricle. In fact, the muscular wall of the left ventricle is typically two to three times thicker. This thick muscle allows the left ventricle to generate the great force needed to force blood through the entire body.

## Pattern of Blood Flow

Now that you've learned about the four chambers of the heart and the major vessels entering and leaving the heart, you should be able to trace the path of blood as it travels through this marvelous

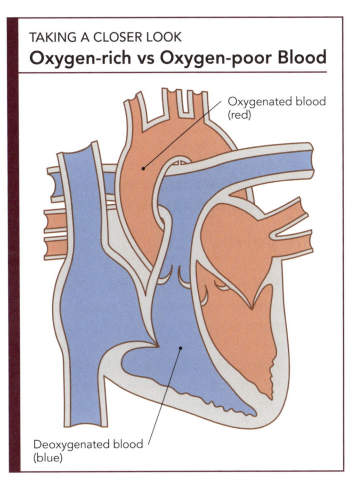

TAKING A CLOSER LOOK
### Oxygen-rich vs Oxygen-poor Blood

Oxygenated blood (red)

Deoxygenated blood (blue)

double-pump. Oxygen-poor blood enters the right atrium from the superior and inferior vena cavae. At the same time, oxygen-rich blood is brought by the pulmonary veins to the left atrium. (There are four pulmonary veins, two from the left lung and two from the right lung.) Blood flows from the right atrium into the right ventricle. At the same time, blood flows from the left atrium into the left ventricle.

After each atrium contracts, pushing that last little bit of blood into the ventricles, the ventricles give a mighty squeeze. Oxygen-poor blood from the right ventricle goes out through the pulmonary artery. The pulmonary artery soon branches to the right and left, and each of these subdivides and branches many times to carry blood to the lungs. At the same time, the left ventricle pushes oxygen-rich blood out of the heart through the aorta. The aorta goes upward, sends off some branches, and then arches downward where it continues as the descending aorta to carry blood to the lower body.

Be sure you understand that the right and left pumps fill and then contract simultaneously. Then see if you can trace the path of a red blood cell as it enters the heart, travels to the lungs, returns to the heart, and is sent out through the aorta. Then see if you can do it without looking at the illustrations. If you don't get it right away, relax. It will be easy for you in no time.

## Heart Valves

You know that most of the rooms in your home have doors. It is obvious why those doors are there. But are there rooms that don't have doors? Those rooms were designed for a reason. The rooms that have no doors allow access in and out much more easily, right? On the other hand, you've probably seen businesses that have one-way doors — separate doors for going in and for going out.

Which design do you think would work best for the heart's "rooms," its chambers? What would happen to the blood in the ventricles when the ventricles squeezed if the heart's rooms had no doors? If you said some blood would go backward into the atria, you see the problem. The ventricles would waste much of their effort if part of the blood went back-

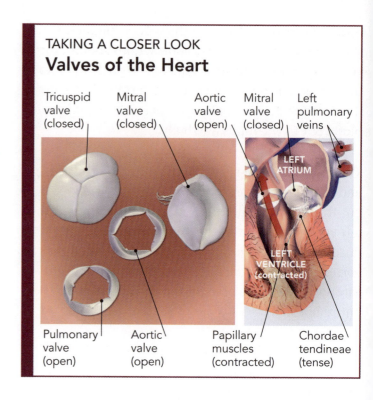

### TAKING A CLOSER LOOK
### Valves of the Heart

ward. To keep this from happening, the chambers are separated by one-way valves. A valve must allow the blood to flow freely in one direction but then shut to stop any back-flow.

Blood passes from the right atrium into the right ventricle through the *tricuspid valve*. Blood passes from the left atrium into the left ventricle though the *bicuspid valve*, also known as the *mitral valve*. Notice that both of these valves have "cusp" in the

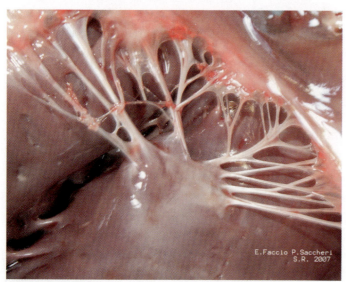

Chordae tendineae

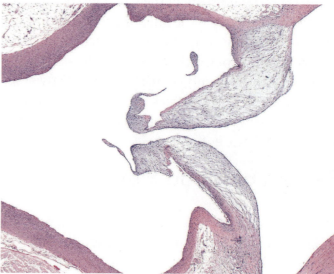

Aortic valve

name. A *cusp* is like a little parachute that fills with blood from the ventricle under pressure, distending the cusp back toward the atrium as the ventricle squeezes. The cusps keep the blood from flowing back into the atria. The tricuspid valve consists of three ("tri") cusps, and the bicuspid (mitral) valve has two ("bi") cusps. The name *mitral* is used for the bicuspid valve because the two cusps look a little like a bishop's headdress, called a miter.

If these cusps were not secured to the walls of the ventricles, the high-pressure blood filling them would push back into the atria. The cusps are therefore tethered to the ventricular walls. The ties that bind these cusps to the ventricular wall are called *chordae tendineae*. This Latin name means "heart strings." As the high-pressure blood distends the cusps, it is kept from being pushed back into the atria by these little tethers.

## Heart Strings

Already you can probably see the great design in this arrangement. But there could be a problem: when the ventricles contract, they shrink. And as they shrink, the chordae tendineae (heart strings) tethering the cusps must somehow get shorter. Otherwise, the cusps would push back into the atria! God designed an amazing feature to keep the chordae tendineae tight as the ventricles shrink. These little cords are attached to the ventricular walls by tiny papillary muscles. As the ventricles contract, the papillary muscles also contract, being perfectly coordinated with the ventricles. These muscles keep the chordae tendinae taut and stabilize the cusps of the valves. (No way this is just a cosmic accident, right?).

The heart's valves do not require a doorman to close them. The pressure of the blood inside the ventricles pushes them shut. We could even say the pressure makes them slam shut. But they make no noise. You've probably heard that the heart makes a "lub-dub" sound with each beat. The "lub" sound comes from the closure of the tricuspid and mitral valves, but it isn't the "slamming shut" that makes the "lub." It isn't even the silent squeezing of the ventricles that makes the "lub" sound. The "lub" comes from the turbulence of the blood rushing against the valves. (Think of the sound a wave makes as it crashes into a beach. Moving liquids, whether water

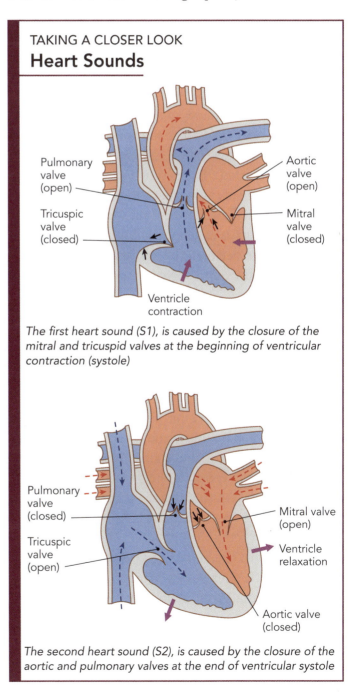

### TAKING A CLOSER LOOK
### Heart Sounds

*The first heart sound (S1), is caused by the closure of the mitral and tricuspid valves at the beginning of ventricular contraction (systole)*

*The second heart sound (S2), is caused by the closure of the aortic and pulmonary valves at the end of ventricular systole*

or blood, are powerful!) Of course, since the "lub" happens when the tricuspid and mitral valves close, it may be easier for you to think of the "lub" as the result of the doors slamming shut.

When the blood leaves the heart through the pulmonary artery and the aorta, another set of valves is needed to keep it from flowing backward into the ventricles. If any blood flowed backward, the ventricles would have to do extra work by pushing it out again with the next beat. Such an arrangement would not be very efficient! (In fact, this very problem happens when valves are damaged, as we will discuss later.)

These valves — the valves guarding the exit from the ventricles — are called *semilunar valves*. As you know already, *lunar* means "moon," so *semilunar* means "half-moon-shaped." Each "ventricular exit" valve consists of three of these crescent-shaped cusps. The semilunar valve between the right ventricle and the pulmonary artery is called the *pulmonary valve*. The semilunar valve between the left ventricle and the aorta is called the *aortic valve*.

The semilunar valves do not have any chordae tendineae. The pressure in the pulmonary artery and the aorta is not high enough to force them backward into the ventricles, so none are needed.

Just as the tricuspid and mitral valves needed no doorkeeper, the pulmonary and aortic valves need no doorkeeper to open or shut them. Fluid pressure does the job. When the ventricles begin to contract, the pressure they generate slams the tricuspid and mitral valves shut. The pressure in the ventricles then quickly rises, forcing the pulmonary and aortic valves to silently open. The blood in the ventricles rushes out through the open valves. When the ventricles have finished their contraction, the semilunar cusps swing closed and balloon slightly toward the ventricles, filling with blood but not leaking backward into the ventricles.

## Heart Murmurs

A doctor often listens to the heart from several locations because the heart sounds transmitted to the chest wall can give a clue about the condition of the different valves. Damaged valves can cause different types of **murmurs**. The location, timing, and type of sound help the doctor know what sort of damage is causing it.

If a valve is damaged and allows blood under high pressure to leak backward, a whooshing murmur may be heard. We say such a valve is **incompetent** because it isn't doing the job a valve is designed for — preventing the back-flow of blood. For instance, were the mitral valve to become incompetent, when the left ventricle contracts, some blood would be pushed back through the valve into the left atrium. The turbulence of the blood passing through the damaged valve would produce a murmur.

If a damaged valve is stiff and does not open normally, the outflow of blood is impeded. This is known as **stenosis**. A whooshing murmur will be heard due to the blood struggling to get through. As an example, if the aortic valve were damaged and became stiff or scarred, it might not open as it should. Then when the ventricle contracts, the blood would not as easily pass into the aorta. Again, the turbulence produced by the forcing of blood through the abnormally small opening would result in a murmur.

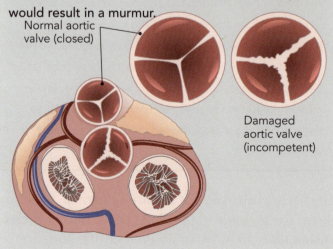
Normal aortic valve (closed)
Damaged aortic valve (incompetent)

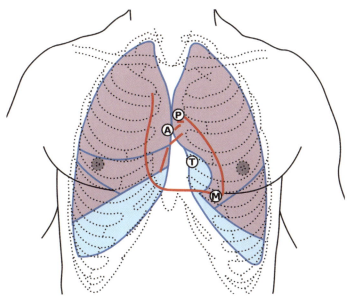

Optimal stethoscope position for listening to heart valves. Heart valves are labeled (Mitral, Tricuspid, Aortic, Pulmonary).

René-Théophile-Hyacinthe Laennec (1781-1826) invented the stethoscope in 1816. The first stethoscope was a simple hollow wooden cylinder. It allowed doctors to listen to the heart and lungs without having to place their ears directly on the patient. Even though that device is primitive by today's standards, it was revolutionary in its day.

If the first heart sound, the "lub," results from turbulence during the simultaneous closure of the tricuspid and mitral valves, what do you think causes the second sound, the "dub"? The turbulence of blood created when the semilunar valves close creates this second heart sound. If you have the opportunity to borrow a stethoscope, you can listen to your own heart's sound. The heart sounds can both be heard at many locations on the chest wall.

## The Cardiac Cycle — What Happens In a Heartbeat

The *cardiac cycle* is the name given to the five steps involved in filling the heart's chambers and pumping the blood. We will now examine this process more closely. All five steps must take place — in just the right order — every time your heart beats.

There are specific terms used to describe what a heart chamber is doing during the different steps in the cardiac cycle. The period of time when a heart chamber is contracting is called *systole* (pronounced "sis-tuh-lee"). The phase during which the chamber is relaxing is called *diastole* (pronounced "dī-as-tuh-lee"). Now let's apply those terms — *systole* and *diastole* — to each of the four steps in the cardiac cycle. (Later we will see that these words help us understand a measurement called "blood pressure." You may have even had yours measured!)

The first step in the cardiac cycle is the "filling phase." While they fill with blood, the atria and ventricles are all in diastole. That is, all the chambers are relaxed. Since the heart muscle is relaxed, the pressure inside them is low. This low pressure allows the atria and then the ventricles to fill with blood. First, blood enters the atria. As they fill, blood pushes the tricuspid and mitral valves open, allowing blood to flow into the ventricles too. At the end of this phase, the ventricles are about 75 percent full.

During this phase what do you think is happening with the heart's "exit-doors" — pulmonary and aortic valves? Since the pressure in the ventricles is low at this point, both of these valves will be closed, right? Otherwise, the blood would flow backward into the ventricles. The pressures in the pulmonary artery and the aorta are keeping the pulmonary and aortic valves closed for now.

# CARDIOVASCULAR & RESPIRATORY SYSTEMS

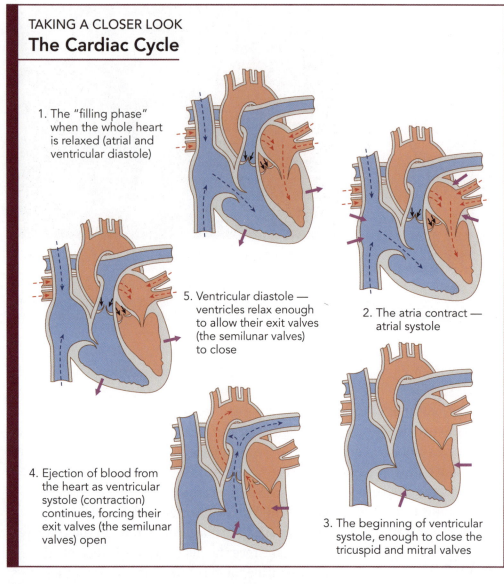

**TAKING A CLOSER LOOK**
## The Cardiac Cycle

1. The "filling phase" when the whole heart is relaxed (atrial and ventricular diastole)

2. The atria contract — atrial systole

3. The beginning of ventricular systole, enough to close the tricuspid and mitral valves

4. Ejection of blood from the heart as ventricular systole (contraction) continues, forcing their exit valves (the semilunar valves) open

5. Ventricular diastole — ventricles relax enough to allow their exit valves (the semilunar valves) to close

During the fourth step of the cardiac cycle, blood is forcefully ejected from the heart. The increasing pressure from the ventricular contraction forces the pulmonic and aortic valves (the semilunar valves) to open, and the blood rushes out into the pulmonary artery and the aorta.

Finally, in the fifth and final step of the cardiac cycle, the ventricles relax. Because of this relaxation, the pressure in the ventricles decreases. The higher pressure in the pulmonary artery and the aorta causes the semilunar valves to close. Thus, blood is prevented from flowing backward into the ventricles. This is *ventricular diastole*, and it is the end point of one complete cardiac cycle.

So, the five steps in the cardiac cycle are:

After this passive filling of the ventricles, the atria simultaneously contract. This is *atrial systole*. The squeezing of the atria pushes more blood into the ventricles to help really fill them up. This atrial "squeeze" is the second step in the cardiac cycle, and it adds another 25 percent or so to the filling of the ventricles.

Next comes relaxation of the atria (*atrial diastole* — the second step) and then contraction of the ventricles, or *ventricular systole*. During this, the third step of the cardiac cycle, the ventricles begin to contract. As a result of this contraction the pressure in the ventricles increases enough to slam the tricuspid and mitral valves shut, causing the "lub" sound.

1. the "filling phase" when the whole heart is relaxed (atrial and ventricular diastole)

2. the atria contract — atrial systole

3. the beginning of ventricular systole, enough to close the tricuspid and mitral valves

4. ejection of blood from the heart as ventricular systole (contraction) continues, forcing their exit valves (the semilunar valves) open

5. ventricular diastole — ventricles relax enough to allow their exit valves (the semilunar valves) to close

# Congestive Heart Failure

The pumping action of the heart is nothing short of amazing. The right side of the heart sends blood to the lungs, and the left side of the heart pumps blood out to the body. Each side pumps the same amount of blood, at the same time, and the process takes place in a coordinated fashion. This precise balance continues day in and day out.

However, we live in a fallen, cursed world. Things go wrong. At times the heart does not function correctly. A heart weakened by disease or heart attack will not be as efficient or pump as powerfully. This is called "heart failure." With heart failure, the heart still works, but one or both of its pumps is weak.

Since the heart consists of two pumps, it is possible for either pump system to function abnormally. If either of the pumps fails to keep up with the amount of blood it is supposed to pump, blood will back up, like cars in a traffic jam. We sometimes say that traffic is "congested," and the same word can be used for blood that backs up due to heart failure. **Congestive heart failure** can be a problem caused by failure of either the right or the left side of the heart to keep up.

If the pump on the right side of the heart fails to pump properly, the blood returning to the heart from the body is not pumped to the lungs efficiently. Then, the vena cavae and other systemic veins that bring blood to them become **congested** with excessive blood. Remember, this is like a traffic jam — traffic congestion — with blood instead of cars. Blood is backed up. Due to this **congestion**, the pressure in these vessels increases. The most noticeable result of this is swelling in the legs and feet. This swelling is called **peripheral edema**. (**Edema** is swelling caused by fluid accumulating in tissues. **Peripheral** means the swelling happens in parts of the body far away from the heart.)

If the pump on the left side of the heart fails to do its job properly, the oxygenated blood returning from the lungs is not adequately pushed out to the body. Now, if the normal amount of blood is being pumped to the lungs by a correctly functioning right heart pump, but the left heart pump cannot keep up with this volume of blood, what do think will happen? The blood will back up into the lungs! This time the "traffic congestion" backs up into the lungs. This problem is called **pulmonary edema** (fluid in the lungs). Pulmonary edema causes patients to be quite short of breath and make it difficult to exercise or even to walk. In its most severe forms, pulmonary edema can lead to death.

The degree of heart failure can be assessed by the severity of the patient's symptoms, such as shortness of breath or how much exercise they can do. Also, it can be quite helpful to obtain a measurement of the patient's ejection fraction — the fraction of the blood ejected during systole. The lower the ejection fraction, the more severe the heart failure is said to be.

Treating heart failure is challenging. Patients are often given drugs that cause the body to get rid of the excess fluid that accumulates in the lungs or other tissues. There are also certain drugs that can help damaged cardiac muscle contract more efficiently and make the heart pump better. However, these drugs can also have serious side effects at higher doses, so they must be used cautiously. In certain very severe cases, a heart transplant may even be considered.

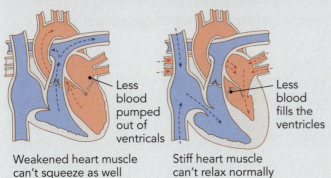

Weakened heart muscle can't squeeze as well — Less blood pumped out of ventricals

Stiff heart muscle can't relax normally — Less blood fills the ventricles

That is what happens every time your heart beats! What do you think happens next? Remember this is a *cycle*, so when the fifth step is completed, the whole cycle begins again. The heart's chambers are all relaxed and the valves are in the right position so that they can fill with blood and the heart can beat again.

## How Empty Is Empty?

When you wring out a washcloth or sponge, is it completely dry? No. It still contains some water. You cannot squeeze it enough to make it dry. Likewise, after your heart's ventricles contract, they still contain some blood. Not every drop of blood gets emptied from the ventricles as they squeeze. In fact, a healthy heart only empties around 60–70 percent of its contents with each beat! This percentage is called the *ejection fraction*. If a person's heart is not working properly, its ejection fraction may be far lower than this. Measuring the ejection fraction can be very important for physicians when they are caring for patients with heart problems.

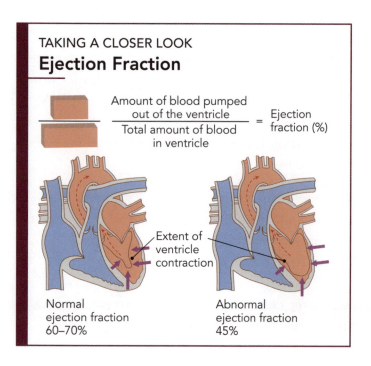

TAKING A CLOSER LOOK
**Ejection Fraction**

$$\frac{\text{Amount of blood pumped out of the ventricle}}{\text{Total amount of blood in ventricle}} = \text{Ejection fraction (\%)}$$

Normal ejection fraction 60–70%

Abnormal ejection fraction 45%

Extent of ventricle contraction

## What the Heart Needs

The heart pumps oxygen-rich blood to every organ in the body. But how does the heart get the oxygen-rich blood it needs? After all, the heart needs a constant supply of oxygen and fuel (in the form of sugar called glucose) in order to keep pumping constantly, day in and day out, for a lifetime! Therefore, God designed the *coronary circulation* — a way for the heart to pump blood to itself.

When something goes terribly wrong with the coronary circulation, a person can have a heart attack. You may know of someone this has happened to. Once you see how the coronary circulation works and why it is so important, you will understand what a heart attack is.

You might wonder why the heart needs its own separate blood supply. After all, the heart is a pump that pumps blood. It is filled with blood most of the time. So why can't it just get the things it needs from the blood in its chambers?

What it comes down to is this: because the heart works constantly, it needs *lots* of oxygen and nutrients. Even though the left ventricle is filled with oxygen-rich blood, the heart wall is just too thick for nutrients to seep into it. A more efficient system is needed to supply the heart muscle — the *myocardium* — with oxygen and fuel.

The *coronary circulation* is a system of arteries and veins that delivers oxygen-rich blood to the heart muscle and carries away deoxygenated blood.

The coronary circulation begins just past the aortic valve. Right after the place where blood exits the heart's left ventricle, two arteries branch from the very first part of the aorta (called the *ascending aorta*). These are the *right and left coronary arteries*. They divert a little of the blood flowing into the aorta toward the heart's muscular walls.

# THE HEART

## TAKING A CLOSER LOOK
### Coronary Circulation

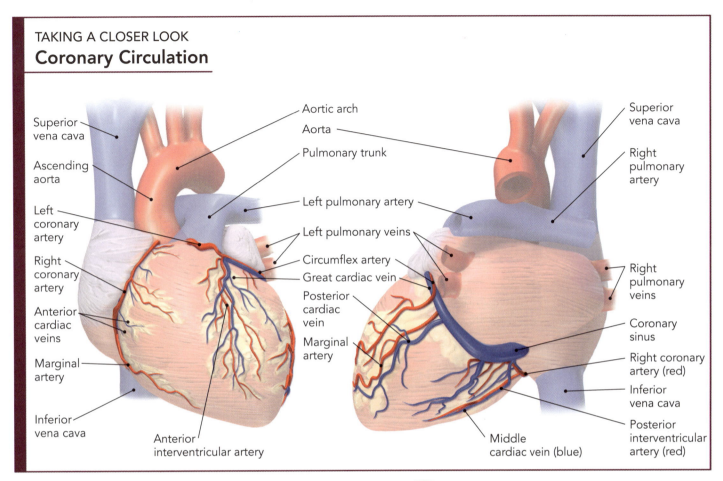

The *right coronary artery* primarily supplies the right atrium and the right ventricle. It divides and divides into many smaller arterial branches to completely supply the right side of the heart.

The *left coronary artery* supplies the left side of the heart. It has two major branches. One of these — the left anterior descending artery, or LAD — supplies the front (*anterior*) walls of both the right and left ventricle as well as the wall of myocardium between the ventricles. (This muscular wall between the ventricles is called the interventricular septum). The other branch — the circumflex artery — brings blood to the left atrium and the left ventricle's back (*posterior*) wall. The two main branches divide and subdivide into many smaller vessels to ensure complete circulation to the left side of the heart.

After supplying the heart's muscular walls with the oxygen and fuel they need, deoxygenated blood returns to the right atrium through several cardiac veins.

### Try This

Squeeze one hand into a tight fist. Then try to push a finger into that fist. You can't! It won't fit if the fist is tightly contracted. Likewise, if your heart is busy squeezing hard — contracting — how can its muscular walls have room to let blood flow through the coronary circulation to bring them the oxygen and fuel they must have to keep working? Well, God is a great engineer. This is the solution He designed: As the heart relaxes during diastole, the pressure in the aorta pushes blood into the coronary arteries to supply the heart. The heart muscle receives most of the oxygen and fuel it needs during the relaxed parts of each heartbeat, enough to keep it going until the next diastole.

## Coronary Artery Disease

It is possible that you know someone who has suffered a heart attack. If not, I expect you have at least heard the term "heart attack." A heart attack can be very serious and is often fatal. Every year over 700,000 people in America have a heart attack!

The primary problem that leads to a heart attack is called **coronary artery disease**. You already know what a coronary artery is. Coronary arteries are the arteries that keep the heart's muscular walls supplied with freshly oxygenated blood. Coronary artery disease, abbreviated CAD, occurs when the lining inside a coronary artery becomes thick. As the lining thickens, the channel inside the artery becomes smaller and smaller. Less and less blood is able to squeeze through the narrowing opening. Severe narrowing is called a "blockage."

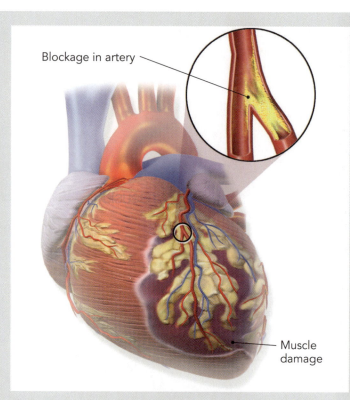

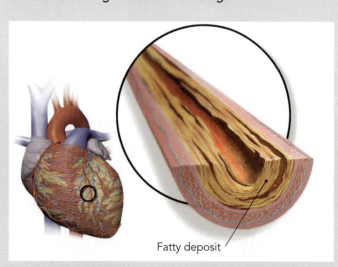

Eventually, the blood flowing through this narrowed artery cannot adequately supply the needs of the myocardium. The situation where adequate oxygen is not delivered to the heart muscle is called **myocardial ischemia**. Coronary artery disease can involve a single "blockage" in only one coronary artery or several blockages in multiple coronary arteries. Obviously, the more blocked arteries, the more serious the situation.

Myocardial ischemia is not the same thing as a heart attack, but it can lead to one. There are degrees of myocardial ischemia. Some people with myocardial ischemia experience episodes of **angina pectoris**, which literally means "strangled chest." A person with angina pectoris has episodes of chest pain, usually described as a tightness or a burning sensation in the chest. Some feel like their chest is in a vise. Often the pain radiates to the left arm, neck, or jaw. Angina pectoris can occur with activity (so-called "stable" angina) or at rest ("unstable" angina). The underlying problem is that due to restriction of blood flow. the heart muscle does not get adequate oxygen to meet its needs, thus resulting in chest pain. However, with angina alone the situation is intermittent, and there is no permanent damage to the heart muscle.

As coronary artery disease worsens, there is increasing danger of myocardial infarction (often called an "MI"). This is commonly known as a heart "attack." Here the disease in the coronary artery (or arteries) has progressed to the point that the myocardium can no longer get the amount of oxygen it needs, and some of the heart muscle dies. Logical, isn't it? If an artery that takes oxygen to a certain part of the heart becomes blocked, then the muscle tissue in that part of the heart is at risk of death. Myocardial infarctions can range from relatively mild to fatal. The severity depends on how much myocardium is damaged and how efficiently the remaining heart muscle functions.

Treatment for coronary artery disease depends on its severity. If it is very mild, a patient may be treated with simple things like exercise, medication, and changes in diet. For more serious blockages, patients may undergo a procedure to open or to by-pass the blockage in order to improve blood flow to the heart.

There are two main sorts of procedures used to deal with a coronary artery blockage. One is called **coronary angioplasty**. (By the way, the author of this book has undergone this procedure.) Here, using special dye and a type of x-ray called fluoroscopy, a tiny wire is threaded through the blockage and a balloon is inflated to open up the artery. Most often, a small mesh device, called a stent, is then put in place in the coronary artery to help keep it open.

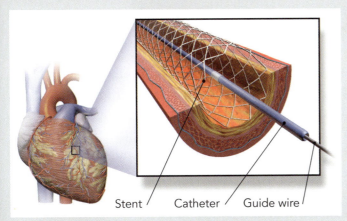
Stent / Catheter / Guide wire

In the most severe cases of coronary artery disease, **coronary artery bypass surgery** is done to route blood around a blockage. In bypass surgery, a section of a vein from the person's arm or leg is removed and used as a bypass graft. One end of the vein is attached to the aorta, and the other end is attached to the diseased coronary artery at a point past the blockage. Thus, the blockage is effectively "bypassed," and blood flow is restored to the heart muscle at risk for damage.

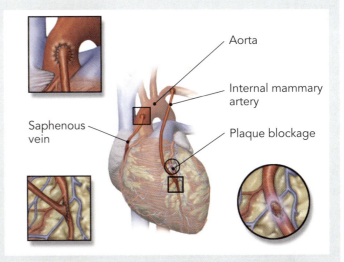

Coronary artery disease is a type of **cardiovascular disease**, a term that includes heart attacks and strokes and other diseases of the heart and blood vessels. Cardiovascular disease is the world's leading cause of death. Heart attacks are the leading cause of death in the United States.

Who is most likely to have a heart attack? Some people are at greater risk than others. A **risk factor** is something that puts a person at greater risk of suffering a particular thing than other people. Some risk factors are beyond a person's control. However, there are some things you can do to lower the risk of ever having a heart attack. There are many risk factors that can lead to heart disease. These include (but are not limited to) smoking, a lack of exercise, obesity, poor diet (especially diets high in fats), high cholesterol, diabetes, and high blood pressure.

We need to take all the steps we can to take good care of our hearts. So make a lifelong practice of getting plenty of exercise (and, no, video games are not exercise), maintain a healthy weight, get in the habit of primarily eating nutritious foods (I'm not saying don't eat hot fudge sundaes, I'm just saying don't make a regular habit of them), and never, ever . . . let me say it again . . . never, EVER, start smoking!

Now is the time to learn a heart healthy lifestyle!

# Beats

As we mentioned earlier, cardiac muscle is involuntary. This means you don't have to think about your heart beating. It happens all on its own. Unlike skeletal muscle, you have no conscious control over the contraction of cardiac muscle. For example, you can willfully make skeletal muscle move... reaching for a glass or throwing a ball. However, you cannot will your heart to beat.

It turns out that just beating isn't enough. Not only must the heart beat (and even be able to speed up when you are running), but both sides must beat simultaneously. Remember, the heart is really two pumps. How does it get the timing right so that both sides pump simultaneously? The answer is electrical. Your heart has a built-in system to produce an electrical signal that triggers the heart muscle in each pump to beat... and to do it over and over and over again.

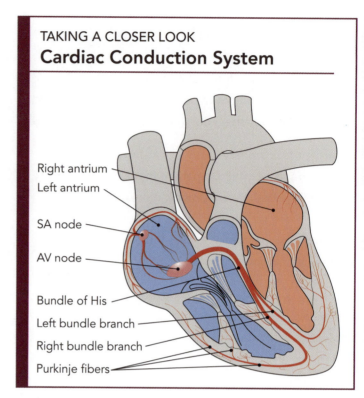

TAKING A CLOSER LOOK
## Cardiac Conduction System

- Right antrium
- Left antrium
- SA node
- AV node
- Bundle of His
- Left bundle branch
- Right bundle branch
- Purkinje fibers

## The Pumping Heart

Most of the heart consists of cardiac muscle cells. The vast majority of these muscle cells are in the business of contracting. They are responsible for the pumping action of the heart. However, about 1 percent of these cells have a very special property and are not primarily involved in heart contraction. These special cells are the ones that *stimulate* the contractions! These cells have the ability to spontaneously generate an electrical signal all on their own. These are called *autorhythmic* ("self-rhythm") cells. They repeatedly produce electrical signals that stimulate the heart to contract.

These autorhythmic cells generate electrical impulses without any outside stimulus from the nervous system. Even if all nerve fibers to the heart were severed, the heart would continue to beat. For example, hearts removed from a body to be transplanted continue to beat for several hours, even though all nerve fibers to the heart have been cut. (This does not mean that the nervous system is not important. Nervous system input can play an important role in controlling the heart *rate* as we will see.)

These autorhythmic cells have two important jobs. First, they function as the *pacemaker* for the heart. That is, they establish and maintain the basic rhythm of the heart. They trigger the start of each and every heartbeat. They set the pace! Second, they are lined up to form a pathway that helps move the electric signals through the heart from muscle cell to muscle cell in a very orderly fashion.

The cardiac conduction system is also called the *intrinsic conduction system*. *Intrinsic* means this conduction system is completely contained within the heart; it does not bring in messages from outside the heart. This intricate network of rhythm-generating cells is designed to distribute signals to the cardiac muscle in an orderly way to ensure that the heart contracts in a coordinated manner. If the heart's chambers did not coordinate their squeezing action, chambers would start squeezing before they filled. Furthermore, just as a toothpaste tube squeezed near the top traps toothpaste in the bottom of the tube, so a heart that doesn't squeeze in a coordinated manner would not empty blood very well. Let's take a closer look at the way the electrical system of the heart is designed to avoid this sort of problem.

## The Cardiac Conduction System

The cardiac conduction system (or intrinsic conduction system) has two "nodes" that set the pace of the heartbeat. The first node to fire signals the beginning of a heartbeat. This pacesetter is the *sinoatrial node*, also called the *SA node*. The SA node is a small group of cells located in the upper portion of the right atrium's wall, near the entrance of the superior vena cava. The SA node is the heart's main pacemaker. The SA node initiates each electrical *impulse* that stimulates the heart to contract. On average, the SA node generates an impulse 72 times a minute. The SA node generates impulses faster than the other pacemaking node. Therefore, under normal circumstances, the SA node controls the heart rate. For that reason, the basic rhythm of the heart is called *sinus rhythm*.

Once generated, the impulse from the SA node travels through the muscle cells themselves. The impulse spreads throughout both atria causing them to contract. Atrial contraction squeezes the blood from each atrium into the ventricles.

At the end of its journey through the atria, the electrical impulse produced by the SA node reaches another group of cells called the *atrioventricular node* (AV node). The AV node is located in the wall (or *septum*) between the right and left atria, just above the tricuspid valve. It is the job of the AV node to send the electrical signal on to the ventricles. That signal makes the ventricles contract. (If, however, the SA node fails for some reason, the AV node can act as a backup system and stimulate the heart to beat.)

Do you see a problem here?

These electrical impulses travel very rapidly. What would happen if the atria and the ventricles all contracted at the same time? The atria would not be able to squeeze their blood into the ventricles because the ventricles would be contracting too. And without getting re-filled with blood from the atria, the ventricles would soon have no more blood to pump out to the body and the lungs. The entire heart would stop pumping blood. Not a good situation at all, right? The beating of the heart must be coordinated, so that the atria both contract before the ventricles do.

God has designed the cardiac conduction system to avoid this problem. When the SA node "fires," both

### How Fast?

As we examined the cardiac conduction system, we saw that the SA node is the primary pacemaker of the heart. The SA node beats at an average rate of 72 beats a minute.

Are there other pacemaker locations in the heart? As it turns out, there are.

If the SA node ceased to function (say, as a consequence of disease or aging), the AV node would take over the pacemaker duties. However, the heart rate generated by the AV node is around 50 beats per minute.

And if both the SA and AV nodes stopped working, the Purkinje fibers also have the potential to act as a pacemaker. Purkinje fibers can only generate a heart rate of around 30 beats as minute. This is certainly not ideal, nor is it as efficient as a properly functioning SA node.

God designed two backup systems to keep the heart beating if its chief pacemaker malfunctions.

atria respond almost instantly, and then the AV node "fires." But when the AV node fires, the ventricles do *not* respond immediately. Instead, the AV node's electrical signal is delayed by about 0.1 second (that's one-tenth of a second . . . not very long at all . . . but long enough). This delay happens because the cells in the fibers near the AV node do not transmit the electrical impulse as rapidly. (They have fewer *gap junctions*, little gateways between cells, and that slows down the passage of the impulse from cell to cell.) Once through the AV node and these signal-slowing muscle fibers, the signal travels normally (that is, very rapidly) through the remainder of the conduction system.

After leaving the AV node, the signal is carried by the *atrioventricular bundle* (sometimes called the *bundle of His*) into the ventricles. You might be thinking, "Wait a minute, we just saw that this impulse was carried through the atria though the muscle cells themselves. Why can't the signal that passed through the atria reach the ventricles the same way?" Good question. The answer is that the atria and ventricles are separated by the connective tissue that makes up the fibrous skeleton of the heart. This fibrous tissue acts as sort of an insulator that stops the electrical signal from passing directly. The only electrical pathway between the atria and the ventricles is the atrioventricular bundle. Here is

# Pacemakers

Even though a healthy heart does have the ability to generate its own conduction signals, there are circumstances when the cardiac conduction system does not function correctly. At times due to aging or illness, the pacemaker center (SA node) may not generate signals rapidly enough to maintain adequate blood pressure. Or perhaps as the consequence of a heart attack, the AV node is damaged and cannot conduct the electrical signals to the ventricles properly.

In many situations like these a patient may require the implantation of a pacemaker. A pacemaker is a small battery-powered device that can help control a patient's heartbeat. The device is attached to a small electrode that is placed in the heart. An electrical signal is sent from the pacemaker to stimulate the heart to beat.

The simplest style of pacemaker has one electrode that is threaded into the right ventricle (under fluoroscopic guidance). The pacemaker itself is usually placed in a small surgically created pocket under the skin just below the left collarbone. The pacemaker can monitor the patient's heartbeat, and if a beat is not detected within a certain period of time, the pacemaker sends an electrical impulse to stimulate the heart. If a normal heartbeat is detected, then the pacemaker would not fire.

Pacemakers have become more and more sophisticated. These devices can be programmed for a wide range of heart rates. Some pacemakers have multiple electrodes and can pace both the atrium and the ventricle. Other pacemakers sense activity levels and can adjust the patient's heart rate to match.

but one more example of the marvelous design of the heart. Without this electrical barrier it would not be possible to control the pumping action of the heart so precisely.

Very soon after reaching the ventricles, the atrio-ventricular bundle splits into two branches, the *right bundle branch* and the *left bundle branch*. These two bundles proceed down through the *interventricular septum* (the wall between the ventricles) toward the apex of the heart. The right bundle branch delivers the impulse to the right ventricle and the left bundle branch signals the left ventricle. In the septum, the bundle branches also to small branches that penetrate deep into the myocardium of the ventricles. These are called Purkinje fibers. Because the *Purkinje fibers* deliver the electrical signals to their final destination, they are vital for maintaining the heart's smooth, coordinated pumping action. Purkinje fibers cause the heart to contract from the bottom up and not from the top down.

Think of squeezing a tube of toothpaste. Is it better to squeeze it from the top or the bottom?

Remember, the heart's conduction system is designed (1) to set its own pace by generating an electrical impulse and (2) to send that signal to all parts of the heart in a coordinated manner that first triggers the atria to squeeze blood into the ventricles and then causes the ventricles to squeeze that blood out from the bottom to the top. See if you can name the parts of the conduction system in the order an impulse travels through them.

## The Electrocardiogram

The electrical impulses transmitted through the heart can be detected on the body's surface. The heart's electrical signals can be measured with an electrocardiograph. The recording that is produced from this is called an *electrocardiogram* (abbreviated ECG or EKG).

To record an ECG, one electrode (called a *limb lead*) is placed on each arm and leg. (This does not hurt.) Then six other electrodes are placed across the front of the chest. These are the *chest leads*. Multiple

### Heart Squeeze

When the AV node's signal is transmitted, the heart muscle cells in the ventricles do not contract at the same time. What would happen if they did? The blood in the ventricles would get a hard squeeze, but it wouldn't move efficiently toward the aorta and pulmonary artery. To avoid this problem, God has designed the heart's conduction system to start responding to the signal from the apex (the sort of pointy part at the bottom of the heart) and move toward the top of the ventricles. The heart's muscle cells are arranged in a spiral so that they contract and efficiently push the blood in the ventricles out, squeezing from the apex upward. So the heart really does squeeze from the bottom up, and that's the most efficient way!

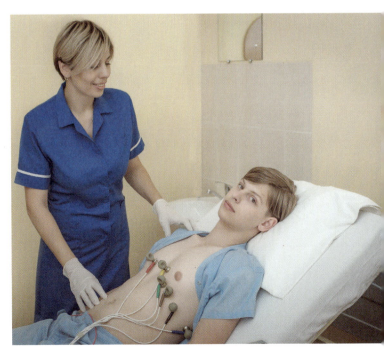

*Getting an Electrocardiogram*

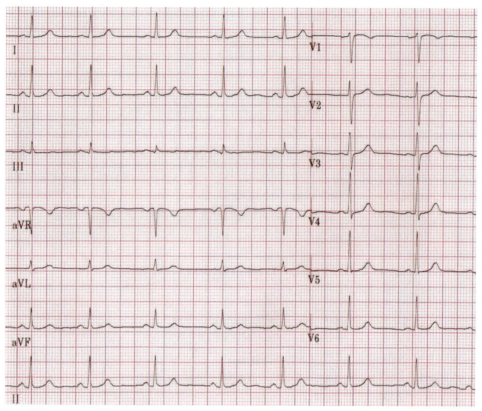

*Normal 12 lead Electrocardiogram*

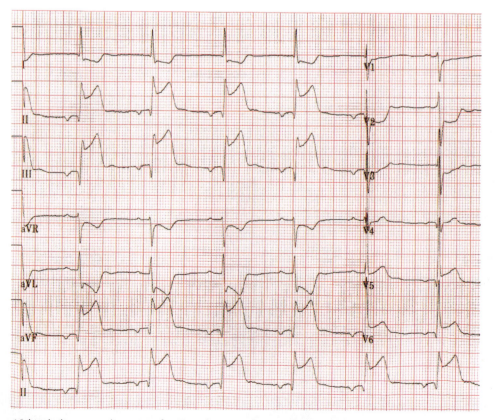

*12 lead electrocardiogram of patient having a heart attack. Note the distinct differences from the normal EKG above.*

leads are necessary in order to measure the electrical signals from many different positions relative to the heart. The electrocardiograph machine amplifies the signals obtained by the various electrodes and prints out the patterns as an electrocardiogram.

The ECG tracing is a reflection of the electrical signal being transmitted through the cardiac conduction system. As your eyes move from left to right along the tracing, you are seeing a measurement of the electrical signal as it signals each part of the heart in turn. The ECG shows us the electrical signal that instructs the heart to beat and reveals how well that signal travels through the heart, but it does not actually show the heart's response to that signal — the squeezing of the muscle. Other techniques, such as the *echocardiogram*, show the actual beating of the heart.

The first major wave seen on the ECG is called the *P wave*. The P wave reflects the electrical signal that begins the domino effect that ultimately makes the heart beat one time. The P wave reflects the movement of the electrical impulse from the SA node through the myocardium of the atria. About 0.1 second after the P wave begins, the atria contract. The flat segment between the P

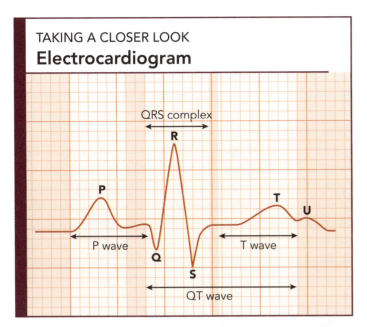

**TAKING A CLOSER LOOK**
**Electrocardiogram**

wave and the beginning of the QRS represents the time after the signal has passed through the atria and is being delayed in the AV node. Remember, it is this delay at the AV node that gives the atria time to squeeze their blood into the ventricles before they contract.

The second large wave seen in a typical ECG is called the *QRS complex*. During the time reflected in the QRS complex the electrical impulse is moving through the ventricles. The QRS complex has a complicated appearance due to the paths that the electrical impulses travel as they move through the ventricular myocardium. This is the time when the ventricles contract. The QRS lasts about 0.1 second.

The last wave in an ECG is the *T wave*. During this time, the ventricle is starting to relax. The ventricles are preparing to receive the next electrical impulse. The duration of the T wave is about 0.16 second, and during this time the electrical system of the heart resets itself in preparation for the next heartbeat.

Then the process begins again.

By understanding the pattern and timing of normal ECGs, doctors can use abnormal ECG patterns to help diagnose and treat patients. In fact, ECGs have become one of the most important tools in modern medicine. Damaged cardiac muscle, for instance, might not transmit the electrical signal properly, and this can be revealed in the ECG. ECGs can be particularly helpful in diagnosing coronary artery disease and cardiac rhythm disorders.

## Cardiac Output

To more completely understand how the heart works, there is another concept you must understand. This is known as cardiac output.

*Cardiac output* (CO) is the amount of blood pumped by the heart in one minute. Cardiac output can vary from minute to minute. For example, when you are running, your leg muscles need more oxygen, right? Of course they do. So what do you think happens to the output of the heart when these muscles need more oxygen? It increases!

When you are asleep, your leg muscles need less oxygen, right? So what happens to the heart output when less oxygen is needed? It is not as high.

Cardiac output is the product of two things: the heart rate (HR) and the stroke volume (SV). *Heart rate* means just what it says, the rate of the heart in beats per minute. *Stroke volume* is the amount (volume) of blood pumped with each heartbeat.

The relationship can be shown this way:

$$CO = HR \times SV$$

So let's calculate an average cardiac output. If the average heart rate is 72 beats per minute, and the stroke volume is 70mL, then

$$CO = 72 \text{ beats/minute} \times 70\text{mL/beat}$$
$$CO = 5040\text{mL/minute}$$
$$CO = 5.04 \text{ liters/minute}$$
$$(CO = 1.33 \text{ gallons/minute})$$

This is a typical cardiac output for an adult at rest.

There are two ways that the cardiac output increases — either the heart rate increases or the stroke volume increases. As you are aware, with exercise, the heart rate increases. You've probably felt your heart beating very fast at the end of a sprint. What you may not realize is that with exercise, your heart's stroke volume — the amount of blood pumped out with each beat — can also increase. If, while running, your heart rate goes to 110 beats per minute and the stroke volume increases to 100 mL (3.4 ounces) per minute, what is the cardiac output?

$$CO = 110 \text{ beats/minute} \times 100 \text{ mL/beat}$$
$$CO = 11{,}000 \text{ mL/minute}$$
$$CO = 11 \text{ liters/minute}$$
$$(CO = 2.9 \text{ gallons/minute})$$

So we see that with only mild increases in heart rate and stroke volume, the cardiac output more than doubles! Soon we'll see how the body can let the heart know that it must pump out more blood — that is, that it must increase its cardiac output.

## Echocardiogram

An echocardiogram is an ultrasound of the heart. Whereas the EKG evaluates the heart's function by measuring electrical conduction, an echocardiogram uses sound waves to see inside the heart. Using a *transducer* placed on the patient's chest, sound waves are painlessly bounced off various parts of the heart. The resulting pictures show the heart's shape, its walls, valves, and even the blood flowing through its chambers.

By using sound waves to see inside the heart and make measurements, doctors can determine if the heart is working normally or not. Do the walls move properly? Are they too thick? Is the heart enlarged? Do the valves close completely, or does blood leak back through them? How much blood do the heart's chambers pump out with each squeeze? The heart's ejection fraction, a valuable way to assess how well it is pumping, can be calculated based on information from the echocardiogram.

Here you can see samples of echocardiographic images.

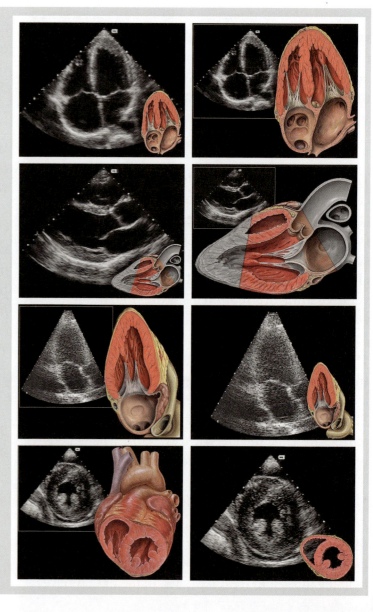

## Cardiac Reserve

Some people's hearts are able to increase their cardiac output more than others. A healthy person who runs regularly, for instance, may be able to increase his or her cardiac output much more than a person with heart disease can. We say their *cardiac reserve* is greater.

*Cardiac reserve* is the difference between the cardiac output at rest and cardiac output during maximal exertion. The average person's heart can increase its output about five times above its resting output. That would be around 24 liters/minute (6.5 gallons/minute). In a highly trained athlete, the maximum cardiac output during heavy exertion might reach 33 liters/minute (9 gallons/minute), or seven times the resting CO.

## Regulation of Stroke Volume

Increasing cardiac output requires an increased heart rate, or increased stroke volume, or both. Let's look at ways the stroke volume can increase.

Stroke volume, remember, is how much blood the heart pumps out during one heartbeat. The heart, no matter how healthy, does not empty itself completely during a beat. There is always some blood left behind. Therefore, stroke volume is the difference between the amount of blood in the left ventricle when it is completely relaxed and the amount of blood remaining in the left ventricle when it has just finished contracting.

The ventricle's time of relaxation and filling is called *diastole*, you recall, so the amount of blood in the ventricle when it is full is known as the *end diastolic volume*. *Systole* is the time of contraction, so the amount of blood left in the ventricle after it contracts is called the *end systolic volume*. We could sum this up like this:

End diastolic volume – End systolic volume = Stroke volume

We've said that the heart can increase its stroke volume in order to supply the body's increased needs, like when you want to run. There are several factors that affect stroke volume, but the two most important are *preload* and *contractility*. Preload depends on how much blood is in the left ventricle before it squeezes. Contractility involves how hard the ventricle squeezes. Let's look at these two things more closely.

Cardiac muscle cells contract most efficiently when they are stretched somewhat before they begin contracting. *Preload* is the amount that cardiac muscle is stretched by the blood in the ventricle before it contracts. The more blood that enters the ventricle, the more its walls are stretched. This stretching helps increase the force of the contraction of the muscle. Imagine blowing up a balloon. The more air you blow into a balloon, the more the balloon stretches. Up to a point, the more the ventricle is stretched (preloaded), the stronger will be its contraction.

Preload depends on the amount of blood that can enter the ventricle before it beats. Let's consider how preload can change. The heart's *rate* can alter its preload. If it beats slowly, there is more time between beats. This allows more time for blood to fill the ventricles and increases stroke volume. The opposite can occur with extremely fast heart rates. A very rapidly beating heart leaves little time between beats to fill the ventricle, and the stroke volume could consequently decrease.

The heart muscle's contractility also helps determine stroke volume. *Contractility* refers to how hard the muscle can contract when it is stretched to a certain point. When you are running, your body can send messages to the heart to increase contractility. Some of the most important chemical messengers in the body are called hormones. *Hormones* travel through the blood stream to deliver their messages to many destinations in the body. The hormones *epinephrine* and *norepinephrine* (also called *adrenaline* and *noradrenaline*) can increase the contractility of cardiac muscle, making the muscle squeeze more forcefully. When the heart squeezes harder, it empties more completely with each beat. Thus, stroke volume increases.

## Stress Testing

A heart suffering from coronary artery disease might have sufficient blood circulation to function normally at rest but not when stressed with exercise. Therefore, one of the most common tests performed to detect coronary artery disease is called an exercise test, or a "stress" test.

An exercise stress test is performed by having the patient walk on a treadmill while connected to an EKG monitor. Every few minutes, the speed and incline of the treadmill are increased, thus demanding more work from the patient's heart. (Those patients unable to walk on a treadmill can be tested using vigorous arm exercises or an exercise bicycle.) The test ends when the patient cannot continue or when a specified heart rate is achieved.

During the stress test certain characteristic EKG patterns may suggest the presence of coronary artery disease. Abnormal heart rhythms also commonly develop during the exertion of the stress test. These rhythms are recorded on the EKG tracings for evaluation.

Although primarily thought of as a test to detect disease, stress tests are also useful in other ways. For example, special types of stress tests are sometimes used to evaluate and monitor the conditioning of healthy athletes as a part of their training regimen.

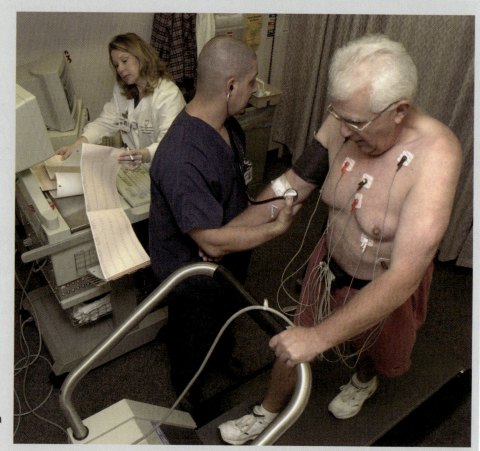

# Regulation of Heart Rate

We said the increased cardiac output requires an increased heart rate, or increased stroke volume, or both. Just as there are factors that regulate stroke volume, there are factors that regulate heart rate. The SA node is the heart's main pacemaker. The SA node is part of the heart's *intrinsic* conduction system — a signaling network *inside* the heart — but it responds to input from the nervous system, hormones, and other stimuli.

Everyone knows that our heart beats faster when we are frightened or excited. This increase in heart rate is due in large part to stimulation of the cardiac conduction system by the nervous system. Nerve fibers from the *sympathetic nervous system* release a chemical (norepinephrine) that binds to special receptors on the heart. Sympathetic nerve stimulation causes the SA node to fire more rapidly, and thus increases the heart rate.

The nervous system can also cause the heart rate to decrease. The *parasympathetic nervous system* has effects opposite to the sympathetic nervous system. (We will learn much more about these two divisions of the nervous system in other volumes of *Wonders of the Human Body*.) Parasympathetic fibers release a different chemical (acetylcholine) to slow the speed at which the SA node fires.

The primary pacemaker of the heart, the SA node, fires at an average of 72 beats per minute. However, the SA node is actually "pre-set" at a rate of nearly 100 beats per minute. The SA node fires at a slower average rate because it is reined in the parasympathetic nervous system's input.

Heart rate can be influenced by other things, such as hormones — chemical messengers that travel though the blood stream. One of these — adrenaline (also called epinephrine) — is made by the adrenal glands when you are exercising and when you are frightened. Adrenaline (epinephrine) increases heart rate. Thyroxine, a hormone produced by the thyroid gland, can also increase the heart rate. Fever can increase the heart rate, and an abnormally low body temperature can lower the heart rate.

# TWO KINDS OF HEARTS

Your heart is designed to beat constantly and to respond to your body's special needs. It pumps in a coordinated fashion to send blood through your lungs to gather oxygen and then on to the rest of your body. If it did not do this, you could not live. The heart can also pump faster or slower or stronger, depending on the signals it receives. Without any conscious thought on your part, it makes all these adjustments. God designed your heart to be dependable and steadfast.

The Bible often refers to a different kind of heart. This sort of heart refers to your character—the emotional, intellectual, and moral being that you are. This kind of heart represents the sort of person you are on the inside.

> *For as he thinks in his heart, so is he.*
> *"Eat and drink!" he says to you,*
> *But his heart is not with you.*
> (Proverbs 23:7)

Here we learn that the thoughts of a person's heart reveal the kind of person he or she is.

> *But the Lord said to Samuel, "Do not look at his appearance or at his physical stature, because I have refused him. For the Lord does not see as man sees; for man looks at the outward appearance, but the Lord looks at the heart."*
> (1 Samuel 16:7)

The Bible says that while people see what we are like on the outside, the Lord sees the heart. Our true character—that invisible kind of heart—is always visible to the Lord, and He has said in that every person is a sinner.

> *For all have sinned and fall short of the glory of God.*
> (Romans 3:23)

God's Word also tells us that we need to truly believe in our hearts in Jesus Christ in order to have salvation from sin. Jesus loves each of us and bought salvation for us when He died on the cross and rose again.

> *That if you confess with your mouth the Lord Jesus and believe in your heart that God has raised Him from the dead, you will be saved. For with the heart one believes unto righteousness, and with the mouth confession is made unto salvation.*
> (Romans 10:9-10)

God designed both kinds of hearts—your physical heart that keeps you alive all the days you live on this earth, and the spiritual heart that must trust in Jesus Christ to receive eternal life with Him.

141
TWO KINDS OF HEARTS

The heart knows, thinks, sees, is wise, speaks, and understands.
Proverbs 15:13–14
Psalm 90:12

The heart is very intentional.
Psalm 27:14
Psalm 119:112

The heart desires, wishes, and envies.
Proverbs 14:30
Psalm 139:23

The heart can be hard, stubborn and calloused.
Proverbs 28:14
Ezekiel 36:26

The heart is emotional. It loves. It feels things good and bad.
Matthew 22:37
2 Thessalonians 3:5

The heart can be wicked and store evil.
Jeremiah 17:9

The heart can be good, pure and holy.
Psalm 51:10
Proverbs 21:2

# BLOOD VESSELS

As incredible as the human heart is, without the body's intricate system of blood vessels, it would serve no real purpose. After all, a pump is no good without "pipes" to carry the blood where it's going. The body's vascular system does just that. And more. . .

You see, the body's vascular system is much more than just a collection of tubes that carry blood away from the heart and then back again. These tubes are able to contract and expand in order to control blood pressure and to divert blood to the places where it is most needed. Parts of the vascular system are strong enough to withstand the high pressures the heart generates. Other parts of the system are thin and delicate enough to allow oxygen and nutrients to diffuse across their walls into adjoining tissues.

Not just any old set of tubes is it? Let's take a closer look.

 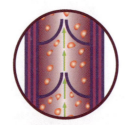 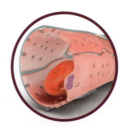 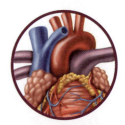

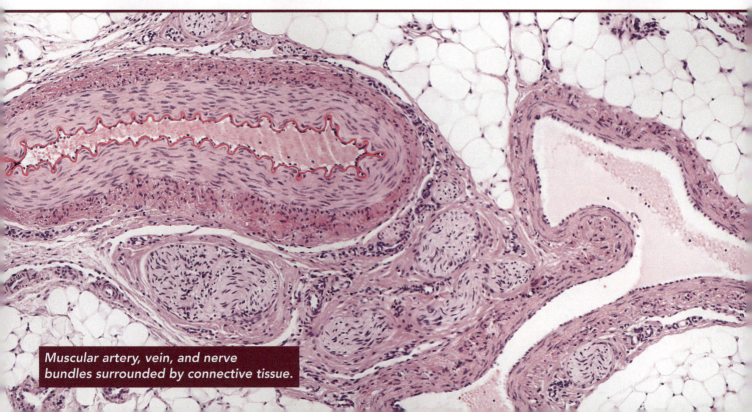

*Muscular artery, vein, and nerve bundles surrounded by connective tissue.*

# Blood Vessels — the Basics

You had probably heard of arteries and veins, even before you began reading this book. We've talked about the difference: veins bring blood toward the heart, and arteries carry blood away from the heart. Now it is time to learn how these vessels branch into other smaller blood vessels and to learn how the differences in the vessels equip them for the important jobs they do.

There are five primary types of blood vessels: arteries, arterioles, veins, venules, and capillaries. Let's review what each type of blood vessel does.

Remember this: *arteries — and their branches — take blood away from the heart*. So it stands to reason that veins and their branches must do the opposite, and that is precisely what they do. *Veins — and their branches — carry blood back to the heart. Capillaries*, which are very small, connect the two. Capillaries get the blood from the arterial system back into the vessels of the venous system.

As we saw earlier, in the systemic circulation (where blood is pumped from the left ventricle into the aorta and out to the body's organs), arteries carry oxygen-rich blood to the tissues and then veins carry the oxygen-poor blood back to the heart. In the pulmonary circulation (where blood is pumped from the right ventricle into the pulmonary artery and out to the lungs), the arteries carry the oxygen-poor blood to the lungs, and the veins carry the oxygen-rich blood back to the heart.

Ultimately, the point of pumping all this blood throughout the body is to deliver oxygen and nutrients to the capillaries where these substances can be made available to the body's tissues. Capillaries are the smallest blood vessels. Their walls are very thin — thin enough to allow the nutrients in the blood to make their way into the surrounding tissues. Waste products produced in the tissues can also cross the

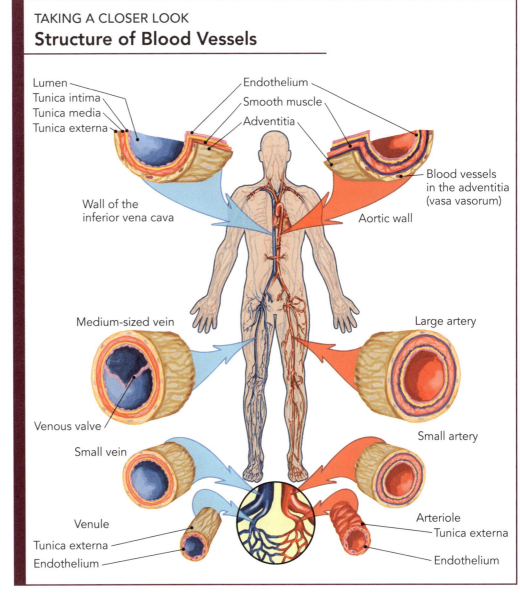

TAKING A CLOSER LOOK
**Structure of Blood Vessels**

capillary walls to be carried away in the blood. (These waste products include carbon dioxide, which eventually leaves the body when you breathe out. We'll talk more about that later in this book.)

Arteries take blood from the heart, and as they make their way to the various organs and tissues, they branch again and again, getting smaller and smaller as they find their way to all parts of the body. (You can think of this like a tree. Imagine that the trunk of the tree is the aorta. See how the branches get smaller and smaller the farther away they get.) This branching ends when the smallest arteries reach the capillaries. After the blood passes through the capillaries, the veins take over to return the blood to the heart. Here the veins progressively get larger and larger as they approach the heart.

But what about those other vessels, the arterioles and the venules? Where do they fit in? Simple really. *Arterioles* are the smallest arteries, and they lead to the capillaries. Blood travels through the capillaries to the venules. *Venules* are the smallest veins. They carry blood from the capillaries to the larger veins. Like the tributaries of a stream, venules and then veins join together to form larger and larger vessels.

## Blood Vessel Structure

Even though arteries and veins have different functions in the vascular system, they do have similarities in their structure. The walls of these blood vessels consist of three layers, each called a "tunic" or *tunica*. These three layers surround the *lumen*. The *lumen* is the open space through which the blood flows.

The innermost layer of a blood vessel is called the *tunica intima*, which means "inner tunic." Tunica intima lines every blood vessel. This lining must be very smooth in order to minimize friction as blood moves through vessels. If the tunica intima were rough, it might trigger blood to clot when it is not supposed to clot. The tunic intima is so smooth because it is mainly made of a smooth layer of tissue called the *endothelium*. Endothelium lines all the blood vessels and the heart itself, where it is the main component of the heart wall's inner layer, called the endocardium, which we discussed earlier. The endothelium lining blood vessels is a continuation of the endocardium of the heart.

The middle of the three layers of a blood vessel is called, not surprisingly, the *tunica media*, which means "middle tunic." It is made of muscular and elastic materials. The smooth muscle and sheets of elastic fibers making up the tunica media are arranged circularly around the blood vessel. This smooth muscle regulates the size of the lumen through which blood flows. When the smooth muscle contracts, the lumen *constricts*, or gets smaller. This is called *vasoconstriction*. (You can see that this word means "vessel getting smaller.") When the smooth muscle relaxes, the lumen *dilates*, or gets bigger (This is called *vasodilation*, meaning "vessel

### TAKING A CLOSER LOOK
### Structure of a Blood Vessel

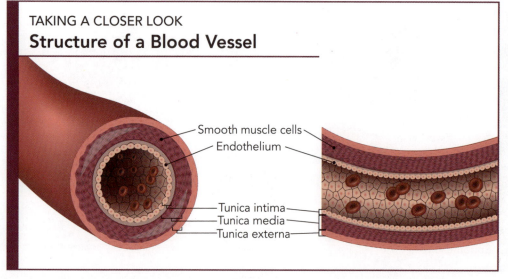

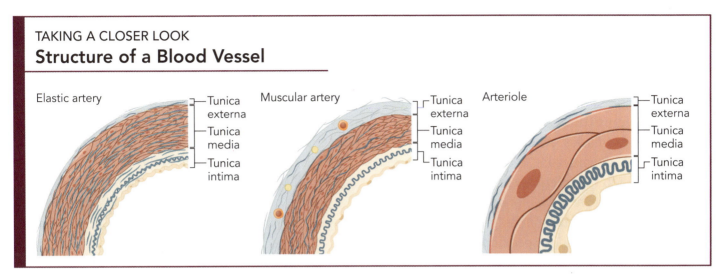

### TAKING A CLOSER LOOK
#### Structure of a Blood Vessel

getting bigger.") The elastic sheets enable blood vessels to stretch. The contracting and relaxing of the muscle in the tunica media is controlled by the nervous system as well as by many different chemicals and hormones. More about that later.

The outermost layer of a blood vessel is the *tunica externa*.[1] This layer is made of collagen and elastic fibers. Elastic fibers enable the vessel to stretch and return to its normal size. Collagen supports and protects blood vessels and anchors them to surrounding tissues. Have you ever heard of a disease called *scurvy*? Scurvy is a severe vitamin C deficiency. It was once very common during long sea voyages because fresh fruits and vegetables were generally unavailable. People with scurvy are unable to manufacture collagen properly. This causes many blood vessels to rupture, leading to spongy, bleeding gums, tooth loss, and many severe problems. The collagen in the outer layer of blood vessels is very important!

## Arteries

Arteries are the blood vessels that take blood away from the heart. These vessels branch more and more the farther they get from the heart in order to get blood to the entire body. As arteries get smaller they take on different roles, and these new roles are very important. Let's explore the three types of arteries: *elastic arteries*, *muscular arteries*, and *arterioles*.

*Elastic arteries* are the arteries that are closest to the heart. These include the aorta and the pulmonary artery along with the largest of the arteries that branch from them. Elastic arteries are the largest arteries in the body. They have a lot of elastic fibers in their walls, so they are very "stretchy." Elastic stretches and returns to its normal size. Because of their great elasticity, these large elastic arteries keep pushing the blood forward during the time the ventricles are relaxing. When the left ventricle contracts, it pushes blood under high pressure into the elastic arteries. This pressure stretches the elastic arteries. Then, when the ventricle relaxes, the elastic arteries recoil and keep blood flowing forward.

Muscular arteries are smaller than elastic arteries. The tunica media in muscular arteries is mainly made of smooth muscle and has fewer elastic fibers. Muscular arteries can contract to decrease the flow of blood to certain organs, diverting it to other places. When the muscle relaxes again, the lumen of the vessel is fully open. Having this ability allows the muscular arteries to regulate the amount of blood that is delivered to different parts of the body at any given time. For example, these arteries might help direct more blood flow to your leg muscles if you are running. On the other hand, if you were sitting outside on a cold day watching your favorite team play, muscular arteries might constrict and direct

---
[1] It is also at times called the *tunica adventitia*.

blood away from your skin and certain muscles to help conserve your body heat.

The smallest arteries of all are the arterioles. The walls of arterioles are mainly composed of smooth muscle with very little elastic tissue. The arterioles lead into the capillaries. When arterioles contract, the capillaries "downstream" receive very little blood flow. When the arterioles relax, the capillaries "downstream" receive more blood. In this fashion, the arterioles are important in helping regulate the precise delivery of blood to specific tissues.

## Capillaries

Capillaries are the smallest blood vessels. They can be found near practically every cell in the body. Capillaries spread throughout tissues. The more capillaries in a tissue, the better it can be supplied with water, oxygen, and nutrients. The brain and kidneys require a lot of energy, and they are supplied with more capillaries than tendons, for instance, which have a lower energy requirement.

Capillaries have a structure that is very different from that of arteries and veins. As you have seen, arteries and veins have walls composed of three layers ("tunics"), and the inner tunic of each is composed of endothelium. The wall of a capillary, however, consists of only a single layer of endothelial cells. This arrangement makes the capillary ideally suited for its purpose, namely, delivering vital substances to the cells surrounding them. Because a capillary wall is so thin, oxygen, water, and nutrients can easily move across it. These substances need only make their way across one cell rather than having to cross all three layers as they would in a larger blood vessel. Because capillaries branch so extensively in tissues, they provide lots of surface area to allow nutrients to quickly move into tissues.

To give you an idea of how small a capillary really is, consider this. The lumen of a capillary is so narrow that red blood cells must pass through single file! (Now that's small!) However, this is still one more example of the incredible design of the body. This single-file passage of the red blood cells through the capillaries may seem inefficient, but it actually allows more contact between each red blood cell and the capillary wall. Thanks to this contact, oxygen moves quickly from the red blood cell through the capillary wall and into the surrounding tissue. See, it wasn't an accident at all.

## Veins

After passing through the capillaries, the blood begins its journey back to the heart. This journey begins as the blood from the capillaries drain into small vessels called venules. Think of venules as the first step of the venous system. Venules merge as they get farther away from the capillaries. Eventually they form veins.

Venules are somewhat porous, although not as porous as capillaries. Interestingly, white blood cells are able to leave the bloodstream and enter the tissues through the walls of venules. White blood cells help the body fight infection.

As venules continue to merge, they form veins. Veins continue to merge,

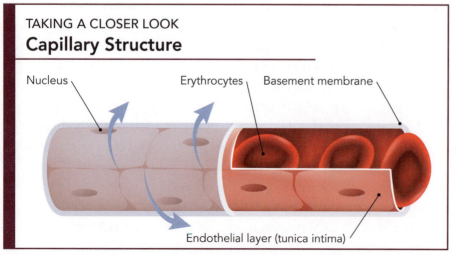

TAKING A CLOSER LOOK
**Capillary Structure**

Nucleus — Erythrocytes — Basement membrane — Endothelial layer (tunica intima)

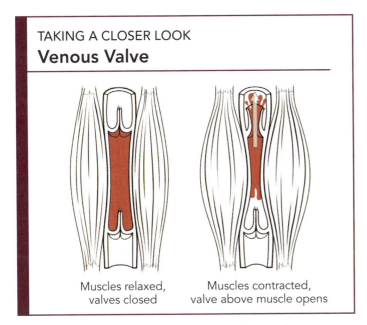

TAKING A CLOSER LOOK
Venous Valve

Muscles relaxed, valves closed

Muscles contracted, valve above muscle opens

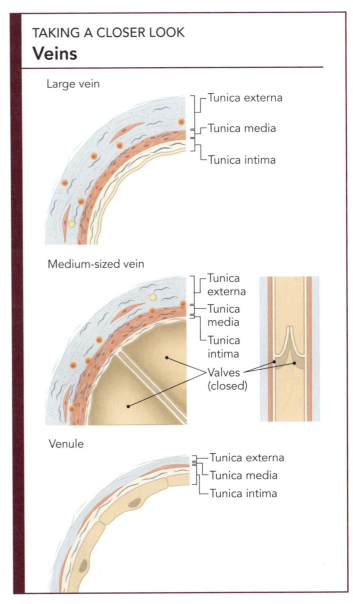

TAKING A CLOSER LOOK
Veins

Large vein — Tunica externa, Tunica media, Tunica intima

Medium-sized vein — Tunica externa, Tunica media, Tunica intima, Valves (closed)

Venule — Tunica externa, Tunica media, Tunica intima

coalescing into larger and larger veins as they get closer to the heart.

Veins have three layers as mentioned previously. Their walls are generally thinner than the walls of arteries. Consequently, the lumens of veins are larger than arteries of comparable size.

The primary function of veins is to return blood to the heart. Doesn't sound like too difficult a job, right? Well, it's not as easy as you might think.

You see, arterial blood is pumped out to the body at a high pressure. However, after passing through the capillaries, the pressure is much lower. So it's not as easy getting the blood *back* to the heart.

Fortunately, our Master Designer has this problem solved.

It's easy to imagine how blood pumped to the head and arms gets back to the heart. Gravity can do some of the work, because the blood mostly has to flow down. But what about the blood that has to get back from the legs and other parts of the body below the heart? Blood in those veins must go against gravity to reach the heart. As it turns out, many veins that come from areas of the body below the heart have small folds of the tunica intima that form *valves*. These valves prevent blood from flowing backward in the veins.

Return blood flow to the heart is also enhanced by the contraction of the muscles in the legs. As we move about, the leg muscles help compress the veins in the legs. This helps push blood back to the heart.

But no matter what part of the venous system we are talking about, more blood is still coming along from the capillaries. Therefore, venous blood is herded along through the venous system, pushed toward the heart by the blood behind it.

# PHYSIOLOGY OF CIRCULATION

Ultimately, the cardiovascular system is just a pump connected to a set of tubes. The pump pushes the blood out into the tubes. The tubes take the blood to its destinations and return it to the pump. This loop repeats and repeats and repeats. . .

But what controls the pump? What determines how much it pumps? How does it know how hard to pump? How do the pump and tubes fine tune their performance to your body's immediate needs? These are great questions. Let's get some answers.

  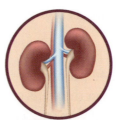 

# Hemodynamics 101

Have you ever felt your own pulse, had your blood pressure checked, or seen someone have their blood pressure checked by a nurse or a machine at the pharmacy? Most people know the term "blood pressure," even if they are not really sure what it means.

Now we are going to take a closer look at these concepts and help you better understand how the cardiovascular system works. We will start with a few simple ideas.

## Your Pulse

Your pulse is a measure of how fast your heart is beating. The heart pushes blood into your arteries every time it beats. You can feel the effect of the blood being repeatedly forced into the arterial system and count it to determine your pulse.

You can feel your pulse at many different places on your body. These are locations where certain arteries run close to the surface of the body. Some of the most common places to check the pulse are the brachial artery (in the bend of the elbow), the radial artery (in the wrist near the base of the thumb), and the *dorsalis pedis* (on the top of the foot). Check the illustration to see many of these locations. If you ever take a first aid course, you may learn to check the pulse by feeling the *carotid artery* in the neck. Many runners also do this. It is *very important* to only check a carotid pulse on one side of the neck. *Never put pressure on both left and right carotid arteries at the same time*, as this can trigger a dangerous reflex in which the heart rate drops.

Pulse rate is recorded in beats per minute. The normal resting pulse rate in an adult is around 70–75 beats per minute, but this can vary. Young children often have faster resting heart rates.

## Blood Flow

The amount of blood that flows through the cardiovascular system in a given time period is called *blood flow*. (At times it may be helpful to consider blood flow through a specific blood vessel or an organ. However, for now, let's just look at the system as a whole.) But wait a minute. Haven't we already studied the output of the heart? Indeed, we have. Remember the term *cardiac output* (CO)? For all practical purposes (when talking about the whole system and not just a single organ), blood flow and cardiac output are the same thing. The heart squeezes and the blood flows.

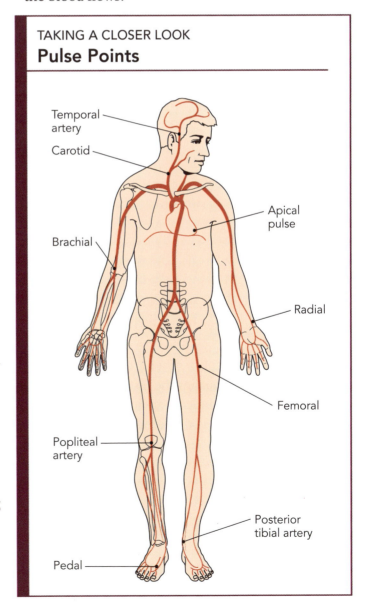

TAKING A CLOSER LOOK
**Pulse Points**

## Too Slow? Too Fast?

A heart rate below 60 beats/minute is called *bradycardia*. It is not uncommon for well-conditioned people to have a resting heart rate in this range. However, in certain circumstances a low resting heart rate can be an indicator of heart disease.

A heart rate above 100 beats/minute is called *tachycardia*. It is quite normal for the heart rate to exceed 100 beats/minute with exertion, and a young infant typically has a faster resting heart rate than an older child or adult. However, in an adult, a resting heart rate above 100 beats/minute most often requires further investigation.

Have you ever stopped to think about *why* blood flows? It's very important to know this. In fact, you really can't understand the cardiovascular system until you get this. Ready? Blood flows from areas of high pressure to areas of lower pressure.

Yes, that's pretty much the story. Blood flows from higher pressure to lower pressure.

The heart muscle contracts, generating a high pressure. This high pressure pushes blood into the aorta. The high pressure of the blood being forced into the aorta's lumen stretches it. Then, after the aortic valve closes, the aorta recoils a bit. This elastic recoil maintains a relatively high pressure in the aorta while forcing the blood into other arteries downstream. That pressure pushes the blood on toward the capillaries, then on into the venules and veins. Notice here that the blood flows toward the lower pressures.

## Blood Pressure

When blood is pushed into a blood vessel it exerts some degree of force against the wall of the vessel. Blood pressure is simply the pressure of blood inside a vessel.

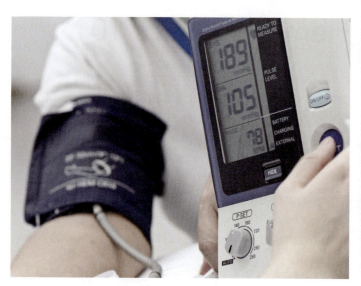

When the left ventricle squeezes (systole) and pushes blood into the arteries, the pressure in the arteries rises. The highest pressure reached in the arterial system at this time is called the *systolic blood pressure* (SBP). The average SBP in a typical healthy adult is 120 mm Hg (blood pressure is measured in millimeters of mercury).

Next, the ventricle relaxes (diastole). Now the walls of the arteries recoil and the pressure decreases as blood is pushed forward. The lowest pressure reached is called the *diastolic blood pressure* (DBP). The average DBP in a healthy adult is 80 mm Hg.

When a blood pressure reading is taken, it is most often recorded like this: 120/80 or 136/74 or 118/62, etc. Those are strange fractions, right? Well, they are not fractions. This is just the way that blood pressures have come to be displayed. The top number is the systolic pressure. The bottom number is the diastolic pressure. No fractions involved.

## Pulse Pressure

The difference between the systolic and diastolic blood pressure is called the *pulse pressure*. In our "average" adult with a blood pressure of 120/80, the pulse pressure is around 40 mm Hg (120 mm Hg - 80 mm Hg). The pulse pressure is greatest in the large arteries near the heart, and it decreases as you get farther and farther from the heart. By the time blood gets to the capillaries, the pulse pressure disappears, but the blood keeps moving.

We mentioned above that "pulse" is a measure of how fast the heart beats. Have you ever felt your pulse? If you have, then you have felt the effect of the pulse pressure. You see, the pulse pressure results in a throbbing sensation in an artery that can be felt at certain points along the paths of arteries that run near the surface of the body.

So if the arterial pressure is higher with systole and lower in diastole, which pressure is important? If the pressures just go up and down all the time, how can we make sense of all this? Actually, there is an easy way to calculate an average pressure, and its called the *mean arterial pressure* (MAP).

$$\text{MAP} = \text{DBP} + \text{pulse pressure}/3$$

Let's use our typical adult and make the calculation:

$$\text{MAP} = 80 \text{ mm Hg} + (40 \text{ mm Hg}/3)$$

$$\text{MAP} = 80 \text{ mm Hg} + 13 \text{ mm Hg}$$

$$\text{MAP} = 93 \text{ mm Hg}$$

Here is the key. The MAP is the pressure that pushes the blood through the blood vessels. It is greatest near the heart and gradually decreases as the distance from the heart increases. For example, near the end of the capillaries, the MAP has decreased to 15-20 mm Hg.

## Hemodynamics 102

So we know more about blood flow and pulse and blood pressure. How can these things adjust to changes in a person's activity? Or illness? Don't these things change a lot in both sickness and health? What controls those changes?

The body functions best when it operates under optimal conditions. The body functions best when the blood pressure is just right and the blood sugar level is just right and the body temperature is just right and the oxygen level is just right. The body's many control systems exist to try and keep things just right. Having these things in the right balance is called *homeostasis*. The same control systems that God designed to maintain the body's homeostasis also enable the body to adjust to changes. Let's consider the controls that keep the cardiovascular

Taking your pulse is easy. Using light pressure with your index finger, locate the throbbing sensation on the inner portion of the wrist just below the base of the thumb. This is your radial pulse.
Using a watch or a timer that indicates seconds, count the number of beats in 15 seconds. Then, multiply that number by four. That is your heart rate in beats per minute!

system in balance and enable it to adjust to meet the body's changing needs from moment to moment.

## Cardiovascular Center

In the brain — specifically, in the *medulla oblongata* — is the cardiovascular center. The cardiovascular center is the part of the nervous system that oversees regulation of the heart and blood vessels. This region of the brain gets input from multiple sources in the body and then responds by sending nerve signals to the heart and/or the blood vessels.

The cardiovascular center gets input from higher regions of the brain, such as the *cerebral cortex*, where our conscious thinking takes place, as well as the parts of the brain that handle our emotions. The cardiovascular center also processes input from special pressure receptors in the arteries. After processing these various inputs, the cardiovascular center sends the appropriate signal to the heart and blood vessels in various parts of the body to achieve the desired response.

What sort of messages might the cardiovascular center send? Well, a nerve signal might be sent to a particular arteriole to relax and dilate slightly in order to decrease the mean (average) arterial pressure. Or a signal could be sent to constrict slightly and increase the arterial pressure. These changes could make blood pressure drop or rise slightly, or return it to normal. If you are anxious about something, a signal from the cerebral cortex to the cardiovascular center also might result in your heart rate increasing. Signals might be sent to certain arterioles to constrict and decrease blood flow to certain organs while at the same time providing increased blood supply to other organs. For instance, if you need to run away in an emergency soon after you eat, blood that was busy supplying your digestive system will be diverted away from your stomach and sent instead to your legs. Digestion will just have to slow down and wait until you are through running to safety!

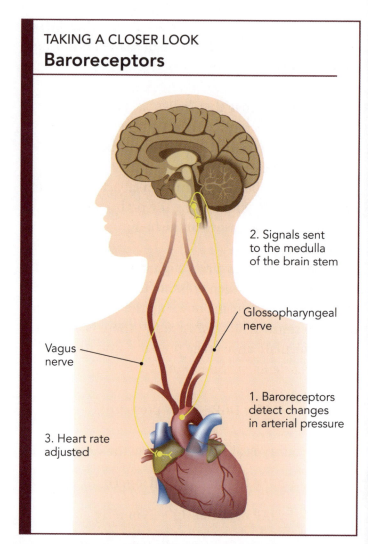

**TAKING A CLOSER LOOK**
**Baroreceptors**

1. Baroreceptors detect changes in arterial pressure
2. Signals sent to the medulla of the brain stem
3. Heart rate adjusted

Vagus nerve
Glossopharyngeal nerve

Now we will examine a few of the sensors that constantly send information to the cardiovascular center in more detail.

## Baroreceptors

Have you ever heard of a barometer or barometric pressure? These terms refer to measurement of the air pressure. Barometric pressure affects weather. The prefix *baro* means "pressure." Well, your body has its own pressure sensors inside your blood vessels.

Located in the aorta and several of the larger arteries in the upper body are pressure-sensitive receptors known as *baroreceptors*. These receptors are necessary for the minute-to-minute control of blood pressure. And remember, we said that the body works best at optimal conditions. The goal of the cardiovascular center is to regulate blood pressure to keep it in the normal range while allowing your body to do all the things it needs to do.

When blood pressure increases, baroreceptors are stretched. This stretching causes the receptors to send more nerve impulses to the brain. This in turn results in the cardiovascular center signaling the arterioles to dilate. Thus, this nerve reflex ultimately causes a decrease in blood pressure, returning the pressure to normal. Stimulation of baroreceptors also can slow the heart rate to some degree and decrease the contractility of the heart. These changes obviously lower the cardiac output, which also lowers the blood pressure. Lower cardiac output and vasodilation lower the blood pressure to appropriate levels.

In the opposite situation, when the blood pressure is low, the baroreceptors are not being stretched much and therefore send fewer signals to the brain. As a result, the cardiovascular center stimulates arterioles to constrict, increasing blood pressure back to normal. Additionally, the cardiovascular center causes an increase in heart rate and myocardial contractility, and with these there is an increase in cardiac output. Higher cardiac output and vasoconstriction raise the blood pressure to appropriate levels.

## Hormones and Blood Pressure

Some hormones have a direct effect on the cardiovascular system. Remember that hormones are chemical messengers that travel through the blood to reach many parts of the body.

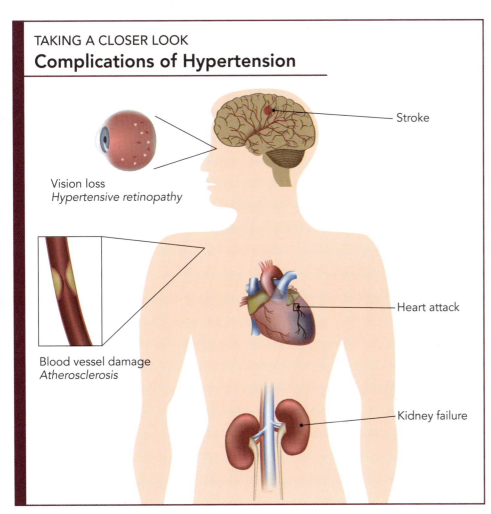

TAKING A CLOSER LOOK
**Complications of Hypertension**

Vision loss
*Hypertensive retinopathy*

Stroke

Blood vessel damage
*Atherosclerosis*

Heart attack

Kidney failure

Stimulation of the adrenal glands by the nervous system — perhaps due to fear, excitement, or anger — can cause the release of the hormones epinephrine (adrenaline) and norepinephrine (noradrenaline). As we have seen previously, these hormones can cause an increase in both heart rate and myocardial contractility, and consequently cardiac output may increase. The blood pressure can go up when more blood is moved more forcefully out of the left ventricle.

The kidneys have a great deal of control over blood pressure. When blood pressure decreases, blood flow to the kidneys is also decreased. When the kidneys detect the low blood flow, cells in the kidney secrete the hormone renin. Renin activates a hormone known as angiotensin

II. Angiotensin II causes vasoconstriction and raises blood pressure.

# Hypertension

The average blood pressure in a healthy adult is 120/80. However, blood pressure varies throughout the day, reacting to the various stresses we encounter. Blood pressure can increase with emotional stress or heavy exertion, but these situations are very short in duration, and blood pressure quickly returns to the normal range. When a person has a blood pressure that is chronically over 140/90 it is known as hypertension. This is commonly called "high blood pressure."

When the left ventricle squeezes, the pressure inside the ventricle rises. When the pressure in the ventricle becomes greater than the pressure in the aorta, the blood is pushed through the aortic valve and out to the body. However, if a person has hypertension, the pressure in the aorta is higher than normal. In this case, the left ventricle has to generate a higher pressure to push the blood out. As the heart is required to constantly pump against

this higher pressure for many months, the wall of the ventricle thickens. Ultimately, this stress on the ventricle results in damage to the muscular wall of the ventricle and the heart is weakened.

Also, exposure to these abnormally high pressures will cause the walls of arteries to thicken and become less flexible. This can lead to damage not only to the arteries of the heart but also to the blood vessels in the brain and kidneys.

Hypertension is one of the most common medical problems in the world. By some estimates, 25 percent of the world's population has hypertension. It is often called a "silent killer," because people can have the problem for many years before they develop symptoms. The damage to the heart, blood vessels, kidneys, and other organs can slowly progress for years before a person has any symptom to make them aware there is any problem at all. For this reason, one of the primary goals in modern medical practice is to identify patients with hypertension and take all necessary steps to control it.

### TAKING A CLOSER LOOK
### Hypertension – Damage to the Heart

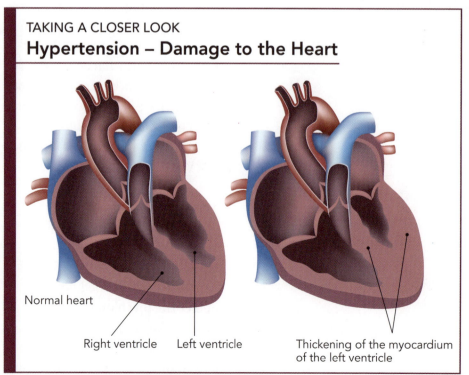

Normal heart

Right ventricle    Left ventricle    Thickening of the myocardium of the left ventricle

# Shock

As we have stated from the beginning, the primary purpose of the cardiovascular system is to deliver oxygen and nutrients to the body's tissues. As a general rule, the cardiovascular system does this extremely well (just like it was designed to do). However, we live in a fallen, cursed world, and things don't always work as planned.

There are situations in which the cardiovascular system cannot deliver adequate blood flow to meet the body's needs. This is called *shock*. There are various forms of shock and people in these situations are critically ill. Immediate medical intervention is required in all cases of shock.

The most common type of shock occurs when a person suddenly loses a great deal of blood. This might occur as the result of an accident, for example. Here, the loss of volume is so great that the body cannot generate adequate blood pressure and blood flow. This is called *hypovolemic* ("low volume") shock. The body's cardiovascular center tries to restore normal blood pressure — to restore homeostasis — by causing the heart rate to increase and many arterioles to constrict. When a person is in hypovolemic shock, the body's efforts to maintain homeostasis are not sufficient. The body cannot generate a sufficiently high blood pressure to supply the needs of the brain and other vital organs. Medical help is needed. The primary method of treating hypovolemic shock is by giving the patient blood transfusions and intravenous fluids. This increases the patient's blood volume, giving the cardiovascular system a larger amount of fluid to move around. With increased blood pressure, circulation can better bring needed oxygen and nutrients to the brain and body. Naturally the cause of the blood loss must also be treated!

There are situations where a person has a massive heart attack, and the resulting damage to the heart is so severe that the heart cannot pump adequately. This situation can result in *cardiogenic* shock. In this case the blood volume of the body is normal, but the heart cannot pump the needed oxygen and nutrients to the tissues. Doctors may give the patient medication to try to increase the heart's ability to contract and to decrease the damage to the heart. A patient in shock is also given extra oxygen to breathe in hopes of getting more oxygen into the blood and to the brain, heart, and other vital organs.

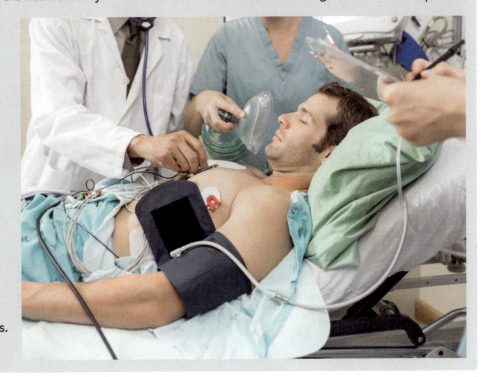

Another form of shock is called *vascular shock*. Vascular shock results from circumstances where blood vessels become too dilated (vasodilation). Here the pumping action of the heart is normal but adequate blood pressures are not obtained because the vasodilation is so extreme. This type of shock can occur during life-threatening infections or severe allergic reactions.

# THE CIRCULATORY SYSTEM

By this time you should be very familiar with the anatomy and physiology of the heart. You have even learned quite a bit about the structure of blood vessels and how they work. Before we finish our exploration of the cardiovascular system, let's spend a little time on the anatomy of the systemic circulation. (We will examine the pulmonary circulation a little more closely later in this book.)

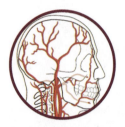
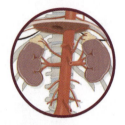
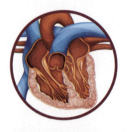

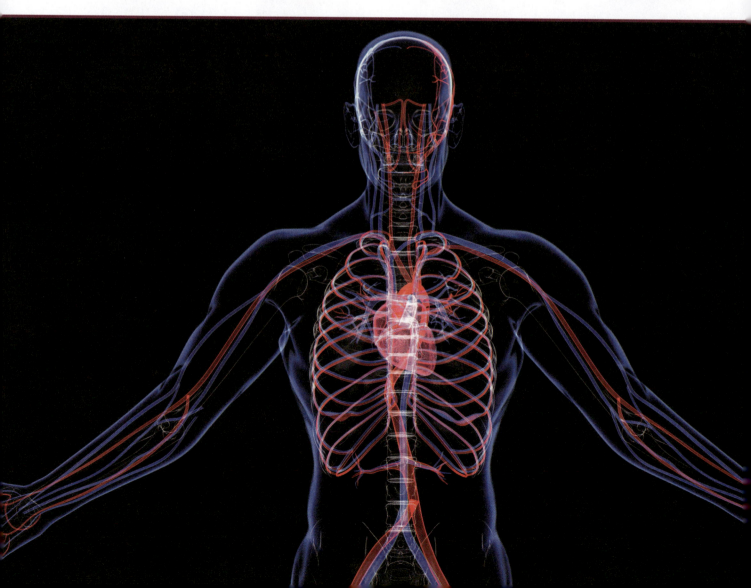

# The Systemic Circulation

The systemic circulation consists of the blood vessels that carry blood from the heart to the body's tissues and back again. It begins with the aorta.

### TAKING A CLOSER LOOK
**The Systemic Circulation**

- Head and neck
- Upper limb
- Abdomen and pelvis
- Lower limb

The aorta is the largest artery in the body with a diameter of roughly one inch. It is considered to have four portions.

## The Aorta

The small section of the aorta when it leaves the heart is called the *ascending aorta*. This small section is noteworthy because the coronary arteries that supply oxygenated blood to the heart muscle begin here. Next, the aorta bends and turns downward. It arches. Therefore, this part of the aorta is called the *aortic arch*. As the aorta proceeds down through the body, it changes names again. The *thoracic aorta* is the portion of the aorta in the thoracic (chest) cavity above the diaphragm, which separates the chest from the abdomen. The *abdominal aorta* is the remaining section of the aorta below the diaphragm. The abdominal aorta is located in the back of the abdomen, near the spine.

### TAKING A CLOSER LOOK
**The Aorta**

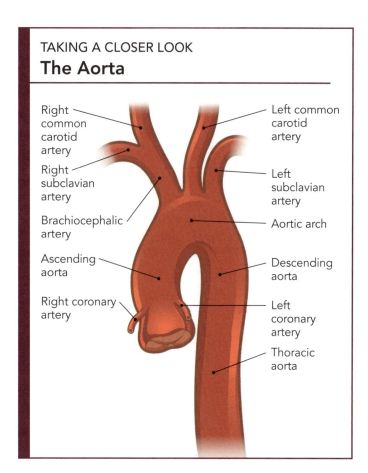

- Right common carotid artery
- Right subclavian artery
- Brachiocephalic artery
- Ascending aorta
- Right coronary artery
- Left common carotid artery
- Left subclavian artery
- Aortic arch
- Descending aorta
- Left coronary artery
- Thoracic aorta

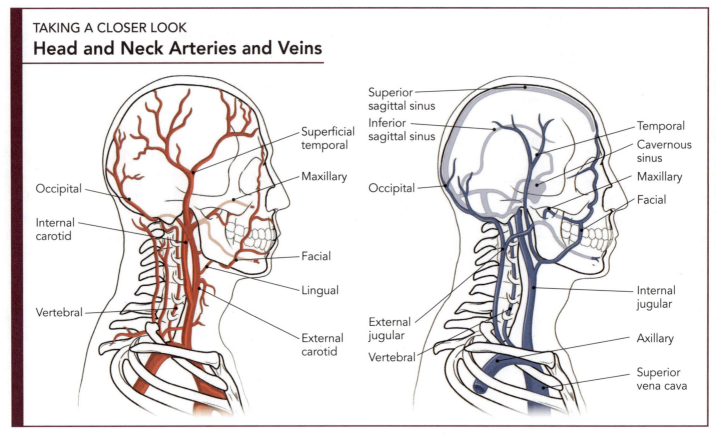

## The Head and Neck

As you examine the arch of the aorta, you will see three major arteries — the brachiocephalic trunk, the left common carotid artery, and the left subclavian artery. As you can see from the illustration, soon after it branches from the aorta, the brachiocephalic trunk gives off the right common carotid artery. The common carotid arteries provide blood supply to the head.

The common carotid arteries each soon divide into internal and external branches. The internal carotid arteries provide the primary blood flow to the brain and the eye. The external carotid arteries supply blood to the other tissues in the head.

The venous drainage of the head brings blood back to the heart through the internal and external jugular veins.

## Upper Limb

Blood is supplied to the upper limbs via the subclavian arteries. The right subclavian artery branches off the brachiocephalic trunk, but the left subclavian artery comes directly from the aortic arch. Each subclavian artery runs down the arm and changes names according to where in the arm it is located. First it becomes the axillary artery and later the brachial artery, in the upper arm. Near the elbow the brachial artery splits into the radial and ulnar arteries which travel down the forearm carrying blood toward the hand.

The venous return is by way of the cephalic, basilic, and brachial veins which ultimately form the subclavian vein.

The final return path from both arms and the head is the superior vena cava, the large veins that drain into the right atrium from above.

### TAKING A CLOSER LOOK
### Upper Limb Arteries and Veins

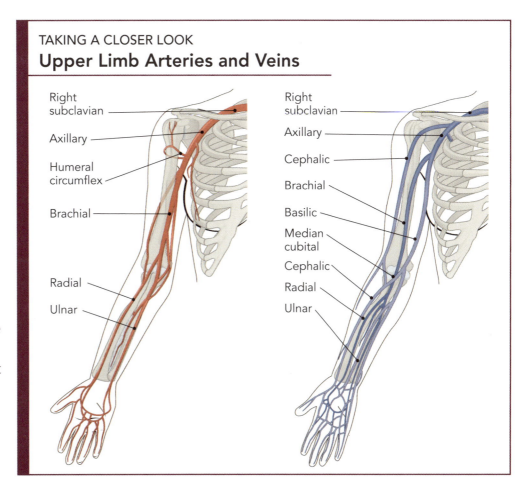

### TAKING A CLOSER LOOK
### Thorax and Abdomen Arteries and Veins

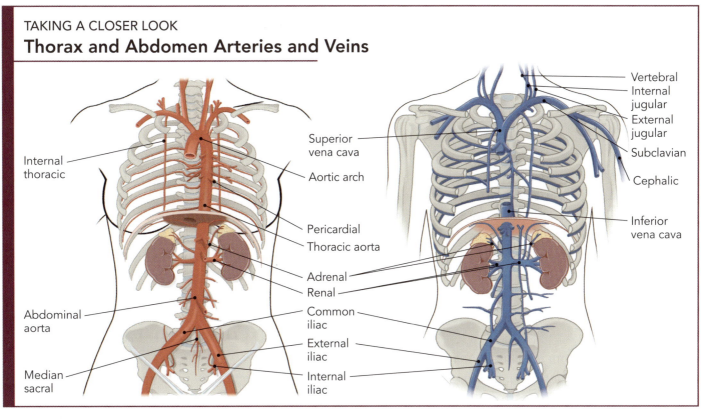

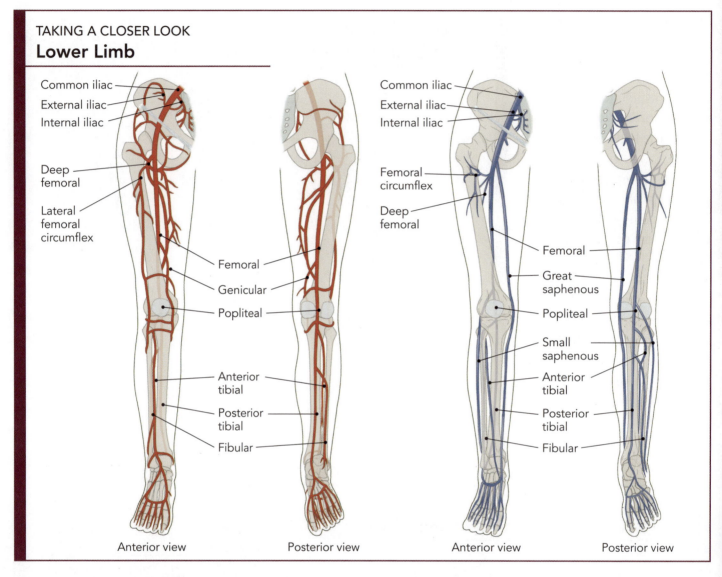

## Abdomen and Pelvis

The portion of the aorta below the diaphragm is the abdominal aorta. As it courses through the abdomen, several arterial trunks arise from the aorta. These include the celiac trunk, the superior mesenteric artery, and the inferior mesenteric artery. The renal arteries also branch from the aorta.

In the pelvis, the abdominal aorta branches into the right and left common iliac artery.

The venous return from the abdomen and pelvis is via the inferior vena cava. The inferior vena cava is formed by the joining of the two common iliac veins.

## Lower Limb

The common iliac arteries supply the lower limbs. The primary arteries in the lower limb are the femoral, the popliteal, and the anterior and posterior tibial arteries.

Venous return from the lower limb is via the tibial, saphenous, and femoral veins.

# Your Spiritual Heart

The heart is commonly thought of as the well-spring of our deepest emotions. It seems to pound when we're scared, and soar upward when we're in love. When we are very disappointed or sad, we might say our heart is broken. Perhaps, if a person is particularly enthusiastic about an activity, we might say about them, "John has a real *heart* for playing the piano" or "Mary has a *heart* for that project."

The heart also represents our genuine thoughts and motives. The Bible warns, "The heart is deceitful above all things, and desperately wicked" (Jeremiah 17:9), letting us know how we as sinful people often deceive ourselves. A person can "harden" his heart (Exodus 8:32) like the Egyptian pharaoh in the time of Moses. Or a person can be sorry for his sins: Psalm 51:17 tells us God honors a "broken and a contrite heart," referring to sincere repentance.

So imagine the surprise people felt when William Harvey in 1628 told the world the heart was a pump!

Nonetheless, the heart is a good metaphor for that invisible part of ourselves that thinks and feels and will even live after our physical bodies die. Jesus promises to "dwell in your hearts through faith" (Ephesians 3:17) if you have accepted the salvation He offers. So His presence in your spiritual heart gives you eternal life, and your physical heart must keep doing its job for you to have physical life

But more than anything else, Jesus wants you to love God with all your heart.

Jesus said to him, "You shall love the LORD your God with all your heart, with all your soul, and with all your mind." This is the first and great commandment. (Matthew 22:37–38).

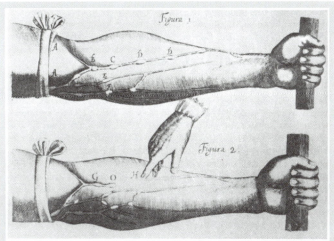

William Harvey (1578-1657) was an English physician who was the first person to describe the circulatory system in detail. He demonstrated that blood flows throughout the body by way of a single system of arteries and veins. This was supported by solid observations and experiments. Before Harvey, most people thought that the arterial and venous systems were separate.

Harvey's famous work on this subject was called *Anatomical Exercise on the Motion of the Heart and Blood in Animals.*

# SECTION 2
# THE RESPIRATORY SYSTEM

It's a fact. In fact, it's a simple fact. We can't live without breathing.

The cells in our bodies need oxygen. Without oxygen, the chemical reactions in our cells that generate energy would quickly shut down. Without energy, the cells would die.

As our cells use oxygen and nutrients to produce energy, carbon dioxide is produced. This carbon dioxide is essentially a waste product. When it is dissolved in liquid, carbon dioxide becomes an acid. Therefore, if carbon dioxide were to remain in the body, it would cause the acid levels in the blood to increase to dangerous levels.

So how do we get oxygen in and carbon dioxide out? Just take a deep breath and you've got your answer.

> And the LORD God formed man of the dust of the ground, and breathed into his nostrils the breath of life; and man became a living being (Genesis 2:7).

# 1

# BREATHING — NO BIG DEAL?

Ever think about breathing? You probably don't. At least not very often.

It is something that the average person does 14 to 16 times a minute, morning, noon, and night, day in and day out, for a lifetime. Ninety-nine percent of the time, you never give it a thought. It's almost totally automatic.

Yet without breathing, you would die.

The respiratory system provides a means to take in oxygen and get rid of carbon dioxide. But it does so much more. . .

 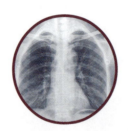 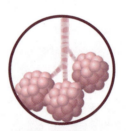 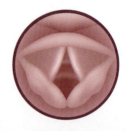

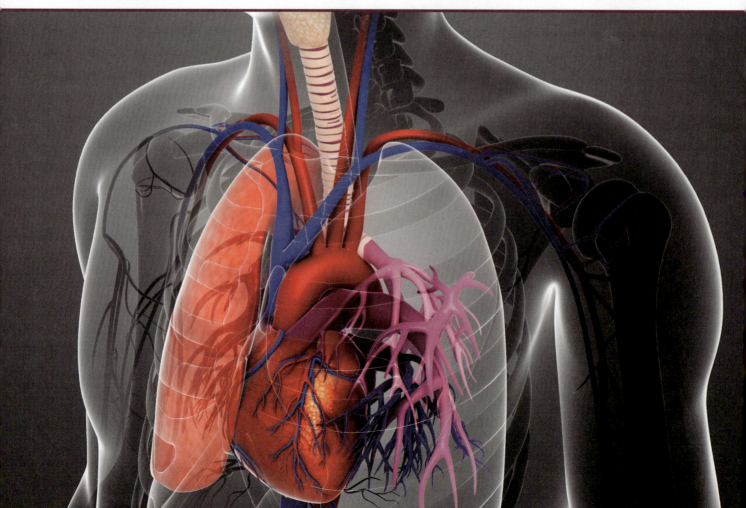

The respiratory system allows us to communicate with others more effectively. When we speak to a friend or shout at a ballgame or laugh at a joke or sing in church, we are relying on our respiratory system. Without it, none of these activities would be possible.

As we will see, the respiratory system also protects us from bad things in our environment. There are lots of bad things in the air we breathe — dust, pollen, and germs. The air we take into our lungs can contain these things. The respiratory system is designed to, at least in some degree, protect us from them.

The respiratory system even protects us from air that is too cold or too dry for comfort. It warms and humidifies the air we breathe even before it reaches our lungs.

Our Creator designed this system pretty well, don't you think?

If you don't think so yet, you will before we're done.

# Anatomy of the Respiratory System

As we begin our exploration of the anatomy of the respiratory system, let's look at its parts and what they do. When you breathe in — or *inhale* — air must travel through the respiratory system from top to bottom. And when you breathe out — or *exhale* — air must follow the same path in reverse.

Inhaled air must first travel through the "upper respiratory system." The upper respiratory system is the part above your chest. It consists of the nose, nasal cavity, the sinuses, the pharynx, and the larynx. *Pharynx* — pronounced "fair-inks" — is the anatomical word for "throat." *Larynx* is pronounced "lair'-inks" and rhymes with pharynx. Larynx is the word for your voice box. Your throat is designed to direct air toward your lungs and food and drink toward your stomach. Air passes through your voice box as it enters and leaves your chest, and you are able to use exiting air to generate the sound vibrations of speech.

Inhaled air continues on through the "lower respiratory system." The lower respiratory system contains the trachea,

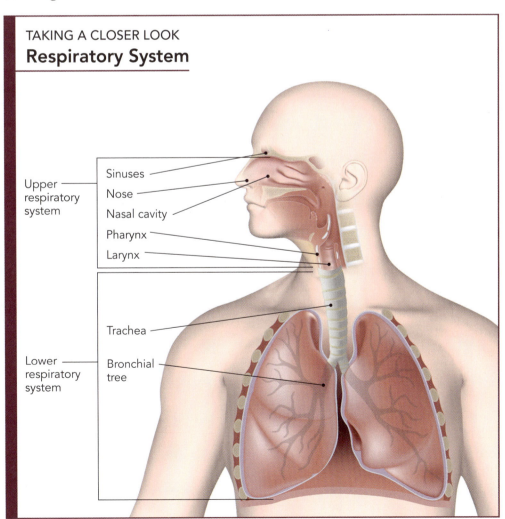

TAKING A CLOSER LOOK
**Respiratory System**

the branching tubes of the bronchial tree, and the lungs. The *trachea* — pronounced "trā'-kē-uh" — is the windpipe. The trachea and the bronchial ("bron'-kē-ul") tree conduct air to and from your lungs and distribute that air through them. The tubes that branch off of the trachea (the *bronchi*) keep on branching into smaller and smaller tubes (*bronchioles* and finally *terminal bronchioles*), just like a tree.

"Upper" and "lower" of course only refer to the general location of the parts of the respiratory system. A doctor might say that a patient has an "upper respiratory infection" to refer to something like a head cold or sinus infection or a "lower respiratory infection" when referring to bronchitis or pneumonia. We'll learn more about these problems later.

Another way to think about the parts of the respiratory system is by describing the sort of function the parts have. Some parts of the respiratory carry air from place to place. Other parts of the respiratory system are designed to get oxygen from the air into the blood and to get carbon dioxide out of the blood and into the air to be exhaled. This is called "gas exchange," because oxygen and carbon dioxide enter and leave the body as gases. During gas exchange in the lungs, oxygen and carbon dioxide sort of trade places.

Many parts of the respiratory system conduct air from outside the body to the areas in the lungs where gas exchange takes place. Inhaled air moves through the nose, nasal cavity, pharynx, larynx, trachea, bronchi, bronchioles, and terminal bronchioles, in that order. The nose and other parts of the upper respiratory system filter, warm, and humidify inspired air long before it reaches the lungs.

After air is conducted to the farthest reaches of the lungs, gas exchange finally takes place. Air enters alveoli, tiny thin-walled sacs, where it is able to be close enough to the blood in capillaries for oxygen and carbon dioxide to move between the blood and air. Oxygen from inspired air is taken into the blood. Carbon dioxide from the blood is released into the alveoli to be removed with the expired air.

## The Nose

The nose is probably the most overlooked part of the respiratory system. After all, it just sticks out from our face, right? I mean, it doesn't do anything, right? Well, it does hold people's glasses, but other than that...

In reality, the nose is very important, and does much more than provide a place for eyeglasses to rest. Let's take a closer look.

The nose is the beginning of the respiratory system. The two openings on the nose are called the *external nares*, or nostrils. Air, of course, can bypass the nostrils when you breathe through your mouth, but that air misses out on some of the important things the nostrils, nasal passages, and sinuses do. The nose is far more than just an opening for air. The nostrils, or nares, provide the ideal entrance to the respiratory system.

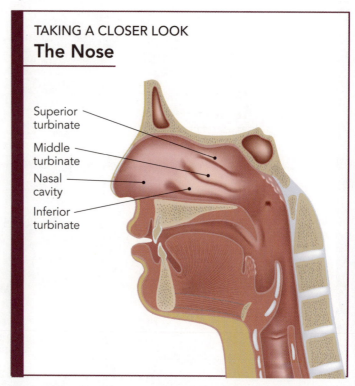

TAKING A CLOSER LOOK
### The Nose

- Superior turbinate
- Middle turbinate
- Nasal cavity
- Inferior turbinate

When the air passes through the nares, it enters the nasal cavity. In the nose are coarse hairs that help filter out dust particles. This cleaning action minimizes the amount of foreign material that finds its way to the lungs.

Inspired air then passes over the lining of the nasal cavity. This lining is not ordinary skin but instead is a "mucous membrane." Mucous membrane is thin and far more moist and supple than skin, and it is rich in small veins and capillaries. Warm blood flowing through these blood vessels helps warm cold air as it passes over the mucous membrane. The moisture produced by the mucous membrane humidifies the air. The gooey mucus, coating the mucous membranes, traps lots of dust and bacteria. Tiny hairs on the mucous membrane, called *cilia*, sweep this mucus and the debris stuck in it toward the

## Sneezing

A sneeze is a very forceful expulsion of air from the nose and mouth. Everyone sneezes at one time or another.

Why do we sneeze?

As it turns out, there is a pretty simple explanation.

Anything that irritates the nasal lining can trigger a sneeze. This might be an irritation due to dust particles or pollen. It could be irritation from a cold virus. Exposure to fumes, say a perfume or a strong household cleaner, can cause enough irritation to trigger a sneeze.

When nerve fibers in the nasal lining detect the irritation, a nerve reflex is triggered. This reflex causes muscles in the chest and throat to contract vigorously. The result is a powerful release of air from the mouth and nose; that is, a sneeze! The purpose of the sneeze is to expel the mucus that contains the things that are irritating the membranes.

A sneeze can fire air out of your body at over 100 miles per hour. Air traveling at that speed can make a really loud sound! One sneeze can shoot out 40,000 droplets containing thousands of germs — and the mucus-containing droplets propelled by a sneeze can travel quite a distance, sometimes ending up on someone else's mucous membranes. Yuck. Simply covering your nose and mouth when you sneeze helps stop germs from spreading.

There are some common misconceptions about sneezing. First of all, the heart does not stop when you sneeze. Doesn't happen. Next, your eyes cannot pop out of your head when you sneeze. Although the natural tendency is for the eyelids to be closed during a sneeze, if the eyelids are held open, the eyeballs won't pop out. Promise.

# CARDIOVASCULAR & RESPIRATORY SYSTEMS

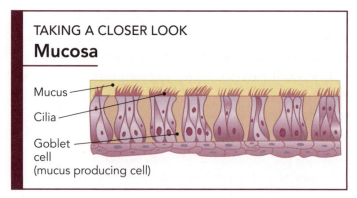

TAKING A CLOSER LOOK
**Mucosa**

- Mucus
- Cilia
- Goblet cell (mucus producing cell)

throat where it can be swallowed and destroyed in the stomach. Already you can see that the nose and nasal cavity are a built-in air purifier and conditioner protecting your lower airways and lungs.

The mucous membranes in the nasal cavity are designed to secrete lots of mucus. By the way, notice the spelling here. *Mucous* membranes make *mucus*.[1] Mucus is a gel-like coating produced by the mucous membrane lining our nose, sinuses, mouth, throat, lungs, digestive tract, and even our eyelids.

Mucus might sound yucky, but it is very important. The mucus that is made in the nasal cavity not only moistens the air but helps trap pollen, dust particles, and bacteria. It is a very important part of the defense system in our respiratory tract. Mucus contains special enzymes, like lysozyme, that help destroy bacteria. Mucus even contains molecules that attract helpful bacteria-eating viruses! (These are called bacteriophages.) God designed mucous membranes and the mucus they produce to clean up and condition air before it enters our bodies.

God's design for cleaning up our inspired air goes beyond this wonderful goo coating the inside of our nose. Our nose and nasal cavity does not provide a straight path for the air passing through. Instead, they contain shelves and barriers that force the air to flow over the surface of a lot of mucous membrane. There are, for instance, three shelf-like *nasal conchae* (pronounced "kong'-kē") sticking out from each side of the nasal cavity. (These are also called *turbinates*.) They disrupt the straight flow of air and thus stir it up. By maximizing air's contact with the mucous membranes in the nasal passages, each breath of inspired air has plenty of opportunity to get warmed or cooled to within a degree of your body's temperature and moistened. The circuitous route air takes protects the upper airway and lungs from harsh, cold, dry air when you are outside on a cold day.

## Sinuses

Have you ever heard somebody say, "If you believe that, you've got a hole in your head"? Actually, you do have holes in your head. In fact, we all do. And it's no joke. We need those holes in our heads!

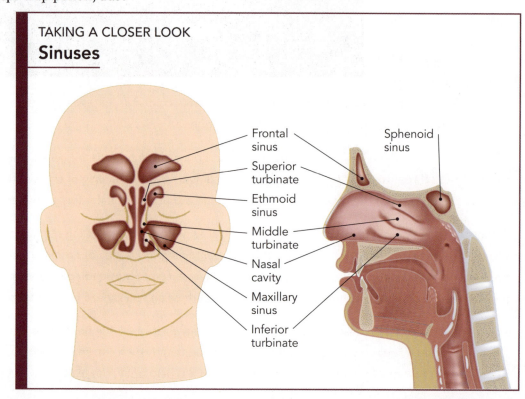

TAKING A CLOSER LOOK
**Sinuses**

- Frontal sinus
- Sphenoid sinus
- Superior turbinate
- Ethmoid sinus
- Middle turbinate
- Nasal cavity
- Maxillary sinus
- Inferior turbinate

---

[1] Mucous is an adjective, but mucus is a noun.

Surrounding the nasal cavity are several air-filled spaces in the skull. Connected to the nasal cavity by small passageways, these air-filled spaces are called *sinuses*. There are four sinuses, and they exist in pairs, one on each side of the nasal cavity. They are the frontal, maxillary, ethmoid, and sphenoid sinuses. You can see these in the illustration. These sinuses are located in hollowed-out spaces in the skull, but they are lined with mucous membrane just like the nasal cavity itself. The frontal sinuses are in your forehead. The maxillary sinuses are below your eyes. The ethmoid and sphenoid sinuses are located deeper behind your face.

The membranes that line the sinuses are continuous with those in the nasal cavities. Mucus that is produced in the sinuses ultimately drains through the passageways connecting them to the nasal cavity and into the nasal cavity itself. Once there, it can be blown out if you blow your nose or drain down your throat to your stomach, where stomach acid can kill any germs that might be left alive.

The sinuses have several functions. First of all, they help decrease the weight of the skull. If the sinus cavities were solid bone, the facial part of the skull would be much heavier. Secondly, as in the nasal cavity, the sinuses defend your body from germs, dirt, and cold, dry, irritating air. The mucus in the sinuses helps trap and kill bacteria. The sinuses warm and humidify the air that passes through them. Finally, the sinuses add some resonance to the voice by adding chambers for vibrating air to bounce around. Consider how different a friend's voice sounds when he or she has a bad cold.

Not bad for a few holes in your head!

## The Pharynx

The next part of the respiratory system is the pharynx. The pharynx is commonly called the throat. It is a funnel-shaped tube that begins at the rear of the nasal cavity and extends down to the larynx (voice box). In order, air passes through three regions of the pharynx: the nasopharynx, the oropharynx, and the laryngopharynx. As you might guess, the prefix on each of these big words simply refers to the nearest landmark — the nose (*naso*), the mouth (*oro*), and the larynx (*laryngo*).

The nasopharynx is the superior (upper) portion of the pharynx. It begins at the rear of the nasal cavity and extends as far as the soft palate. The *soft palate* is the rear portion of the roof of the mouth. When you swallow, the soft palate moves enough to close off the nasal cavity from the nasopharynx. This prevents food from getting into the nasal cavity. If you've ever managed to inadvertently get some of your soda past your soft palate and up into the back of your nose while coughing or laughing, you know that you should thank our Lord for giving you a soft palate to make that a very rare occurrence!

The nasal cavity and the soft palate mark the upper and lower boundaries of the nasopharynx, but there is another way for air and small amounts of fluid to enter and leave it. The nasopharynx communicates

---

**TAKING A CLOSER LOOK**
**The Pharnyx**

- Eustachian tube opening
- Nasopharynx
- Oropharynx
- Laryngopharynx
- Larynx

with each middle ear via a tiny tube. These *pharyngotympanic tubes* are also known as *Eustachian tubes*, named after the 16th-century anatomist who discovered them, Bartolomeo Eustachi. These small tubes drain any excess mucus or other fluid that might collect in the middle ear and empty it into this portion the pharynx. The Eustachian tubes also help equalize the pressure between the atmosphere around you and the middle ear.

If you have ever been in an airplane or on a tall mountain and felt your ears "pop," you understand the value of the Eustachian tubes. You might have had to yawn to encourage the tiny Eustachian tubes to open enough to equalize the air pressure outside your body (and therefore inside your mouth and throat) with the air pressure inside your middle ear. Didn't your hearing suddenly improve once your ears "popped"? You see, the middle ear is the part of your ear where the vibrating eardrum (*tympanic membrane*) transfers sound vibrations to three tiny bones, making it possible for you to ear. If the air pressure inside your middle ear is different from the air pressure outside of your body, the eardrum cannot vibrate freely and your hearing is temporarily impaired. By equalizing the air pressure between your throat — the *pharynx* — and the eardrum — the *tympanic membrane* — the *pharyngotympanic tubes* (see how they got that big name?) restore your hearing to normal.

The *oro*pharynx is the middle part of the pharynx. It is the part of the throat that is behind your mouth, or *oral cavity*. The oropharynx extends from the soft palate down to the level of the hyoid bone, which is located just below the level of the chin. The *hyoid bone* is a small U-shaped bone that supports your tongue. The hyoid bone does not have any joints with

## Ear Infection

Have you ever had an ear infection? Middle ear infections are called *otitis* (meaning "ear inflammation") *media* (meaning "middle"). Acute otitis media causes a great deal of pain, sometimes with fever. Though anyone can develop an ear infection, it is more common in babies and young children. Acute otitis media is associated with fluid accumulation in the middle ear as well as bulging and redness of the eardrum.

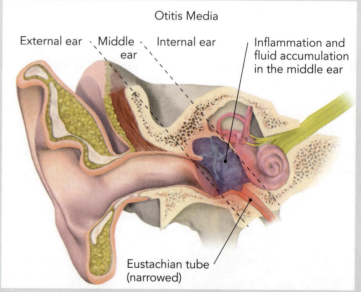

Acute otitis media is usually caused by bacteria. How, you might well wonder, do bacteria manage to get into the middle ear? Typically, they move up through malfunctioning Eustachian tubes. When a sore throat, allergy symptoms, or exposure to tobacco smoke causes inflammation of the nasopharynx, the Eustachian tubes may swell and fail to drain properly. Eventually, because the air pressure in the middle ear cannot easily equalize with atmospheric pressure, negative pressure develops in the middle ear. This negative pressure can then pull bacteria-containing fluid — even the milk a baby sucks from its bottle — up the Eustachian tube and into the middle ear.

Ear infections are more common in the young because their Eustachian tubes are shorter, smaller in diameter, and more horizontal than those of adults. Therefore, their Eustachian tubes swell shut more easily and, being shorter, bacteria-laden fluid only has to make a short trip to reach the middle ear.

other bones but is held in place by ligaments and muscles that help move your tongue and voice box.

The nasopharynx and the oropharynx are not just tubes through which air passes. They are also the home of your tonsils. The *adenoid tonsils* are located in the back of your nasopharynx. The *palatine tonsils* and the *lingual tonsils* are located in the oropharynx. The lingual tonsils are located at the back of your tongue. The palatine tonsils look like two pink mounds on each side of the oropharynx. Because they are exposed to germs entering your body, the tonsils are an important part of your immune system. Like policemen guarding this gateway to your respiratory system, the tonsils contain white blood cells that help your body recognize and fight infection. Sometimes tonsils are surgically removed with a *tonsillectomy* if they become persistently swollen or severely infected.

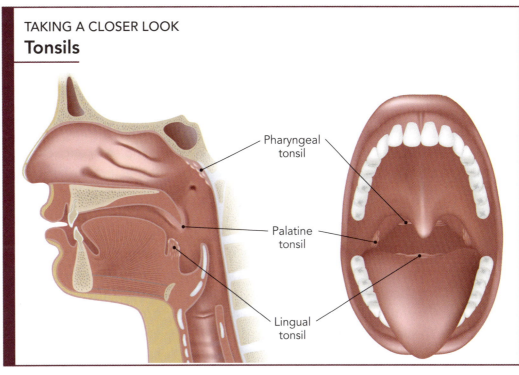

TAKING A CLOSER LOOK
## Tonsils

- Pharyngeal tonsil
- Palatine tonsil
- Lingual tonsil

# Snoring

Snoring is very common. Snoring is caused by a partial obstruction of airflow through the mouth, nose, or throat. Vibration of the obstructing structures or the squeezing of air through a narrowed passage produces the low-pitched noise that may disturb the snoring person's sleep or keep others awake.

Swelling in the nasal passages due to allergies, enlarged tonsils and adenoids, a deviated septum (a crookedness in the wall separating the nostrils), an unusually long uvula, and excessive relaxation of the tongue or throat muscles are just some of the causes of snoring. Men are more likely to snore than women, and overweight people with a lot of bulky tissue in the neck are also more likely to snore.

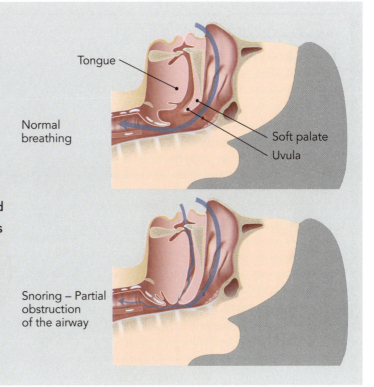

Normal breathing

- Tongue
- Soft palate
- Uvula

Snoring — Partial obstruction of the airway

The third part of the pharynx, the laryngopharynx, begins at the level of the hyoid bone and ends at the larynx. Upon leaving the laryngopharynx, inspired air is permitted by an anatomical gate called the epiglottis to continue on its path to your lungs. The epiglottis slams shut, however, if food or liquids are passing by, and directs them toward your esophagus and on to the stomach. We'll see in the next section how the epiglottis is designed.

## The Larynx

The larynx is the portion of the airway that connects the laryngopharynx to the trachea. It is often called the "voice box." The larynx protects the trachea from foreign material entering from above and transforms exhaled air into speech and song.

Several pieces of cartilage form the larynx. The three largest are called the thyroid cartilage, the cricoid cartilage, and the epiglottis. The epiglottis is like a trap door at the top of the larynx, and below it is the most prominent cartilage, the *thyroid cartilage*. The thyroid cartilage, commonly called the "Adam's apple," is visible on the front of your neck and moves up and down as you swallow. You can feel a little notch on the top of the thyroid cartilage with your finger. It is called the "thyroid" cartilage because the thyroid gland is draped over part of it. Below the thyroid cartilage is the *cricoid cartilage.* The cricoid cartilage is shaped like a signet ring, and it completely encircles the top of the trachea, to which it is attached. The broader part of this signet ring forms the lower rear wall of the larynx and the thin part of the cricoid ring is located in front.

Supported within these larger cartilages forming the outer walls of the larynx are smaller ones that support the vocal folds, tissues that vibrate to produce sounds as air passes over them. These are called the arytenoid, cuneiform, and corniculate cartilages.

The epiglottis guards the entrance to the larynx from above, protecting it from the things you eat and drink. The epiglottis is a leaf-shaped flap of elastic cartilage. The "stem" portion is anchored to the anterior rim of the thyroid cartilage and acts like a hinge for this doorway to the airways. The broader (or "leaf") end of the epiglottis is unattached, allowing the epiglottis to swing up and down like a trap door. When we swallow, the larynx moves upward and, at the same time, the epiglottis flaps down to close off the airways below. This movement of the epiglottis helps direct food and drink into the esophagus and keeps them out of our lungs.

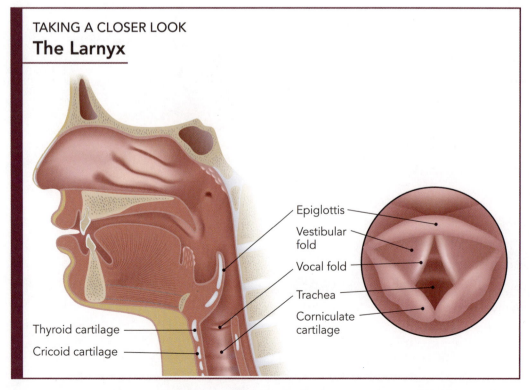

TAKING A CLOSER LOOK
**The Larnyx**

Just below the attachment point of the epiglottis, the mucosa (mucous membrane) lining the airway forms two sets of folds. The superior and thicker ones are the *vestibular folds*, or "false vocal cords." As you might guess from this name, this pair of folds does not participate in sound production, but they do help close off the glottis while you speak.

Inferior to the vestibular folds are the *vocal folds*, or "true vocal cords." Muscles in the larynx can tighten or relax the vocal folds as needed to produce higher and lower sounds as air passes over them and to move the cartilages supporting them to begin "shaping" the sounds into words. The area of the larynx that contains the vocal cords is known as the *glottis*. (The epiglottis is located *above* the glottis: the prefix *epi* means "above.")

Let's look more closely at the way God designed human anatomy to allow you to speak and sing.

> **The Voice**
>
> Some people don't talk much.
> Some people talk way too much.
> Most of us are somewhere in between.
>
> Our Creator has given us a great gift. This is the gift of speech. The ability to speak allows us to communicate and interact with one another in ways that other creatures cannot.

So how does it work?

When we speak or sing, we force air from our lungs through the glottis and past the true vocal cords. As the air passes the vocal cords, the vocal cords vibrate. Muscles in the larynx can increase or decrease the tension on the vocal cords, causing a change in the sound. When vocal cords are more relaxed, they vibrate slower, resulting in a lower pitch. Alternatively, when vocal cords are tighter, they vibrate most rapidly, resulting in a higher pitch.

As a general rule, men have lower voices than women. This is because due to the effect of androgens (male sex hormones) men have slightly longer and thicker vocal cords.

But it is not just the vocal cords alone that produce speech. The nasal and oral cavities, as well as the sinuses, contribute to the resonance of our voices. The movement of the tongue, lips, and pharynx help mold sound into recognizable speech. The force with which air is pushed from the lungs past the vocal cords plays a very important role in the loudness of our voice.

## Laryngitis

*Laryngitis* is an inflammation of the larynx. It can result in hoarseness or even a complete, though temporary, loss of voice.

Inflammation of the vocal cords (vocal folds) can cause swelling and pain, keeping them from moving normally during speech. In severe cases, the swelling can be so extensive as to almost prevent the vocal cords from vibrating.

Laryngitis can be caused by viral or bacterial infections, allergies, or inhalation of chemical fumes. Irritation of the larynx can also result from chemical burns caused by stomach acid that sometimes percolates up the esophagus and slips over into the larynx. This is called *acid reflux disease*. Overuse of the voice can lead to inflammation of the vocal cords. Irritation caused by cigarettes or alcohol can also cause laryngitis.

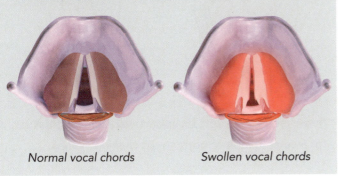

*Normal vocal chords*     *Swollen vocal chords*

# The Trachea

The *trachea*, or windpipe, begins at the larynx and runs down into the chest. It is a tube about 5 inches long and 1 inch in diameter. The trachea branches into the two main *bronchi*, one *bronchus* for each lung.

The wall of the trachea has four layers: mucosa, submucosa, hyaline cartilage, and adventitia.

Like the other airways in the respiratory system, the trachea is much more than a tube. The trachea, as well as most of the airways we are discussing, is lined with a mucous membrane that produces dust-catching mucus and sweeps it away with microscopic cilia. The inner layers of the trachea — the mucosa and and submucosa beneath it — are packed with mucus-producing *goblet cells*. The mucosa is also loaded with *ciliated cells*. Cilia — hairlike projections from the ciliated cells — continually move mucus containing dust and debris toward the pharynx.

## TAKING A CLOSER LOOK
### The Trachea

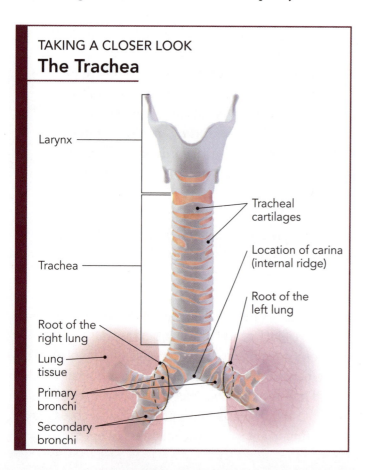

- Larynx
- Trachea
- Tracheal cartilages
- Location of carina (internal ridge)
- Root of the left lung
- Root of the right lung
- Lung tissue
- Primary bronchi
- Secondary bronchi

## Heimlich Maneuver

Occasionally a person may have his or her airway completely blocked by a piece of food. This is obviously a medical emergency. The obstruction must be removed immediately.

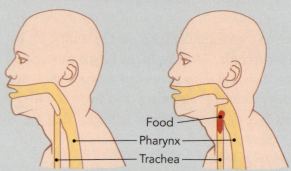

- Food
- Pharynx
- Trachea

In recent years, a procedure known as the Heimlich maneuver was developed to enable laymen with some knowledge of first aid to help in these emergencies, sometimes providing life-saving help long before medical professionals could arrive. This maneuver involves exerting sudden pressure on the victim's upper abdomen just below the diaphragm. This sudden compression would ideally push air into the trachea with enough force to dislodge the obstruction and open the airway. However, it is easy to injure someone with this maneuver (such as fracturing a rib or the end of the sternum), so care must be taken when attempting to perform this on a person. The point at which pressure should be applied in the Heimlich maneuver varies in children and adults.

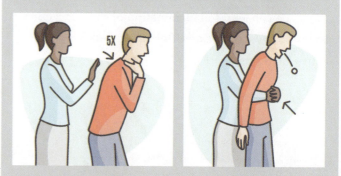

Some medical experts favor back slaps or chest thrusts (or a combination of both) to aid choking victims.

## Emergency Tracheotomy

While the Heimlich maneuver is a procedure that anyone can learn to perform safely, an emergency *tracheotomy* — despite what you might have seen on television — is far riskier. (*Trache* means "trachea" and *otomy* means "opening.") If a choking person is completely unable to move any air at all past an obstructed trachea, and despite repeated attempts to clear the obstruction with the Heimlich maneuver and abdominal thrusts, is still unable to even gasp or cough, it may become necessary to open the trachea surgically to allow life-saving air to enter the lungs.

Actual emergency tracheotomies are extremely difficult to perform without proper medical training and equipment. Therefore, in a life-threatening situation outside a hospital, the procedure normally used as a last resort is not a tracheotomy but a *cricothyrotomy* (though many people still call this a tracheotomy). An actual tracheotomy is placed between tracheal rings, but a *cricothyrotomy* is a surgical opening placed higher on the trachea, between the cricoid cartilage and the thyroid cartilage (Adam's apple). This opening is below the place ordinarily plugged by a misdirected bit of food or other foreign body, and once a small tube is placed in the surgical opening in the cricothyroid membrane, air can enter the lungs.

Despite the apparent ease of this procedure in television drama, even a cricothyrotomy in the hands of an untrained person is extremely risky. A thorough knowledge of anatomy and the techniques involved in the procedure are needed to safely navigate the many major blood vessels and nerves in the neck.

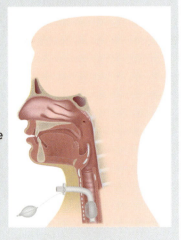

While the cilia in the nasal cavity direct mucus down toward the pharynx, cilia in the trachea sweep mucus upward away from the lungs and toward the pharynx. In this manner, the lining of the trachea works like the lining of the nasal cavity in protecting and cleaning the airway. God has designed every part of our bodies with the details it takes to protect us.

Below its mucosa and submucosa, the trachea is supported by 16 to 20 rings of hyaline cartilage. Without these sturdy "C" shaped rings of vertically stacked cartilage, the trachea could collapse and air movement would stop. Dense connective tissue holds the cartilage rings together. The openings of each "C" face posteriorly (to the back) toward the esophagus. (The esophagus is the tube connecting the throat to the stomach, and it is right behind the trachea.) These rings of cartilage keep the airway open despite the pressure changes that occur in the trachea during inspiration and expiration.

The outer layer of the trachea is adventita. This layer is composed of connective tissue that helps secure the trachea to surrounding structures.

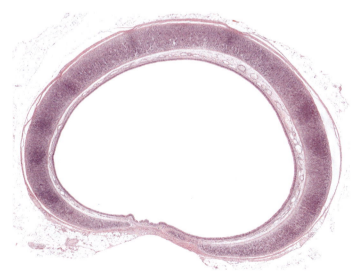

*Cross section of the trachea, showing C-shaped tracheal rings of hyaline cartilage. Ciliated mucosa lines the trachea.*

# Bronchi and Bronchioles

The trachea ends by dividing into two *bronchi*. The *right primary bronchus* goes to the right lung, and the *left primary bronchus* goes to the left lung. As the bronchi continue into the lungs, they branch more and more, much in the same way that blood vessels branch the farther they are from the main blood vessels. These airways are at times called the "bronchial tree."

The primary (first) bronchi divide to become the *secondary bronchi*. There is one secondary bronchus for each lobe of the lung, so secondary bronchi are also called the *lobar bronchi*. Since the right lung has three lobes, and the left lung has two lobes, there are three lobar bronchi on the right and two lobar bronchi on the left.

The secondary bronchi divide to form the *tertiary bronchi* which supply air to segments of the lung (and are therefore also called *segmental bronchi*). Tertiary bronchi then divide to form bronchioles. Bronchioles are much smaller in diameter than bronchi and do not have any cartilage in their walls to keep them open. Instead, they are held open by elastic fibers that attach them to the surrounding lung tissue. The bronchioles continue to divide, getting smaller and smaller, until they become *terminal bronchioles*. The terminal bronchioles are the end of the air-conducting part of the respiratory system.[2]

The structures past the terminal bronchioles are involved in gas exchange. Terminal bronchioles branch to form *respiratory bronchioles*, tiny airways studded with gas-exchanging *alveoli*. Respiratory bronchioles divide into *alveolar ducts*, which are also studded with outpocketing alveoli. Each alveolar duct ends in an open space into which grape-like clusters of alveoli open. A person develops new alveoli until he or she is about eight years old, by which time the average person has about 300 million of them. Do you get the idea that alveoli are important?

# The Lungs

The lungs are the organs in the body responsible for both ventilation and respiration. Respiration is the process in the lungs that gets oxygen into the blood and carbon dioxide out of the blood. Here we are referring specifically to external respiration, which is gas exchange in the lungs. *Internal respiration* is gas exchange between capillaries and the cells and tissues of the body. *Cellular respiration* is the process by which cells obtain energy from nutrients like glucose.

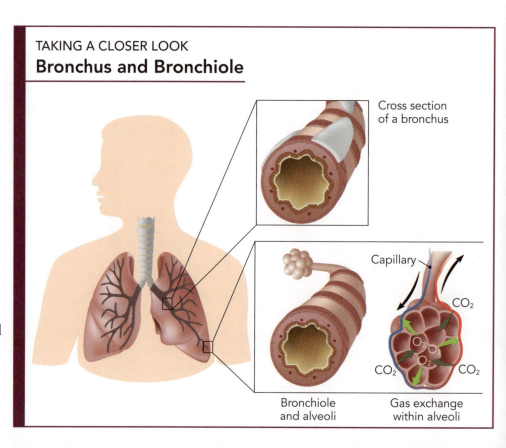

TAKING A CLOSER LOOK
**Bronchus and Bronchiole**

Cross section of a bronchus

Capillary

$CO_2$

$O_2$

$CO_2$  $CO_2$

Bronchiole and alveoli

Gas exchange within alveoli

## Asthma

Do you have a family member or a friend who has asthma? There is a good chance that you do. Though it varies in severity, about 8 percent of the population in the United States suffers from asthma. That works out to about 1 person in 12, and that's significant.

*Asthma* is a chronic inflammatory disease of the airways. It is characterized by intermittent spasm of the airways. That means that parts of the airways constrict for varying periods of time, making it difficult to exhale or inhale. As you can imagine, this is very uncomfortable and frightening, and it can sometimes become dangerous. Symptoms typically include shortness of breath, cough, and wheezing. Wheezing is the noise produced when air squeezes through a tight space. Wheezing can occur during expiration and during inspiration. The symptoms are often intermittent. In other words, patients can be without symptoms one minute and then suddenly develop severe shortness of breath.

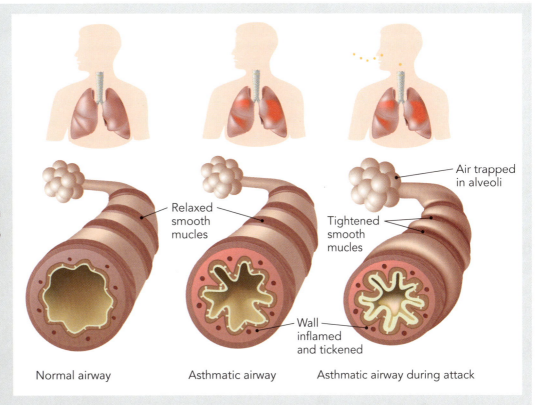

Our understanding of the cause of asthma has improved greatly in recent years. In the past it was thought that asthma was caused primarily by the contraction of the smooth muscle in the airways. This contraction narrowed the airways and made movement of the air into and out of the lungs very difficult. More recently, the role of inflammation in asthma has been much better recognized. This improved understanding of the cause of asthma has led to better treatment for the disease. In the past, the goal was to relieve the airway constriction. Today, the primary goal is to relieve, or at least minimize, the inflammation in the airways.

Asthma can be triggered by many things. Infections, either viral or bacterial, are an obvious trigger. Exposure to cold air, exercise, allergies, and exposure to irritants, such as perfumes, household chemicals, or cigarette smoke can also exacerbate asthma.

Treatment of asthma is focused primarily on minimizing the inflammation in the airways. Long-term management usually includes the use of inhaled corticosteroids (to minimize the chronic inflammation), and often includes long-acting medications called bronchodilators (medication aims at relieving the construction of the airways). These medications are often administered by small hand-held inhalers. Sudden attacks of asthma are often treated with fast-acting bronchodilators designed to help quickly relieve airway obstruction.

Air must reach the lungs in order for respiration to happen. *Ventilation* gets air into the lungs. Ventilation just means "breathing." It is the process of moving air from the environment into the lungs (inspiration, or inhalation) and then moving air back out (expiration, or exhalation). Without lungs, air would not be drawn in through our airways, and with no way to get oxygen into our blood we would die in minutes.

## Gross Anatomy

The human body has two lungs, one on each side of the chest cavity. The lungs are separated from each other by the heart and large blood vessels in the *mediastinum*. Each lung is somewhat cone shaped, broader at the bottom and rounded at the top. The inferior portion of the lung is called the base. The base of the lung sits on the diaphragm. The superior portion of the lung is called the apex.

On the medial surface of each lung is the hilum. The *hilum* is where the bronchi and blood vessels enter the lungs.

The right and left lungs are not identical. The right lung is slightly larger than the left. The right lung has three lobes, and the left lung has only two. Can you think of why? There has to be room for the heart! The left lung has a concave depression that accommodates the heart.

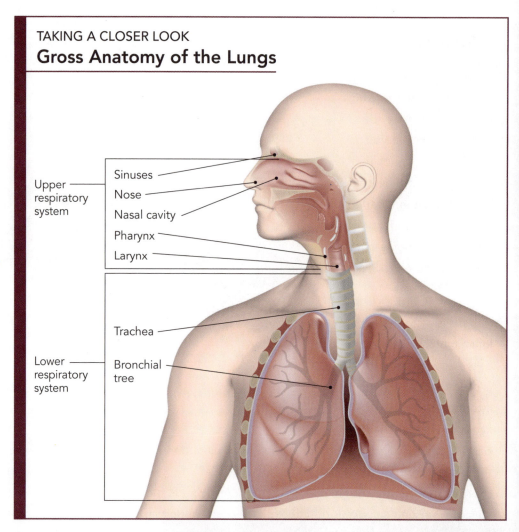

TAKING A CLOSER LOOK
### Gross Anatomy of the Lungs

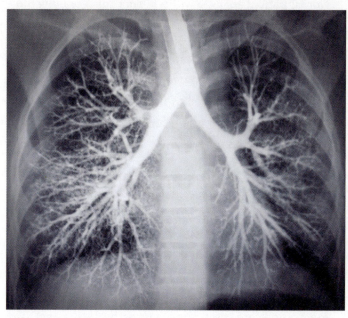

X-ray demonstrating the extensive branching of the bronchial tree.

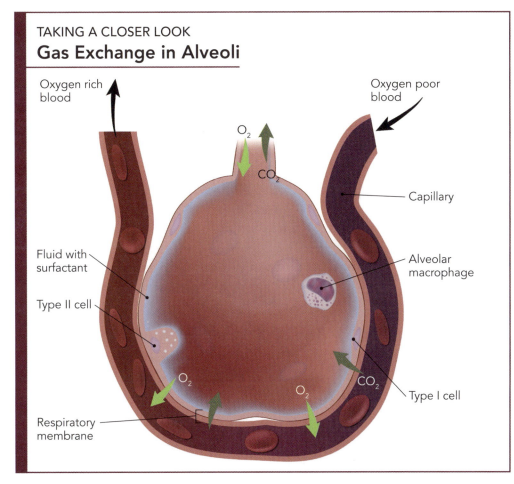

**TAKING A CLOSER LOOK**
**Gas Exchange in Alveoli**

## Alveoli

Gas exchange happens across the walls of little sacs called *alveoli*. Alveoli are like tiny sacs that pooch out from the walls of respiratory bronchioles and alveolar ducts, the smallest airways in the bronchial tree. Some estimates suggest that there are as many as 2 million alveolar ducts in each lung! The alveolar ducts end in clusters of *alveoli* called *alveolar sacs*. Alveolar sacs look like microscopic clusters of grapes.

One *alveolus* (the plural form of alveolus is *alveoli*) is a microscopic air sac. It is the endpoint of the respiratory system. Alveoli are the "grapes" that make up alveolar sacs. Gases are exchanged between the air in the respiratory tree and the blood in capillaries draped on the alveoli. Oxygen and carbon dioxide move across their thin walls and the thin walls of the capillaries.

An alveolus can, in some ways, be compared to a balloon or a bubble. It is a small cavity that is made of a single layer of squamous epithelium. These epithelial cells are known as type I alveolar cells. Type I cells are responsible for the structure of the alveolar wall, making up 95 to 97 percent of the alveolar lining. These cells are extremely thin. This allows the most efficient gas exchange possible.

The outer surface of the alveolus is covered by a mesh of capillaries. At this point, the air in the respiratory system and the blood in the circulatory system are brought very close together. As you recall, capillary walls are also very thin. A thin capillary wall adjacent to a thin alveolar wall is the best possible design for the efficient movement of gases in and out of the bloodstream. Remember, you have at least 300 million alveoli in each of your lungs. That's a lot of alveoli, giving you a HUGE surface area for gas exchange!

Is this another example of the incredible design of the human body, or did this happen by chance? Does this cause us to stop and give praise to our wonderful Creator, or do we merely accept that this is the result of chemicals randomly banging together over millions of years? That's not really a hard choice, is it?

In addition to the type I cells in the alveolus, there is another important cell found there. Incredibly enough, these are called type II alveolar cells (clever, huh?). The type II cells secrete alveolar fluid. This fluid keeps the surface of the alveolus moist. The most important component of this alveolar fluid is surfactant. *Surfactant* is a special kind of molecule that has some detergent-like properties. It aids in reducing the surface tension of the alveolar fluid, and that helps prevent the alveolus from collapsing. Surfactant helps alveoli stay open like the soap bubbles children blow through the air for fun.

There are also macrophages in the alveolus. These cells can move along the surface of the alveolus. Macrophages are like cellular vacuum cleaners. They keep the alveolus free from dust and debris.

## Blood Vessels

Blood enters the lungs by two different arterial pathways. One of these brings blood to the lungs to pick up oxygen for the body to use. The other brings blood for the lungs themselves to use.

The first pathway by which blood reaches the lungs we have already studied, namely, the pulmonary arteries. As you remember, deoxygenated blood enters the lungs via the right and left *pulmonary arteries*. These arteries enter the hilum of each lung and then branch to get smaller and smaller until they become the capillaries that surround the alveoli.

## Pneumonia

Pneumonia is an inflammatory condition of the lungs. It primarily affects the air sacs, the alveoli. The most common cause of this inflammation is infection, either bacterial or viral. Other causes include autoimmune disease or exposure to certain drugs. A particularly dangerous type of pneumonia occurs when stomach contents are aspirated into the lungs (so-called aspiration pneumonia), causing acid damage to the lining of the airways.

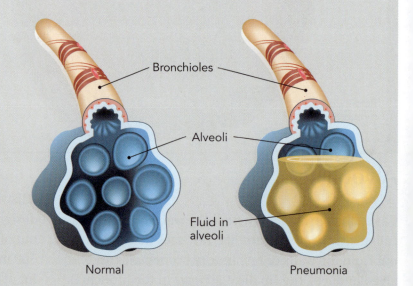

Pneumonia is characterized by cough, fever, and shortness of breath. Chest x-rays are very helpful in diagnosing pneumonia. The inflamed areas of the lung are often seen clearly on the x-ray film. If the cause of the pneumonia is bacterial, it can be helpful to obtain samples of the patient's sputum for testing. This can help determine the type of bacteria causing the infection.

The severity of pneumonia ranges from relatively mild to life threatening. The most common treatment for pneumonia is antibiotics. Management of pneumonia can also at times include supplemental oxygen, breathing treatments, and intravenous fluids. In the most severe cases, patients can be placed on a ventilator (a machine designed to move air in and out of the lungs).

Pneumonia is the eighth leading cause of death in the United States. It is most common in very young children (under age 5) and in adults over age 75.

Risk factors for pneumonia (other than age) include a history of smoking, alcoholism, liver or kidney disease, and illnesses that suppress or weaken the immune system, such as cancer.

After the blood is oxygenated, it exits the lungs via the pulmonary veins on its way back to the heart. The heart, of course, sends this oxygenated blood out to the brain and the rest of the body.

So why, you might wonder, would the lungs need an additional way to receive blood? You probably remember that the heart is full of blood but still has to have a special set of coronary arteries to supply the heart muscle itself with oxygen and nutrients. The same is true of the lungs. Though they are filled with air containing up to 21 percent oxygen as well as miles of capillaries devoted to capturing some of this oxygen for the body to use, the lungs need their own dedicated supply of oxygenated blood.

That other important arterial pathway bringing blood to the lungs involves the *bronchial arteries*. Like the coronary arteries that supply the heart muscle, bronchial arteries arise from the aorta and, like the pulmonary arteries, enter each lung at the hilum. Bronchial arteries supply *oxygenated* blood and nutrients to the lung tissue. Most of the blood from the bronchial arteries returns to the heart through the pulmonary veins, and some returns via the bronchial veins that empty into the superior vena cava.

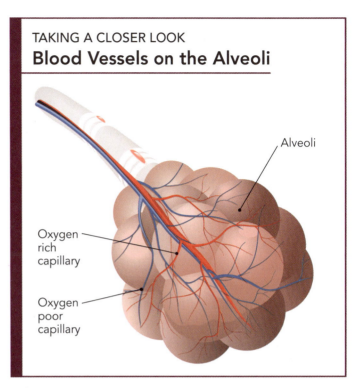

TAKING A CLOSER LOOK
### Blood Vessels on the Alveoli

- Alveoli
- Oxygen rich capillary
- Oxygen poor capillary

## Pleura

Each lung is enclosed by its own double-layered membrane called the *pleura*. The layer of the pleura that covers the surface of the lung is called the *visceral pleura*. The pleural layer adjacent to the interior of the chest wall is called the *parietal pleura*.

## Surfactant

You may have heard of premature babies that have difficulty breathing when they are born. While a baby is inside the mother's womb, the baby does not have to use lungs to exchange gases. Gas exchange takes place between the baby's blood and the mother's blood through the placenta. But as soon as a baby is born, he or she must take a breath to expand the alveoli in the lungs. From then on, it is the job of the baby's lungs to obtain oxygen and remove carbon dioxide from the blood. If there were no surfactant in the alveoli, the baby would have difficulty breathing because many of the alveoli would soon collapse.

God has designed a baby's lungs to begin making surfactant just a few weeks before the time to be born. If the baby is born too soon, he or she may have difficulty breathing because the surfactant is not yet present. Intensive care facilities today can assist a premature baby to breathe and even provide artificial surfactant until the baby's lungs can catch up with production.

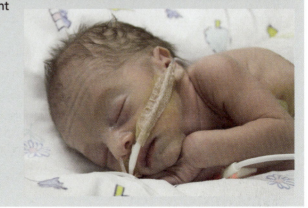

Recall the example we used when illustrating the pericardium. If you take a balloon partially filled with air and slowly press your fist into it, you get the idea. Your fist will be inside a two-layered membrane. This illustration helps describe the pericardium as well as the pleura.

Cells in the pleura produce a small amount of liquid that helps lubricate the potential space between the visceral and parietal pleural surfaces. This is called *pleural fluid*. It reduces friction between the pleural layers and allows the lungs to expand and contract more easily. Even though the pleural layers glide across each other easily, the surface tension produced by the pleural fluid makes it very difficult to actually separate the layers. This helps keep the lungs expanded inside the chest cavity.

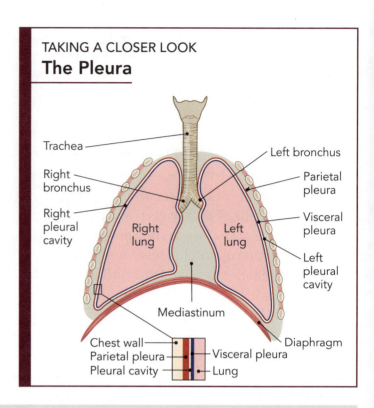

TAKING A CLOSER LOOK
**The Pleura**

## Pleural Problems

As we described, the visceral layer of the pleura covers the surface of the lung, and the parietal layer of the pleura covers the inner surface of the chest wall. These layers glide across one another, but due to surface tension they rarely separate. However, on occasion the pleural space can be compromised due to illness or injury.

One of the more common problems occurs when the pleural membranes become inflamed. This is known as *pleurisy*. Pleurisy can result in severe pain when taking a breath. The pain results from the inflamed membranes rubbing against each other. (Think of how uncomfortable it is when something rubs against your skin when you have sunburn, and you'll get the idea.) Pleurisy is often the result of an infection, for example, pneumonia. However, other diseases can cause it.

At times, fluid can leak into the pleural cavity. When blood collects in the pleural cavity, it is called a hemothorax. This can be the result of a traumatic injury to the lung.

In other situations, watery (serous) fluid accumulates in the pleural cavity. This is called a *pleural effusion*. Effusions can have many causes such as infection, malignancy, heart failure, liver disease, and kidney disease, to name just a few.

Regardless of the cause, when blood or other fluid accumulates in the pleural cavity, problems arise. If fluid prevents the lung from expanding properly, shortness of breath can result, as well as a significant decrease in the amount of oxygen that makes it into the bloodstream. The goal in these situations is to reduce the amount of fluid compressing the lung. This might be accomplished by medication, but often it requires drainage of the blood or fluid in the chest.

A pneumothorax occurs when air enters the pleural cavity. This most often results from lung disease, such as emphysema or asthma, but can also be due to malignancy, connective tissue disease, or trauma.

# Smoking

If I could suggest two things to help keep you healthy, there would be one "do" and one "don't."

First, DO exercise regularly. Exercise helps you feel better. It increases your endurance. It is good for the health of your heart and lungs. Many people say that they even concentrate better when they are exercising regularly. You don't have to run a marathon every week. Just get off the couch, get out from behind your computer, and get active!

Now for the DON'T. Never, ever, ever, under any circumstances, start smoking. There is absolutely nothing positive about smoking. It's not good for you. It's not cool (no matter what some people might say!). It makes your breath and clothes smell bad. It costs money (which then just goes up in smoke). And it can damage your body in ways that time can never undo. Let me repeat, just DON'T!

Smoking is one of the leading causes of death worldwide. Estimates place the number of smoking-related deaths in the United States at about 500,000 per year. Compared to nonsmokers, the life expectancy of smokers is reduced by about 13 years. If you stopped reading right here, that statistic alone should be enough to discourage you from ever taking up this terrible habit! In a nonsmoker, the risk of dying of lung cancer is 1 percent. The risk of dying of lung cancer is around 20 percent for a male who smokes and around 12 percent for a female who smokes. And although lung cancer is very, very serious, it's not only thing related to smoking.

Smoking increases the risk of heart attack, peripheral vascular disease, cancer of the larynx and mouth, emphysema, pancreatic cancer, bladder cancer, and stomach cancer. I honestly cannot think of a single positive thing when it comes to smoking. Let me repeat, just DON'T!

I would conclude this section with a painful personal note. My own mother died in her early 50s. She died of lung cancer. She was a two-pack-per-day smoker. Before she died, she begged her grandchildren to always remember what she went through as a direct result of smoking. She told them never to be so foolish as to start.

Just DON'T!

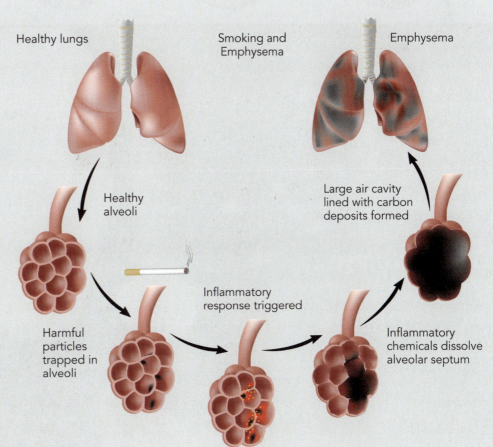

## HOW WE BREATHE

We breathe in. We breathe out. We do this on average 12 to 16 times a minute. Although we can consciously decide to take a breath, to breathe faster or slower, to hold our breath, or to take an extra-deep breath, the vast majority of the time we breathe without giving it a thought. And that's a good thing too! Right? Just imagine trying to get some sleep if you had to think about every breath.

So it's automatic when we need it to be, and yet we can have control when we want it. Our Creator thought of everything.

Now that we have examined the parts of the respiratory system, let's take a close look at how we breathe.

# Breathing Basics

At first glance, breathing doesn't seem all that complicated. It has only two parts — in and out.

The first phase of breathing is called *inspiration* (or *inhalation*), when air is taken into the lungs. The second phase is called *expiration* (or *exhalation*), when air flows back out of the lungs. Air goes in, air goes out. Nothing to it.

But when you look closer at "how" and "why" the air moves the way it does, you begin to see the wonderful complexity. It's not simple at all.

We will begin by seeing how we get air in. (We'll get it back out later...)

## Inspiration

Taking air into the lungs is known as inspiration. It is also called inhalation, inhaling, or just "breathing in." It is such a familiar thing, isn't it? You expand your chest and you feel the air move through your nose and down into your lungs.

Here is how it works.

Recall our discussion about blood flow. How does blood flow? It flows from higher pressure to lower pressure. The higher pressure in the larger arteries causes blood to flow to the smaller, lower pressure arteries downstream.

Air flow works pretty much the same way. Air moves from areas of higher pressure to areas of lower pressure. When the lungs expand or contract, the pressure in the airways changes. This results in air flowing into and out of the lungs.

Imagine the lungs at rest in the thoracic cavity. No inspiration or expiration happening. No pressure changes are occurring. No air is moving.

Now let's take a breath.

Inspiration begins with the contraction of the inspiratory muscles, the *diaphragm* and the intercostal muscles. The diaphragm is the large, dome-shaped muscle located below the lungs. The bases of the lungs rest on the diaphragm. When the diaphragm contracts, it flattens and loses much of its domed shape. As it contracts and flattens, the height of the thoracic cavity increases, making the space in the chest much larger. Air rushes into the enlarged thoracic cavity. Contraction of the diaphragm accounts for about 75 percent of the air movement in a typical breath.

The *intercostal muscles* are the muscles between the ribs. When these muscles contract, they elevate each

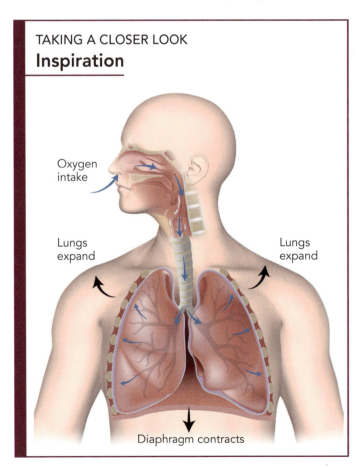

TAKING A CLOSER LOOK
**Inspiration**

rib like the handle on a bucket. This movement of the ribs causes an increase in the size of the thoracic cavity, not only front to back (anteroposteriorly) but also side to side (laterally). The reason for this has to do with how the ribcage was designed. The ribs are attached posteriorly to the vertebral bones and anteriorly to the sternum. When the intercostal muscles contract, the ribs move much in the same way that a bucket handle moves when it is lifted. It moves both up and out at the same time. About 25 percent of the air movement in a basic breath is due to intercostal muscle movement.

As the thoracic cavity enlarges, each lung expands. This happens because the pleura help the lungs "stick" to the chest wall. The parietal pleura is in contact with the chest wall and moves outward with it. Because there is a lot of surface tension between the parietal and visceral pleura, the visceral pleura and the attached lung expand too.

When the diaphragm and intercostal muscles contract, the thoracic cavity expands. As the thoracic cavity expands, the lungs expand as do the airways in the lungs. This lung expansion causes a decrease in the air pressure in the alveoli. When the pressure in the alveoli drops below the pressure in the surrounding environment (atmospheric pressure), air moves into the lungs. This is a simple inspiration.

In some circumstances, when a deeper or more forceful breath is required, there are so-called accessory muscles that come into play. These muscles are not important during normal inspiration, but are used, for example, during exercise. These accessory muscles include the sternocleidomastoid muscles, the scalene muscles, and the pectoralis minor muscles.

# Expiration

Getting air back out of the lungs is called expiration, also known as exhalation, exhaling, or "breathing out."

Isn't expiration just the opposite of inspiration? That seems reasonable, but there is a significant difference. You see, expiration is not ordinarily an active process. It is a passive process.

At the end of inspiration, the diaphragm and intercostal muscles simply relax. As they relax, the diaphragm and ribs return to their original position. In addition, due to their natural elasticity, the lungs contract back to their original shape. As this recoil/relaxation takes place, the air pressure in the alveoli increases. When the pressure in the alveoli increases above the pressure in the surrounding environment, air moves out of the lungs. This is a simple expiration. As you see, no active muscle contraction is needed for breathing out.

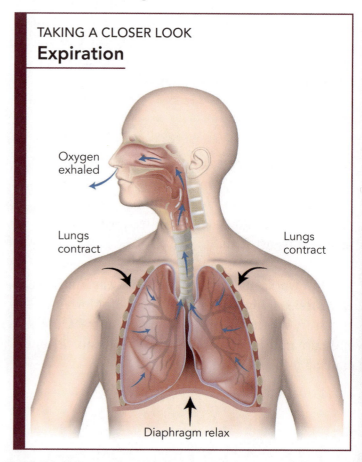

TAKING A CLOSER LOOK
**Expiration**

There are times when we need a more forceful expiration, like when you are blowing up a balloon or playing the trumpet. In these situations, the abdominal wall muscles can assist expiration by contracting and pushing the abdominal organs upward against the diaphragm. The internal intercostal muscles can also help by pulling the rib cage downward.

## Lung Volumes

The *total lung capacity* of an average male is around 6 liters. This is the total amount of air in the lungs at the end of a deep inspiration. We do not, however, move nearly that much air with each breath.

In an adult, a normal breath moves about 500 mL of air into the lungs during inspiration and back out during expiration. This is called the *tidal volume*. The tidal volume is the amount of air in one breath.

Fortunately, we do have the capacity to take in far more than just 500 mL of air in a breath. During heavy exertion we need to take in much more oxygen. We have a great deal of lung capacity in reserve. To see this for yourself, try this: take a normal breath, and stop. Don't exhale. Now start breathing in again. Breathe as deep as you can. The amount of air taken in *beyond* the normal tidal volume is called the *inspiratory reserve volume*.

Now again, take the deepest breath you can. Hold it for a second or two and then exhale as much air as you can. The amount of air you just exhaled is called the *vital capacity*. This is the maximum amount of air that can be exhaled after a maximum inspiration.

Now if you exhale as completely as possible, can you get all the air out of your lungs? Of course not. And it's a good thing too. There needs to be some air left in the lungs at the end of expiration or the alveoli in the lungs would collapse. The amount of air that remains in the lungs after a maximum expiration is called the *residual volume*.

The amount of air that is moved into and out of the lungs in one minute is called the *minute ventilation*.

### TAKING A CLOSER LOOK
### Lung Volume

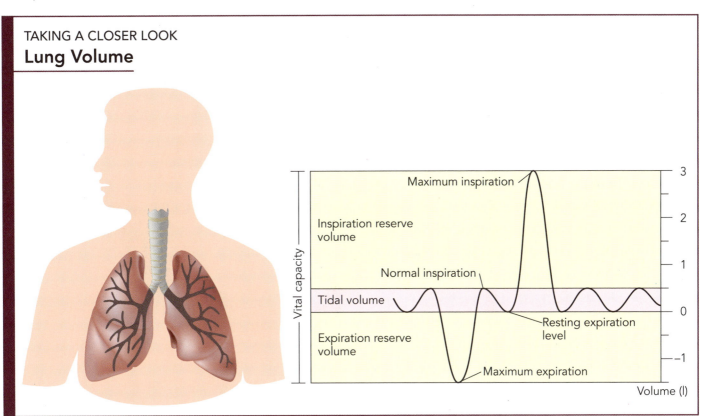

It is easy to calculate. If a person at rest has a tidal volume of 500 mL and breathes 14 times a minute, what do you think the minute ventilation is?

500 mL/breath X 14 breaths/minute

7000 mL/minute

Minute ventilation = 7 liters/minute

This is a calculation of minute ventilation at rest. During exertion, the minute ventilation will be much, much higher.

One other measurement that is often quite helpful is called the *forced expiratory volume in 1 second* (or $FEV_1$). This is, as it would seem, the amount of air that can be forced out of the lungs in one second after a maximal inhalation. Patients with asthma or chronic obstructive pulmonary disease (COPD) often have an abnormal $FEV_1$ and its assessment can be helpful in treating those patients.

## Gas Exchange

The ultimate purpose of the respiratory system is to get oxygen into the bloodstream so that it can be delivered to the body's tissues and to remove the carbon dioxide generated by the body tissues and expel it from the body. Even as incredible as the anatomy of the respiratory system is, it would be a very poor design if it didn't accomplish its mission. But it does work. It works very well indeed!

Oxygenated blood is pumped from the left ventricle out to the body. When its gets to the capillary beds, the oxygen in the blood moves into the tissues where it is used in metabolic processes. As a consequence of these metabolic activities, carbon dioxide is produced. Carbon dioxide is a waste product and would do serious harm to cells if it were allowed to accumulate. Fortunately, the carbon dioxide that is produced in the tissues moves into the blood in the capillaries to be taken away and disposed of.

The blood leaving the capillary beds has a lower oxygen content than the blood entering the capillary beds. This blood is said to be deoxygenated.

Deoxygenated blood returns to the right atrium and is sent to the right ventricle. The right ventricle pumps the deoxygenated blood to the lungs via the pulmonary artery. This blood makes its way to the capillary beds adjacent to alveoli in the lungs.

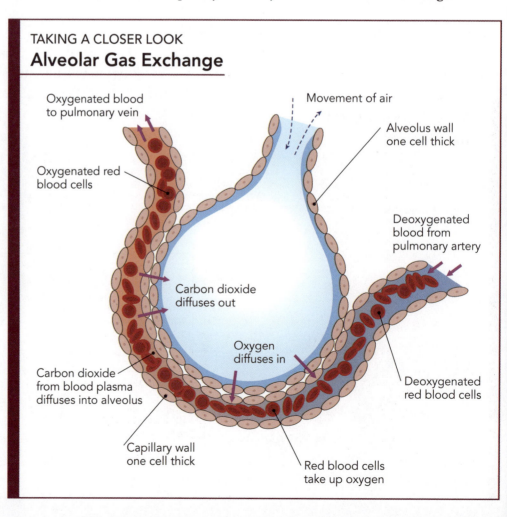

TAKING A CLOSER LOOK
**Alveolar Gas Exchange**

The alveolus, as we said before, is where gas exchange takes place, and this is the ideal place for this process to occur. The very thin wall of the alveolar cell, adjacent to the thin wall of the capillary, makes a perfect environment to promote the movement of gases into and out of the blood.

After all, it makes sense that it's easier for gases to diffuse through something thin than something thick, right? That is why there is no gas exchange farther back up the bronchial tree. Those structures (bronchi and bronchioles) are too thick to allow for proper gas exchange. The alveoli are perfectly designed for gas exchange.

The total surface area for gas exchange provided by the alveoli is staggering! With at least 300 million alveoli in our lungs, if you spread all the alveoli out on a flat surface, it would cover an area a little more than 750 square feet! That is a lot of area for gases to diffuse through. This is another evidence of the amazing design of the human body. It didn't just happen by accident.

## Oxygen Transport

Step one of gas exchange is to get the oxygen in the alveoli into the blood in the alveolar capillaries. In other words, the "deoxygenated" blood needs to be "oxygenated." Seems simple, but how does it work?

Well, doesn't the oxygen in the alveoli just move into the blood on its own? After all, we said these membranes were thin enough for the oxygen to move through them easily. That's true enough, but as usual, there's more to it than that.

When oxygen reaches the blood in the capillaries, it doesn't just dissolve in the blood plasma and get carried away to the tissues. In fact, when you look closely, you will find that blood plasma does not make a very good carrier of oxygen. You see, oxygen does not dissolve well in water. Blood plasma only carries about 2 percent of the oxygen found in the bloodstream — so there must be something else in the blood to "do the heavy lifting," so to speak. There must be something in the blood that can carry oxygen efficiently. And that thing is called hemoglobin.

The erythrocytes (red blood cells) in the blood contain a protein called hemoglobin. Hemoglobin is composed of four polypeptide chains and four heme groups. Each of the four heme groups contains an iron atom. Each iron atom in the hemoglobin molecule can bind to one oxygen molecule, so every hemoglobin molecule can bind with a maximum of four oxygen molecules.

Hemoglobin binds oxygen molecules very efficiently, and the vast majority of the oxygen in the blood (98 percent or so) is bound to hemoglobin. When hemoglobin is bound to one or more oxygen molecules, it is called *oxyhemoglobin*. When the hemoglobin is not bound to any oxygen molecules, it is called *deoxyhemoglobin*.

When four molecules of oxygen are bound, the oxyhemoglobin is "fully saturated." If one, two, or

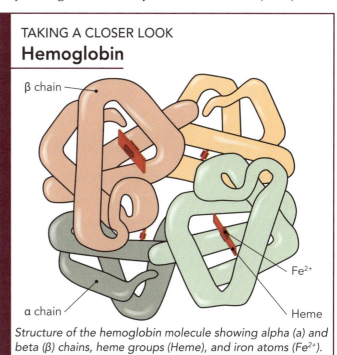

TAKING A CLOSER LOOK
### Hemoglobin

*Structure of the hemoglobin molecule showing alpha (α) and beta (β) chains, heme groups (Heme), and iron atoms ($Fe^{2+}$).*

three oxygen molecules are bound, the oxyhemoglobin is only "partially saturated."

The presence of hemoglobin is the reason that our blood is red. Oxyhemoglobin is bright red in color. Deoxyhemoglobin is a much duller shade of red. This is why arterial blood (oxygenated) looks redder than venous (deoxygenated) blood.

Each red blood cell contains around 270 million hemoglobin molecules. So, at maximum saturation, every red blood cell can carry one billion oxygen molecules! That's a LOT of oxygen molecules.

Oxygen contained in inspired air is delivered to the alveoli, and it moves into the alveolar capillaries where it binds to hemoglobin. In yet another testimony to our Creator, hemoglobin has an incredible property. After the first oxygen molecule binds to hemoglobin, the hemoglobin molecule alters its shape slightly to enhance the binding of other oxygen molecules! (No way that happened by chance. Our Creator thought of everything!) This is a very efficient system.

The now-oxygenated blood is carried back to the left side of the heart by the pulmonary veins. The left side of the heart then sends it out to the body's tissues. Upon arrival in the capillary beds of the body's tissues, the oxygenated blood finds itself in an entirely different situation. Whereas in the alveoli the oxygen level was relatively high, oxygen levels in the tissues surrounding the capillaries are relatively low. So what happens to the oxyhemoglobin

# Carbon Monoxide Poisoning

Carbon monoxide is a colorless, odorless gas that consists of one carbon atom and one oxygen atom. It is generated by the burning of carbon-based substances like wood, coal, and gasoline.

People can be poisoned by exposure to carbon monoxide, and a number of deaths are caused each year by this type of poisoning. This can occur when combustion of things like wood or gasoline takes place in poorly ventilated areas. In circumstances like these, people may not realize they are being exposed until significant symptoms develop. Symptoms can include headache, shortness of breath, and nausea. If exposure continues, more severe symptoms can follow, such as confusion, vomiting, and loss of consciousness. In the worst cases, death can occur.

Many homes contain carbon monoxide detectors to alert sleeping people about carbon monoxide leaks in the home. Since carbon monoxide poisoning makes people sleepy, people would not awaken to save themselves without this sort of warning. In a house with a carbon monoxide furnace leak, such a detector can save the lives of everyone in the home.

Carbon monoxide binds very effectively to hemoglobin. In fact, it binds to hemoglobin much more efficiently than oxygen. Carbon monoxide is particularly dangerous because it competes with oxygen for binding sites in the heme group. Plus, hemoglobin's binding to carbon monoxide is 200 times greater than for oxygen.

People exposed to carbon monoxide require prompt medical attention. Treatment for those patients is therapy with 100 percent oxygen until the carbon monoxide clears from the body.

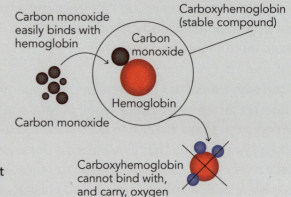

here? If you think it now releases its bound oxygen molecules, you are correct! The released oxygen molecules then move out of the capillaries and enter the body tissue cells where they are needed.

As efficiently as it binds oxygen in conditions of high oxygen concentration, hemoglobin is equally efficient at releasing its oxygen when the surrounding oxygen levels are low. Even at that, not all the oxygen is released from all the hemoglobin molecules in the systemic capillaries. That would not work well at all. Here's why.

When we are at rest, the hemoglobin is almost completely saturated. In fact, it is around 98 percent saturated. That means the hemoglobin in oxygenated blood carries 98 percent of the maximum amount of oxygen it could theoretically hold. After the blood passes through the systemic capillaries we call it "deoxygenated," yet the hemoglobin is still 70 to 75 percent saturated! You see, the hemoglobin doesn't give up all its oxygen on every trip through the body. There's still plenty in reserve. And that's as it should be.

Under normal conditions while at rest, the body only uses about 25 percent of the available oxygen in the blood. But we are not always at rest, are we? During periods of exertion, the body's tissues need much more oxygen to meet their metabolic needs. In situations like this, the body is able to obtain a higher percentage of the oxygen bound to the hemoglobin than it does at rest. But wait a minute, how does hemoglobin release more oxygen at some times and less at others? Hemoglobin can't "know" anything. It doesn't "know" when someone is exercising. It's merely a molecule in the blood. The answer is that there are conditions in the body itself that can cause hemoglobin to release oxygen more readily.

First of all, tissues that are metabolically very active, such as muscles during exertion, can produce acids as a byproduct of that activity. As a result, the acidity of the blood in the region near these tissues increases. Tissues at rest produce far less of these substances, and thus the acid level in the blood near resting tissues is lower. Hemoglobin releases oxygen more readily in a more acidic environment. So the higher the tissue activity, the higher the acidity near these tissues, and ultimately, the more oxygen released to these active tissues.

Second, active tissues produce more carbon dioxide than tissues at rest. Carbon dioxide also binds to hemoglobin, and when it does, it causes the hemoglobin to unload its oxygen more easily. So again, the conditions near active tissues induce hemoglobin to give up its oxygen.

Third, higher temperatures cause hemoglobin to release oxygen. Again, this is another of the conditions found near active tissues. The more active the tissues are, the more heat is generated, and then, the higher the temperature in the capillary beds serving these tissues. So once again, here is a condition expected near tissues requiring more oxygen that leads to more oxygen being released.

# CARDIOVASCULAR & RESPIRATORY SYSTEMS

It seems almost unfair to say that the respiratory system is designed to deliver oxygen to the tissues. It is really over-designed when you think about it. There are so many features working so well together: the design of the alveoli, the hemoglobin molecule itself, and the ability of hemoglobin to alter its binding to oxygen in response to the precise conditions produced by active tissues. It makes no sense to think these things could all be some sort of chemical accident occurring over millions of years.

The entire process of oxygen delivery speaks to the abilities of the Master Designer.

## Carbon Dioxide Transport

Not only does the respiratory system take oxygen to the tissues, it has to remove the carbon dioxide produced by these same tissues. There are three main ways that carbon dioxide is carried by the blood.

Carbon dioxide can dissolve in the blood plasma. It dissolves in plasma much better than oxygen does. About 10 percent of the carbon dioxide in the blood is carried in this fashion.

Hemoglobin not only transports oxygen, it can also transport carbon dioxide. However, the carbon

## The Placenta

We mentioned earlier that a baby in the womb does not have to use its lungs to breathe. Yet a baby of course needs to get oxygen into his or her blood cells and to release carbon dioxide to be carried away. God designed a wonderful way for a baby's gas exchange to take place. He designed a placenta. In the placenta, the mother's blood flows through capillaries right next to capillaries containing the baby's blood. The baby's blood is carried there in the umbilical cord and, once oxygenated, returns to the baby the same way. The mother's and baby's blood do not mix. Instead, oxygen is released from the hemoglobin in the mother's red blood cells. The oxygen crosses from the capillaries containing the mother's blood into the capillaries containing the baby's blood. There, the baby's hemoglobin molecules grab the oxygen molecules and carry it back to the baby.

You might wonder, since the mother's and the baby's blood both contain hemoglobin, how the mother's hemoglobin "knows" to release oxygen for the baby's hemoglobin to pick up. Here God designed a particularly wonderful system. He made a special kind of hemoglobin just for unborn babies. Fetal hemoglobin has an even greater affinity (attraction) for oxygen than ordinary hemoglobin. The unborn baby's hemoglobin is able to overcome the attraction of the mother's hemoglobin and carry away all the oxygen the baby needs from the placenta. Once a baby is born, red blood cells begin to be made with adult hemoglobin, and generally by about six months of age the fetal hemoglobin has been replaced.

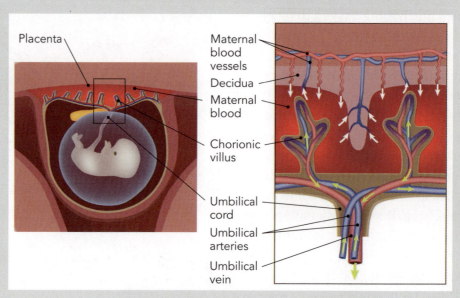

dioxide carried by hemoglobin is not bound to iron but instead is attached to amino acids. These amino acids are part of the polypeptide chains that make up the globin part of hemoglobin. When hemoglobin is carrying carbon dioxide it is called *carbaminohemoglobin*. Approximately 20 percent of the carbon dioxide in the blood is transported by hemoglobin.

The remaining 70 percent of the carbon dioxide is transported in the blood plasma as bicarbonate ions ($HCO_3^-$). After carbon dioxide enters the plasma, it soon finds its way to the red blood cells. Once inside the RBC, the carbon dioxide combines with water and is turned into carbonic acid. This reaction occurs very rapidly due to the presence of a special enzyme called *carbonic anhydrase*. The carbonic acid quickly breaks down into a hydrogen ion and a bicarbonate ion. The bicarbonate ion is the form in which the carbon dioxide is carried to the lungs.

In the systemic capillaries where the carbon dioxide levels are high, the tendency is for more carbonic acid to be produced and then, ultimately, more bicarbonate ions. The bicarbonate ions are then carried to the lungs by the blood. In the lungs, the levels of carbon dioxide are relatively low, so the reaction here is reversed. The bicarbonate combines with the hydrogen ions to again form carbonic acid. Then, with the help of the carbonic anhydrase enzyme, the carbonic acid is broken down into carbon dioxide and water. The carbon dioxide then diffuses out of the blood and into the alveoli. It is then taken away with the expired air!

## Control of Respiration

We breathe in and out, in and out. Thousands of times a day. And we really never give it a thought. It just happens. Automatically.

But stop and think a minute. We do have control over our breathing — a lot of control when we need it. If we didn't have the ability to voluntarily hold our breath, we could not go swimming. If we could not precisely control our respiratory system on command, we would not be able to talk or shout or sing.

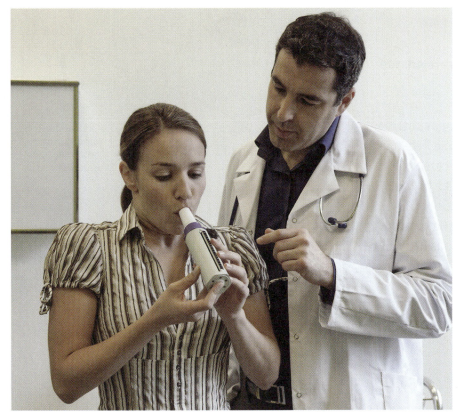

The respiratory system has both voluntary and involuntary controls. And that's a very, very good thing.

The primary control of respiration takes place in the brain, in the medulla oblongata. There are two locations in the medulla that make up what is known as the medullary respiratory center. The first of these is the ventral respiratory group (VRG), and the second is the dorsal respiratory group (DRG). From these two centers, nerve signals are sent to the intercostal muscles and the diaphragm to stimulate them to contract. The contraction of these muscles (as you should recall) triggers inspiration. These impulses are generated for about two seconds, and then they cease. When the nerve signals stop, the intercostal muscles and the diaphragm relax. The natural recoil of these structures leads to expiration. Remember, expiration is generally a passive process. The cessation of nerve stimulation lasts 2–3 seconds, and then stimulation occurs again.

Inspiration is triggered by impulses from both the VRG and the DRG. However, the VRG does have another function. There are times when more forceful expirations are needed. In these circumstances, the VRG sends nerve impulses to the internal intercostal muscles and the muscles of the abdominal wall. Contraction of these muscles helps decrease the size of the thoracic cavity and assists with a forceful expiration.

Another important area is the pneumotaxic area located in the pons. It is sometimes referred to as the pontine respiratory group (PRG). The pneumotaxic area coordinates the switch between inspiration and expiration. The PRG helps regulate how much air is taken in with each breath. It can send signals to turn off the VRG and DRG to help limit inspiration. This can keep the lungs from getting too full of air.

The operation of the respiratory centers is regulated by many inputs from the body.

First of all, we do have significant voluntary control over the respiratory system. This is possible because there are nerve pathways connecting the cerebral cortex to the respiratory centers. This gives us the ability to alter our inspiration and expiration. Of course, this control is within limits. For example, we cannot hold our breath for extended periods of time. Usually, after a minute or so, involuntary control of the respiratory system takes over and causes us to breathe whether we want to or not. These involuntary controls help protect the body and maintain proper levels of oxygen and carbon dioxide.

Further control of the respiratory system is based on input from special sensory cells called chemoreceptors. These special cells are sensitive to changes in the levels of certain chemicals or substances in the body. There are two main groups of chemoreceptors that regulate the respiratory system — the central chemoreceptors that are located in the brain stem, and the peripheral chemoreceptors found in the arch of the aorta and the carotid arteries. These cells help monitor blood levels of oxygen and carbon dioxide, as well as the acidity of the blood itself.

What is the most important thing that the respiratory system does? If you said that taking in oxygen and getting it into the bloodstream is most

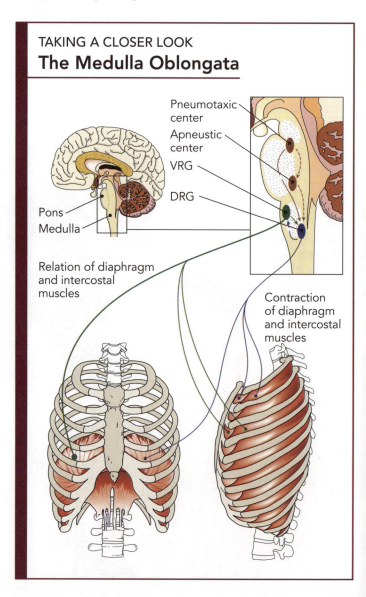

### TAKING A CLOSER LOOK
### The Medulla Oblongata

important, you would be correct. Oxygen is the most important thing. A person can live only a few minutes without oxygen, so it is vital to keeping our bodies alive.

From this fact, you probably would assume that the primary thing that the chemoreceptors monitor is oxygen, right? Well, as it happens, that's not the case. The chemical in the body that has the most influence on respiration is actually carbon dioxide, not oxygen! And this makes sense when you give it some thought. Recall that the body uses only about 25 percent of the oxygen bound by hemoglobin, so under normal circumstances, there is plenty of oxygen in reserve. If the chemoreceptors primarily checked for oxygen, then we would have little stimulus to breathe because there is always so much oxygen in the blood. Our brains wouldn't signal the need to breathe until we were nearly out of oxygen with no reserve. If mindless evolutionary processes "designed" us we might have been made that way, but God is far wiser. He designed us to breathe before we become in desperate need of oxygen to survive.

It makes much more sense to monitor the levels of carbon dioxide, and this is what the central chemoreceptors do. They monitor the levels of carbon dioxide (in addition to monitoring the acidity of the blood). When the receptors sense the levels of carbon dioxide going up, what do you think happens? Correct! The chemoreceptors send signals to the respiratory center to increase respiration and remove the excess carbon dioxide. The opposite occurs when the level of carbon dioxide gets a little too low. The chemoreceptors trigger the respiratory center to slow respiration and allow the carbon dioxide level to return to normal. This control mechanism is very precise. It not only maintains control of the respiratory system, it also helps keep the level of carbon dioxide within very tight limits.

The level of acidity in the blood is also monitored by the central chemoreceptors. An increase in acid levels is usually accompanied by an increase in respiration.

While oxygen is not the primary controller of respiration, its level is also monitored by chemoreceptors. The peripheral chemoreceptors keep a close watch on the oxygen level in the blood. However, the oxygen level has to get very low before the peripheral chemoreceptors trigger an increase in respiration. It is very unusual for the peripheral chemoreceptors to be the primary stimulus for respiration.

# IS THIS "DESIGN" JUST AN ACCIDENT?

The design of the body seems obvious at every turn. But not to everyone it seems. You see, many people would have you believe that the marvelous complexity of the body is merely the result of chance. They want you to believe that the body came into existence by means of a process called "evolution." The basis of this belief is that billions of years ago, matter just appeared. Out of nothing. From nowhere. Then the chemicals in the universe proceeded to form stars, planets, and galaxies. On their own. Without direction.

And then...

These chemicals randomly bouncing off one another formed our sun and the planets in the solar system, including, of course, the earth. In the earth's vast oceans, chemicals continued to bang together, and somehow the first living cell formed. By itself. Without direction. By chance.

Over the next three billion years or so, this one-celled organism became more and more complex, and new types of living things came into existence. More and more complex creatures evolved. By chance. On their own.

Next, there arose an ape-like creature that is said to be our ancestor. Over the last few million years this ape-like ancestor evolved into both the modern apes and humans. By chance. On its own. Without direction.

To those who accept evolution, the incredibly complex machine that is the human body is the result of random chance processes. Ultimately, the body is just a cosmic accident. Nothing marvelous about it at all. Just a cosmic accident.

Curious...

You might perhaps be asking yourself just where all the matter in the universe came from in the first place. Excellent question. The evolutionist has no answer. Maybe you are wondering where all the information came from to allow a one-celled organism to increase in complexity, that increase in complexity resulting in human beings. Another excellent question. Again, the evolutionist has no answer.

I have an answer, and it's painfully simple. Evolution didn't happen. Period.

There is no way that the complexity in our world can merely be written off as the result of chance. Whether it's a simple one-celled organism (actually to suggest that one-celled organisms are simple is ridiculous; they are extremely complex), a jellyfish, a flower, a butterfly, or a giraffe, everything in our world is obviously the work of a Master Designer. One would have to suspend reality to believe that these complex things "just happened."

The Word of God refers to how the amazing things that God made should direct a person's attention to his or her Creator. Romans 1:20–22 tells us:

*For since the creation of the world His invisible attributes are clearly seen, being understood by the things that are made, even His eternal power and Godhead, so that they are without excuse, because, although they knew God, they did not glorify Him as God, nor were thankful, but became futile in their thoughts, and their foolish hearts were darkened. Professing to be wise, they became fools.*

As you explore and consider the human body, I pray you will not lose sight of the greatest wonder of all: the awesome power of Jesus Christ, our Creator and Savior.

He is indeed the Master Designer. As we read in Colossians 1:16–17:

*For by Him all things were created that are in heaven and that are on earth, visible and invisible, whether thrones or dominions or principalities or powers. All things were created through Him and for Him. And He is before all things, and in Him all things consist.*

Let us praise Him for the great things He has done!

# INTRODUCTION

Think for just a moment about the things you do every day. You wake up, walk to the bathroom, comb your hair, and brush your teeth. You sit at your desk and read a book. You take a walk with your dog. You stand in church, and sing a beautiful worship song (hopefully in the right key). You have a conversation with your parents. You go to bed and go to sleep.

How do just the correct muscles know how to contract in just the right way to allow us to walk? How can we control the movements of our hands in a very precise fashion so that we can brush our teeth? How can we decipher those funny marks on a printed page, understand that they are letters and punctuation marks, and make sense of them? How can we hear others singing and make our voices match theirs? How can we understand others' speech? What makes us fall asleep and then wake up again?

Somehow we just "know" how to do these things. Or at least we remember "learning" how to do them. How is this possible? These remarkably complex tasks seem simple because of the remarkably complex human nervous system.

# Functions of the Nervous System

The nervous system processes an amazing amount of information. Sometimes this processing is relatively simple, but often it is incredibly complicated. However, as we explore your master control system in more detail, you will notice that all its processes follow the same basic pattern.

This basic pattern is simply this: information comes into the nervous system, this information is recognized and processed, and then a signal is sent out instructing an organ (or organs) to respond in some manner. If you think of the nervous system functioning in this fashion, things won't seem complicated at all.

Let's look at the three parts of this pattern in more detail.

The first step is sensory function. A vast number of sensory receptors throughout the body provide input to the nervous system. There are receptors designed to detect internal changes, such as blood pressure or acid levels in the blood. Other receptors detect external stimuli, such as heat or cold on the skin, or the sensation of a splinter's sharp point. All these receptors send signals to the nervous system. These signals are the sensory inputs.

The next step is called integration. The nervous system integrates all this incoming information. It must recognize, analyze, and process all the various

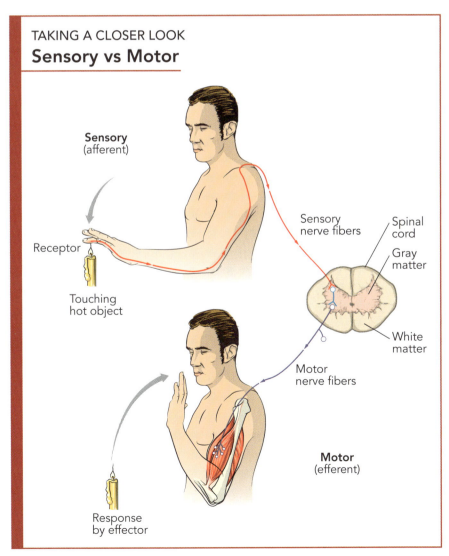

TAKING A CLOSER LOOK
**Sensory vs Motor**

sensory inputs, often comparing what is sensed in the present to what has been experienced in the past. Then the nervous system comes up with an appropriate response, sometimes filing the information away for future use and often creating an instruction to be sent out to deal with the information.

As an example, let's say that you are riding your bike down a steep hill. You feel the wind on your face and sense the speed of the bike increasing. While processing these sensory inputs, you also remember that last month you were going too fast down this hill, wrecked your bike, and sprained your wrist. The processing of sensory inputs is often dependent on your past knowledge and experiences. As your nervous system integrates all this information, you realize that you need to slow down.

The last step is motor output. The word *motor* implies movement or some sort of action. Motor output is simply what the body is told to do as the result of all this information input and processing. In our example, this step causes you to use the muscles in your legs or hands to put some pressure on your coaster brakes or hand brake, and you slow down to a safer speed.

Input, integration, output. Using these three steps, the nervous system controls the complex activities of the human body.

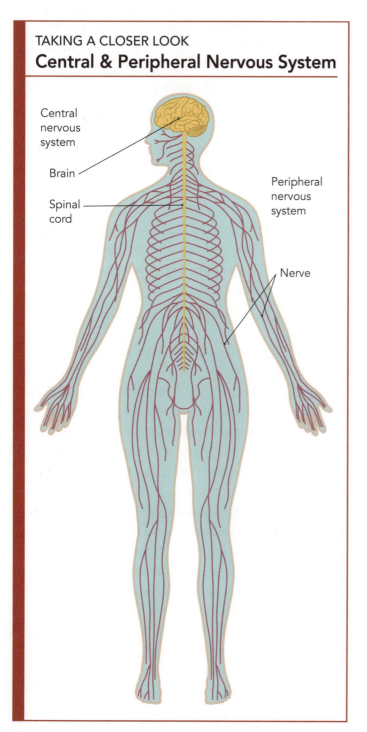

TAKING A CLOSER LOOK
**Central & Peripheral Nervous System**

- Central nervous system
- Brain
- Spinal cord
- Peripheral nervous system
- Nerve

# Overview of the Nervous System

We will begin our tour of the nervous system by taking a broad look at its two major divisions, the central nervous system (CNS) and the peripheral nervous system (PNS). Even though these parts work together as a highly efficient, integrated unit, breaking it down into these two parts can be very helpful as we try to understand how the nervous system works.

The central nervous system is composed of the brain and the spinal cord. The brain is the most recognizable part of the CNS. It is the master control center of the nervous system, containing hundreds of millions of neural connections. Our perception of the world around us, our movements, our intellect, our memories—all are controlled and regulated by the brain. The spinal cord extends from the base of the brain down to the lower levels of the spinal column. It provides a pathway for nerves to and from the brain.

The peripheral nervous system is the portion of the nervous system outside of the central nervous system. It consists of the cranial nerves that extend from the brain, and the spinal nerves that extend from the spinal cord. The peripheral nervous system in effect allows all the other organ systems and body parts to connect and interact with the central nervous system.

The PNS has two basic functions: carrying sensory information to the CNS and transmitting instructions out to the various part of the body. Based on these functions, we can divide the PNS into two divisions, the sensory division and the motor division.

The sensory division carries information from the skin and muscles as well as from the major organs in the body to the central nervous system, where all the sensory input is processed ("integrated"). The sensory division is sometimes called the afferent (meaning "bringing toward") division because it carries nerve impulses "to" or "toward" the CNS.

The motor division, on the other hand, carries instructions from the CNS out to the body. (This is the motor output function of the nervous system discussed earlier.) The motor division is sometimes called the efferent (meaning "carrying away") division because it carries instructions "away from" the CNS.

Some instructions carried by the motor division are taken to muscles that we can consciously control. For example, we can consciously control the muscles we use to hold a glass or throw a ball. This aspect of the motor division is called the somatic nervous system.

### TAKING A CLOSER LOOK
## Sensory vs Motor Nerves

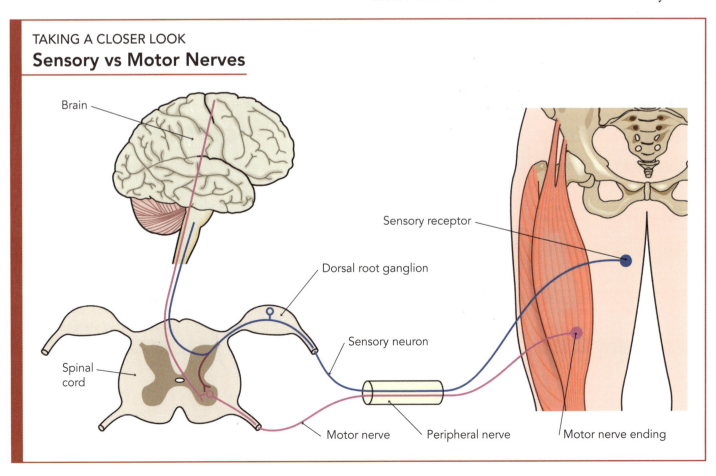

# THE NERVOUS SYSTEM

## TAKING A CLOSER LOOK
## Automatic Nervous System

### PARASYMPATHETIC

**Eye**
Constricts pupil

**Salivary & Parotid Glands**
Stimulates saliva production

**Blood Vessels**
Constricts blood vessels in skeletal muscles

**Sweat Gland**
Inhibits sweat secretion

**Lungs**
Constricts bronchi

**Heart**
Slows heart beat

**Liver**
Inhibits glucose release

**Gallbladder**
Stimulates bile

**Pancreas**
Stimulates pancreas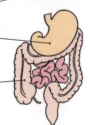

**Stomach**
Stimulates stomach motility & secretions

**Intestines**
Stimulates intestinal motility

**Kidneys**
Decreases renin secretion (lowers blood pressure)

**Bladder**
Stimulates urination

Brain
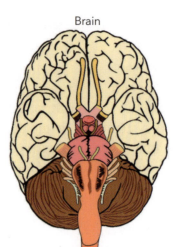
Spinal cord

### SYMPATHETIC

**Eye**
Dilates pupil

**Salivary & Parotid Glands**
Inhibits saliva production

**Blood Vessels**
Dilates blood vessels in skeletal muscles

**Sweat Gland**
Stimulates sweat secretion

**Lungs**
Dilates bronchi

**Heart**
Accelerates heart beat

**Liver**
Stimulates glucose release

**Gallbladder**
Inhibits bile

**Pancreas**
Inhibits pancreas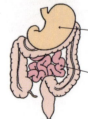

**Stomach**
Inhibits stomach motility & secretions

**Intestines**
Inhibits intestinal motility

**Kidneys**
Increases renin secretion (raises blood pressure)

**Bladder**
Inhibits urination

Somatic means "body," so the somatic nervous system allows us to control our body's movements.

Another equally important part of the nervous system's motor division controls involuntary activities. Involuntary activities—like making sure we breathe and adjusting our heart rate—are vital to survival but not under voluntary control. Such activities continue 24 hours a day whether we think about them or not, and that is a very good thing. Imagine having to think about every single breath you take! What would happen when you slept? It is a good thing the nervous system takes care of this for us. The part of the motor division that controls these involuntary functions is called the autonomic nervous system. (Autonomic sounds a lot like "automatic," so you should be able to remember this easily!)

Let's make sure you have all these divisions and subdivisions straight so far. The nervous system has two parts: the central nervous system and the peripheral nervous system. The central nervous system consists of the brain and spinal cord. The peripheral nervous system brings information to the central nervous system with its sensory nerves, and it transmits instructions from the central nervous system with its motor nerves. Somatic motor nerves instruct skeletal muscles to move voluntarily. Autonomic motor nerves carry instructions for involuntary functions, like breathing and adjustments of the heart rate.

The autonomic nervous system also consists of two parts: the sympathetic nervous system and the parasympathetic nervous system. Both control our involuntary functions, but they have opposite effects on the body. The sympathetic division is more active when we are stressed or exercising. Think of the sympathetic nervous system as the part of you that triggers your "fight or flight" responses to danger. The parasympathetic division does the opposite. The parasympathetic nervous system promotes less-demanding activities like digestion, things that your body needs to do while not busy running or expending lots of energy on other highly active pursuits. Both sympathetic and parasympathetic functions are important for the body to operate properly.

If at this point you are feeling a little overwhelmed with all this, don't worry. Everyone feels that way the first time they encounter all these "divisions." Just keep sight of the big picture and everything will soon fall into place. Remember the three basic functions of the nervous system? They are sensory input, integration, and motor output. No matter how bewildering all these divisions seem to be right now, it all comes down to the basic three functions.

As we examine the nervous system in more detail, you will see just how sensibly it is organized. And you will be amazed at how it works as it assists and controls complex activities throughout your body.

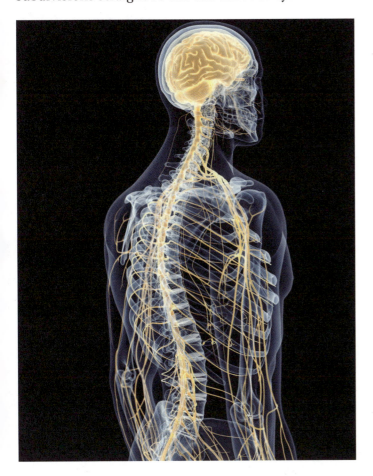

# STRUCTURE OF NERVOUS TISSUE

The nervous system is composed primarily of nervous tissue. Nervous tissue is one of the four basic tissue types that we examined previously in Volume 1 of *Wonders of the Human Body*.

Nervous tissue consists of two primary types of cells: neurons and neuroglia.

 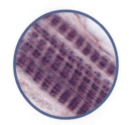  

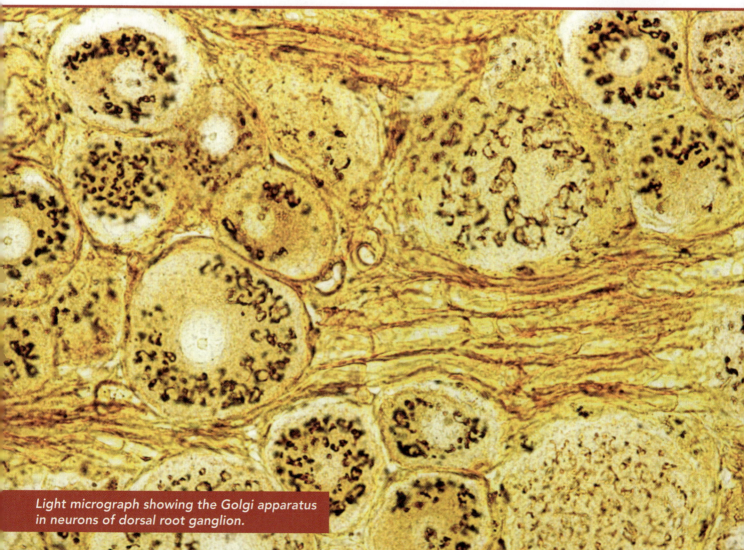

*Light micrograph showing the Golgi apparatus in neurons of dorsal root ganglion.*

# STRUCTURE OF NERVOUS TISSUE

## Tissue Types

### Epithelial Tissue

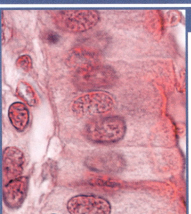

Epithelial tissue (or epithelium) lines body cavities or covers surfaces. For example, the outer layer of skin is epithelium. The sheet of cells that line the stomach and intestines, as well as the cells that line the heart, blood vessels, and the lungs, is epithelial tissue.

### Connective Tissue

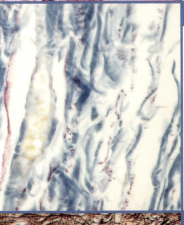

Connective tissue helps provide a framework for the body. It also helps connect and support other organs in the body. Further, it helps insulate the body, and it even helps transport substances throughout the body. This tissue can be hard or soft. Some connective tissue stretches. One type is even fluid. Connective tissue is comprised of three parts: cells, fibers, and ground substance.

### Nervous Tissue

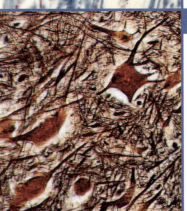

Nervous tissue is the primary component of the nervous system. The nervous system regulates and controls bodily functions.

Nerve cells are incredible. They are able to receive signals or input from other cells, generate a nerve impulse, and transmit a signal to other nerve cells or organs.

### Muscle Tissue

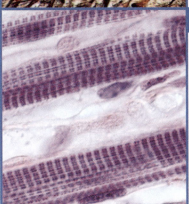

Muscle tissue is responsible for movement. There are three types of muscle tissue: skeletal muscle, smooth muscle, and cardiac muscle.

---

Neurons are the excitable nerve cells that transmit electrical signals.

What starts such an electrical signal? Some type of change in the environment acts as the stimulus that excites a neuron, triggering an electrical signal called an action potential. The electrical signal transmitted by a neuron is also called an impulse. An impulse travels like a wave along the nerve cell membrane from one end of the neuron to another. We will soon study this in depth.

The other cells in nervous tissue are called neuroglia. There are several types of neuroglia cells. They help protect and support the neurons.

Let's examine the neuron in greater detail.

## Neurons

The neuron is often called a nerve cell because it is the cell type that does the primary work of the nervous system. You have neurons in your brain, in your spinal cord, in your peripheral nervous system, and even in specialized sensory organs like your eye, nose, and ear.

A neuron doesn't look like a typical cell. If you have seen sketches of "typical" cells before, you will notice that, while the neuron still has a cell membrane, cytoplasm, and a nucleus, it has an unusual shape. The neuron is a very specialized type of cell that is designed to transmit electrical impulses (nerve impulses) rapidly to various parts of the body.

The neuron is composed of three parts: the cell body, dendrites, and the axon.

The cell body contains the typical organelles we discussed at length in Volume 1 of *Wonders of the Human Body*. The cell body contains a nucleus surrounded by cytoplasm. The cytoplasm contains plenty of protein-building organelles like rough endoplasmic reticulum dotted with ribosomes and free ribosomes. An extensive Golgi apparatus processes the proteins made by these ribosomes. Neurons require a lot of energy to build the substances they require, so lots of energy-generating mitochondria are also found in the cell body. Energy provided by these mitochondria fuels the building of the substances neurons need to do their job. Some of the most important substances synthesized in the neuron's cell body are neurotransmitters. As we will soon see, neurotransmitters are the chemicals that transmit an electrical impulse from one neuron to the next.

Extending from the cell body are numerous projections, or processes. Neuron cell bodies have two kinds of processes protruding from them, dendrites and axons. Dendrites are designed to receive signals. Axons are designed to carry signals away.

Some dendrites resemble the branches of a tree. Others have more thread-like branches, and some have branches covered with tiny spines. The reason for this branching design is simple. Remember, dendrites are the parts of neurons that receive inputs (signals). The branching pattern covers an extensive area, allowing the neuron to receive an enormous number of inputs. When an input is received by a dendrite, an electrical signal is generated and transmitted toward the cell body.

The axon is the portion of the neuron that carries a nerve impulse away from the cell body. The axon begins at a cone-shaped axon hillock on the cell body. The hillock narrows to form the more thread-like axon. The axon can be very short or up to several feet long. The axon of a motor nerve to the muscle that enables you to curl your big toe has to travel a long way, all the way from your spinal cord to your foot.

A neuron can have multiple dendrites

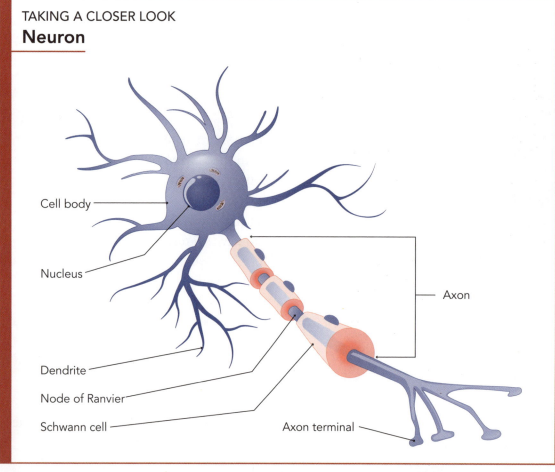

TAKING A CLOSER LOOK
**Neuron**

- Cell body
- Nucleus
- Dendrite
- Node of Ranvier
- Schwann cell
- Axon
- Axon terminal

but only one axon. Axons end in small branches called axon terminals. At the axon terminal, neurotransmitters are released to carry the neuron's signal on to the next cell in line. You will learn more about this shortly.

## Neurons — The Lowdown

There are hundreds of millions of neurons in the human body. And that's a really good thing. Why? Unlike most cell types in your body, neurons cannot be routinely replaced. Once neurons mature, with only rare exceptions, they are no longer able to divide. The neurons you have, once your nervous system matures, are all the neurons you will ever have.

So...when neurons are damaged by drugs, disease, or injury, the loss of function is often permanent. Neurons are designed to last a lifetime, but we need to take care of them. For instance, we must be vigilant about what we put into our bodies, as many illicit drugs destroy these precious messengers. A lifetime of poor eating habits and lack of exercise can increase the risk of a stroke in later life, which can destroy many neurons in the brain. Riding your bicycle without a helmet puts the irreplaceable neurons in your brain at risk right now. Following the rules for safety in contact sports may prevent a tragic accident that could leave you paralyzed. Operating power tools unsafely may lead to permanent loss of peripheral nerve function in an injured body part, even if you do not lose the body part itself. Habitually exposing your ears to loud music or explosive noise without ear protection may destroy the specialized neural structures in your ears and impair your hearing. Looking directly at the sun can permanently damage your retina, the very specialized extension of your brain that enables you to see.

God only gave you one body, and there are no do-overs when it comes to neuron damage. While many diseases and conditions that damage neurons in this sin-cursed world are not preventable, you should take care to avoid those that are.

Further, neurons require lots of oxygen and glucose to function properly. Neuron cells can be quickly damaged by lack of these essentials. Loss of oxygen for as little as four minutes can permanently damage neurons. For this reason, many people take courses in basic CPR and water safety, so that they will be able to help others avoid permanent damage or loss of life.

*Performing CPR (cardiopulmonary resuscitation) on someone who has stopped breathing.*

# Types of Neurons

There are several types of neurons. We can classify them according to how they look or according to how they work. Each type of classification can help us understand how the nervous system works.

One method of classifying neurons is based on the number of processes they have. Remember, processes are dendrites and axons, the projections sticking out from the cell body.

Most neurons have one axon and multiple dendrites. These are called multipolar neurons. This is by far the most common type of neuron in the body.

Bipolar neurons have only two processes: one axon and one dendrite. These are only found in special sensory organs, such as the eye, ear, and nose.

Unipolar neurons have a more unusual configuration. They have only one process extending from the cell body. This process looks like a "T." The dendrite and the axon form the arms of this "T."

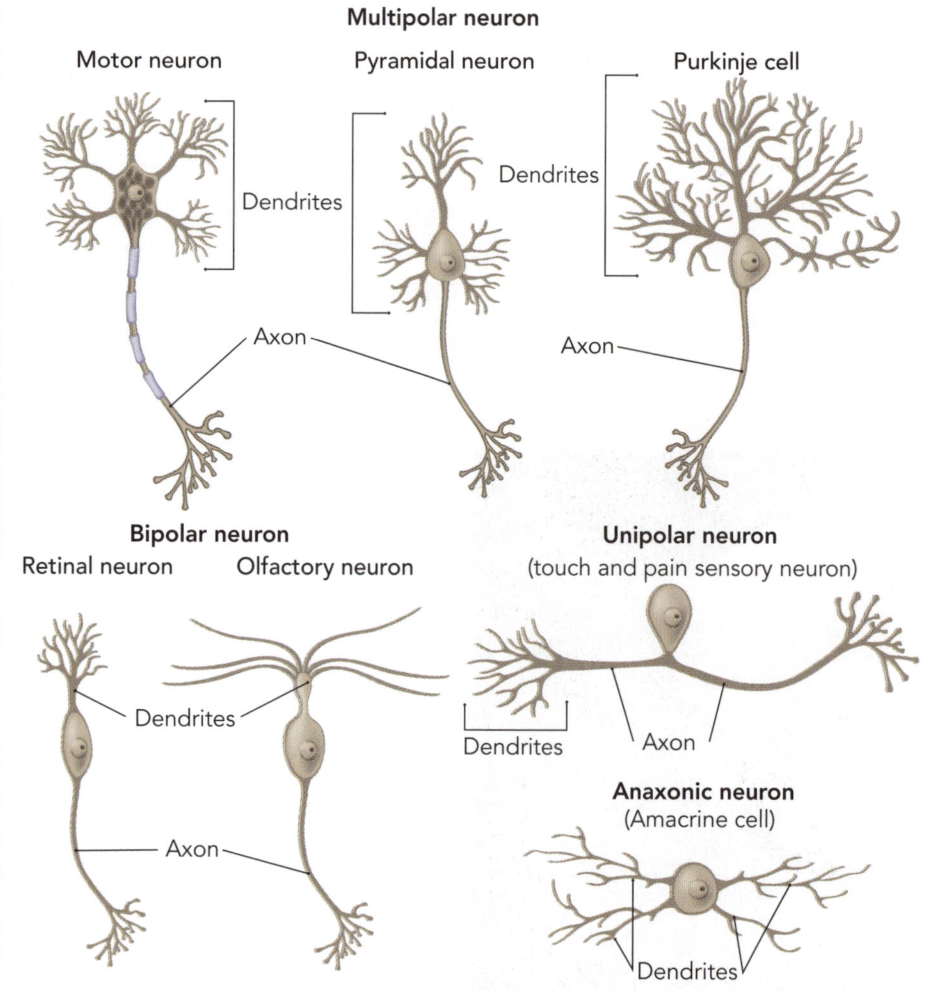

Neurons are also classified according to the direction they carry nerve impulses. Some neurons carry instructions from the central nervous system, and others bring information to the central nervous system.

Neurons that transmit impulses away from the central nervous system are called motor or efferent (remember "carrying away" or "carrying outward") neurons. These impulses contain instructions to muscles or to glands in the body. Most motor neurons are multipolar.

Sensory or afferent (remember "bringing toward") neurons carry impulses triggered by sensory receptors toward the central nervous system. Most sensory neurons are unipolar.

### TAKING A CLOSER LOOK
## Types of Neurons

**Motor neuron**

**Multipolar neuron**
Pyramidal neuron

Purkinje cell

**Bipolar neuron**
Retinal neuron — Olfactory neuron

**Unipolar neuron**
(touch and pain sensory neuron)

**Anaxonic neuron**
(Amacrine cell)

Yet one other class of neurons carries impulses from one neuron to another within the central nervous system. These connectors are called interneurons, a word that obviously means "between neurons." Interneurons make up the vast majority of the neurons in the body. Some estimates are as high as 99 percent. Interneurons are located in the brain and spinal cord, forming connections between sensory and motor neurons. Signals from sensory neurons are delivered to the interneurons. The interneurons pass the impulse on to the appropriate motor neurons. If you recall the basic functions of the nervous system, this is the integration step we discussed, a step in which inputs are processed and passed on to generate suitable output.

# Neuroglia

Neurons are not usually alone. They are generally surrounded by several types of smaller cells in the nervous system. These other cells are known as neuroglia, or glial cells. Neuroglia are found both in the central nervous system and the peripheral nervous system. Neuroglia have various functions depending on their cell type and location.

We will first examine the neuroglia in the CNS.

Astrocytes are the most numerous of the neuroglial cells in the CNS. Astro means "star," and cytes means "cells." Astrocytes are therefore glial cells with

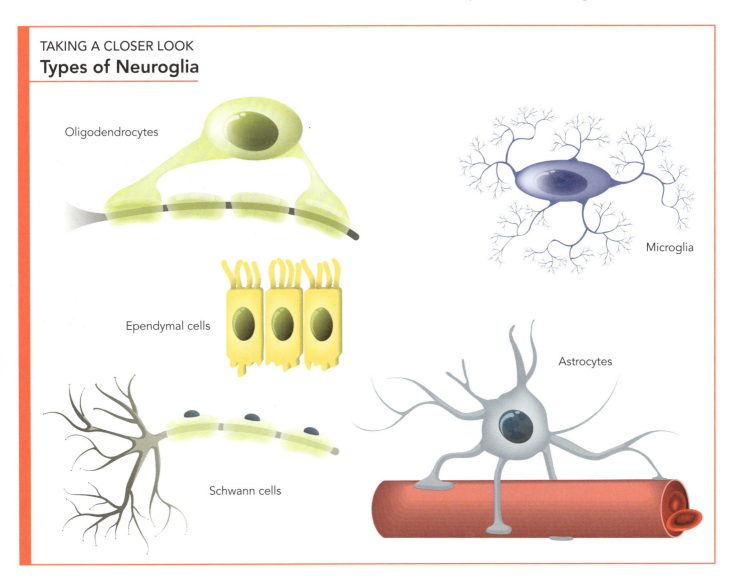

**TAKING A CLOSER LOOK**
**Types of Neuroglia**

Oligodendrocytes

Microglia

Ependymal cells

Astrocytes

Schwann cells

many star-shaped processes. These cells anchor and support the neurons associated with them. They help the neurons pass on impulses efficiently. Astrocytes also protect their neurons. They monitor nearby capillaries, ensuring that harmful substances in the blood do not reach the neuron. Astrocytes help maintain the correct level of ions, such as potassium ($K^+$), and other nutrients around the neurons. They contain a readily available supply of glucose that they supply to neurons when lots of energy is needed. They even help recycle neurotransmitters released from their neurons.

Microglia are small cells with long slender processes. (Micro means "small," so this is a good name.) Microglial cells "keep watch" over neurons in their vicinity. If they detect damage to a neuron or invading bacteria, they transform into a cell that can remove damaged nerve tissue or engulf and destroy the bacteria.

Ependymal cells line the ventricles of the brain and the spinal canal. The ventricles in the brain, like the canal surrounding the spinal cord, are filled with cerebrospinal fluid. Ependymal cells produce much of the cerebrospinal fluid that fills these cavities. Cerebrospinal fluid doesn't just sit still; it circulates through these fluid-filled spaces in the CNS. Cilia on the ependymal cells help move this fluid around.

Oligodendrocytes resemble astrocytes, but they are smaller. Oligodendrocytes produce and maintain a special covering (called a myelin sheath) around neuronal axons. This myelin sheath is made of lipids and protein. We will be learning much more about myelinated axons shortly.

Okay, now you know there are four types of glial cells in the central nervous system—astrocytes, microglial cells, ependymal cells, and oligodendrocytes. There are two types of neuroglial cells in the peripheral nervous system, satellite cells and Schwann cells.

Satellite cells surround the cell bodies of neurons in the PNS. They provide structural support and also control the extracellular environment around the cell bodies. Thus, the satellite cells function in the PNS much in the way astrocytes do in the CNS.

Schwann cells form the myelin sheaths around axons in the PNS. Therefore, Schwann cells function in the PNS the way oligodendrocytes do in the CNS. Let's explore myelination in more detail next.

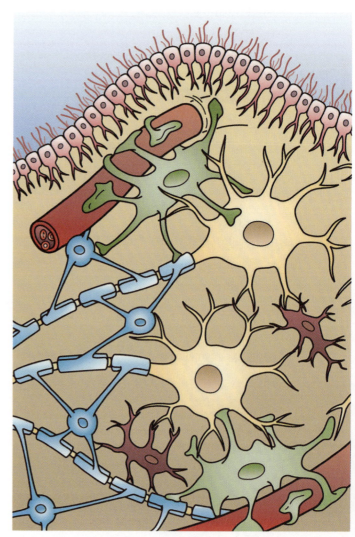

This image shows the four different types of glial cells found in the central nervous system: Ependymal cells (light pink), Astrocytes (green), Microglial cells (red), and Oligodendrocytes (functionally similar to Schwann cells in the PNS) (light blue).

# Myelination

Myelination is a process in which long axons are covered by a myelin sheath. The myelin sheath is a spiral wrapping of the modified cell membranes of the Schwann cells or oligodendrocytes responsible for forming the myelin. Axons having this myelin covering are said to be myelinated. Axons not having this covering are called nonmyelinated.

The myelin sheath provides electrical insulation for the axon. It also increases the speed a nerve signal can travel.

In the PNS, myelination is carried out by Schwann cells. These cells initially indent to receive the axon, and then wrap themselves repeatedly around the axon. Ultimately, this wrapping has the appearance of tape wrapped around a wire or gauze wrapped around a finger. At the end of the wrapping process, there may be several dozen layers of wrapping to the sheath.

Each of the Schwann cells wraps only a small length of a single axon. Other Schwann cells wrap the remaining length of the axon, like so many hot dogs in buns laid end to end. However, Schwann cells do not touch each other. There are small gaps between adjacent Schwann cells. These gaps are called nodes of Ranvier. (They were discovered by—you guessed it!—French anatomist Louis-Antoine Ranvier in the 19th century, and his name is pronounced ron'- vee-ay.)

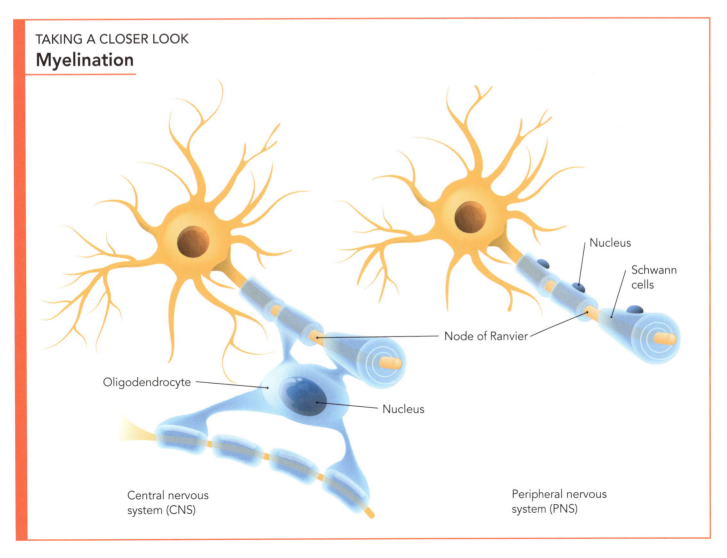

TAKING A CLOSER LOOK
**Myelination**

It should be pointed out here that a Schwann cell can enclose a dozen or more axons without wrapping them. These axons are nonmyelinated even though they are in contact with a Schwann cell.

In the CNS, it is the oligodendrocyte that is responsible for myelination. Because an oligodendrocyte has many processes, it can wrap around numerous axons rather that only one, as in the case of the Schwann cell.

The amount of myelin in the body is very low at birth and increases as the body develops and matures. Thus the number of myelinated axons increases from birth throughout childhood until adulthood. Myelination increases the speed of nerve impulse conduction through the axon. Faster conduction

## Multiple Sclerosis

Multiple Sclerosis (MS) is an autoimmune disease that results in the destruction of myelin sheaths in the central nervous system. (In autoimmune diseases, the body's immune system turns against its owner's own tissues.) In multiple sclerosis, the body's immune system attacks myelin proteins, creating hardened lesions called scleroses. These lesions commonly occur in the optic nerve, the brain stem, and the spinal cord.

As the myelin loss increases, conduction of nerve impulses becomes progressively slower. Short circuits develop and interfere with the proper functioning of the neurons. That this disease is so debilitating shows the importance of the myelination of nerve fibers to proper functioning of the nervous system.

MS primarily occurs in people under 50 years of age. Symptoms include double vision, weakness, loss of coordination, and paralysis.

One form of MS is characterized by periods of active disease alternating with periods of minimal symptoms. Another form of MS is slowly progressive, without the symptom-free periods.

Although in recent years much progress has been made in our understanding of multiple sclerosis, at present there is still no cure.

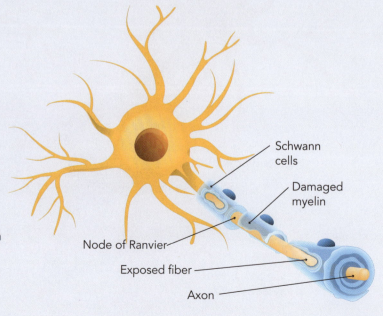

makes those nerves work better, more efficiently, as an individual matures.

Think of a newborn baby. It has very little control of its body in the beginning. It cannot hold its head up or sit up or walk. As more axons become myelinated, it has more and better control of its muscles. Compare this to a teenager. After years of development, the teenager has much better control and coordination of the body. Much of this improvement of due to increased myelination both in the central and peripheral nervous systems.

# Nerves

What are nerves? They not the same thing as neurons.

A neuron is a nerve cell. Neurons have dendrites and axons. The neuron is the cell that transmits electrical impulses in the nervous system. Thus, the neuron, not the nerve, is the basic unit of nervous tissue.

So what is a nerve? Well remember that axons, even though they are part of nerve cell, can be very long. Some reach from your back to your foot. A nerve is made of bundles of axons located in the peripheral nervous system. These bundles of axons are not alone in the nerve. The nerve also contains the Schwann cells associated with the axons, as well as blood vessels, connective tissue, and lymphatic vessels. This cross section shows the various components.

Before we go further, let's see how some of these things fit together.

Individual axons and their associated Schwann cells are covered by a very thin layer of connective tissue known as the

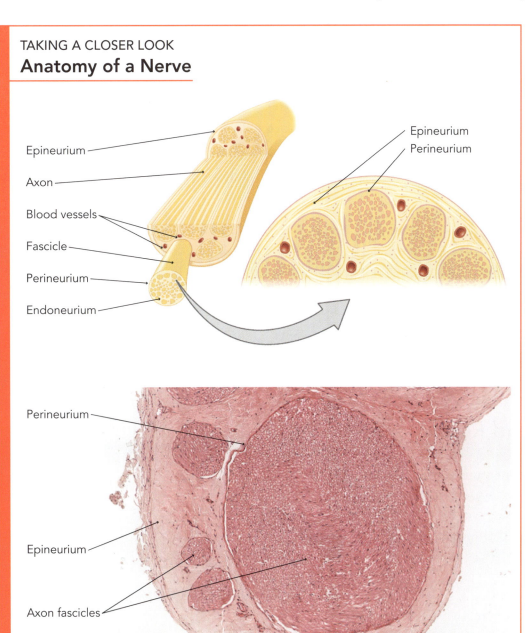

TAKING A CLOSER LOOK
**Anatomy of a Nerve**

endoneurium (endo- meaning "inner," and neurium meaning "nerve"). Next, many such endoneurium-covered axons running parallel to each other are grouped in bundles called fascicles. Each fascicle is then covered by another connective tissue layer known as the perineurium (peri- meaning "around"). Lastly, numerous fascicles, blood and lymphatic vessels are bound together by yet another connective tissue wrapping called the epineurium (epi- meaning "over"). This epineurium-wrapped bundle of bundles—containing axons, neuroglia, blood vessels, lymphatic vessels, and layers of connective tissue—is known as a "nerve."

Remember that neurons can be classified by the direction they carry electrical impulses. Motor neurons carry impulses away from the central nervous system, and sensory neurons carry impulses toward the central nervous system. Nerves can be classified the same way.

Motor nerves carry signals away from the CNS. Sensory nerves carry impulses toward the CNS. But motor nerves and sensory nerves are very rare. The most common type of nerve by far is called a mixed nerve. Even though an individual neuron can only carry an impulse in one direction (remember, from dendrite to cell body to axon), mixed nerves possess both motor and sensory fibers. Mixed nerves have two-way traffic. They carry impulses both toward and away from the CNS.

## Nerve Damage and Repair

With rare exceptions, mature neurons do not divide to reproduce themselves. The mature nervous system is not designed to replace damaged nerve cells. The neurons you have now are pretty much all you are going to get. Because of this, damage to the nervous system is serious.

However, there is a bright spot here. In the peripheral nervous system, there can be regeneration of a nerve after an injury. Recall that a nerve does not contain whole neurons, but instead consists of bundles of axons and their supporting tissues.

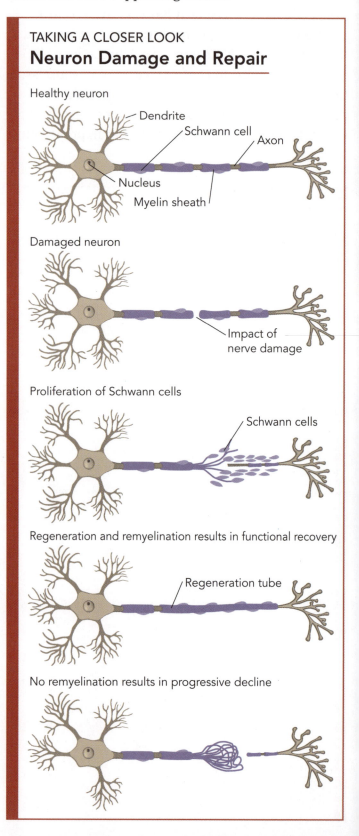

TAKING A CLOSER LOOK
### Neuron Damage and Repair

Healthy neuron
- Dendrite
- Schwann cell
- Axon
- Nucleus
- Myelin sheath

Damaged neuron
- Impact of nerve damage

Proliferation of Schwann cells
- Schwann cells

Regeneration and remyelination results in functional recovery
- Regeneration tube

No remyelination results in progressive decline

When a nerve is badly damaged, proteins and other vital substances produced in the neuron cell bodies cannot be transported all the way out to the ends of their axons. The distal (farther away) portions of the axons—the part beyond the injury—begin to break down without these nutrients. This is known as Wallerian degeneration. However, the Schwann cells near the injured area multiply and begin to form a protective tube. This "tube" helps align the damaged ends of the axons as they regenerate. Further, the Schwann cells secrete growth factors to promote axon regeneration. Therefore, nerve damage in the PNS does not always result in permanent loss of function.

It is a different story in the CNS. Recall that myelination in the CNS is due to the presence of oligodendrocytes. Unlike the Schwann cells in the PNS, oligodendrocytes do not have the capability of supporting regeneration of a damaged axon. For this reason, damage to the brain or spinal cord is more serious and more likely to be permanent than peripheral nerve injury.

---

**SO SIMPLE YET SO COMPLEX — Designed by the Master**

The foundation of our thinking in every area of our lives should be the Word of God.

How we understand the world, how we approach our daily tasks, how we view and treat our fellow man — these things should be based on the principles we find in the Bible.

Unfortunately, too many people are so strongly influenced by the views of the world that they reject the direct teaching found in God's Word. These people view the world around them as just a chemical accident. Matter somehow just came into existence all on its own billions of years ago. Then everything in our world just created itself. Millions of years of chemicals banging together resulted in something as incredibly complex as the human body.

Even though we've only just begun our study of the nervous system, I'll bet you are already getting the idea of how complex just this one body system truly is. Do you really think it could have just created itself, all on its own? No, neither do I.

In the Book of Genesis, we are told

> In the beginning God created the heavens and the earth (Gen. 1:1).

There is an all-powerful God who indeed created all things. The earth, the living creatures, the sun and moon, the planets, the stars in the sky—these things did not come into being as the result of an accident. They are not the result of time and chance. They are the work of our wonderful Creator.

Even more, you and I are not the products of chance. We are special creations.

> Then God said, "Let Us make man in Our image, according to Our likeness" (Gen. 1:26a).

As we continue our study of the human body, we need to always remember that the complex systems we study bear the unmistakable mark of the Master Designer. The enormous complexity of the body should remind us constantly of God's wisdom and creativity. We should also be reminded of His boundless love for us that He should take such care in our creation.

# NERVE SIGNALS

A fundamental characteristic of nervous tissue is that it is excitable. That is, it is capable of being stimulated to produce an electrical signal. The stimulus is some sort of change in the environment, a change in temperature for a temperature-sensitive nerve cell in the skin, or a change in light for the light-sensitive cells in the eye, or a change in the neurotransmitters around the dendrites of sensory neuron. The stimulus triggers an electrical signal that then travels down the nerve cell membrane.

We will now begin to explore this amazing phenomenon in depth.

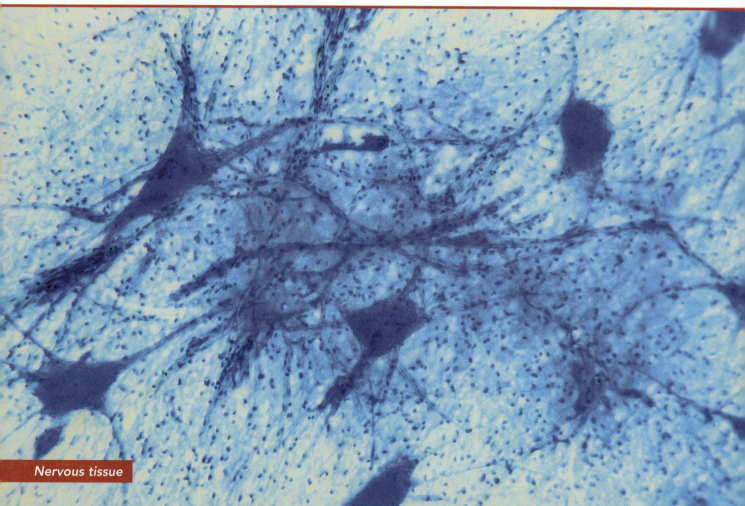

Nervous tissue

## Basic Principles

For a neuron to function properly, it must be able to produce an electrical impulse and then transmit it along the length of its cell membrane. Production of this traveling electrical impulse depends on the movement of charged particles into and out of the cell. As charged particles—called ions—move across sequential parts of the neuron cell membrane, the electrical impulse moves along the length of the neuron. Let's see how these electrical impulses are generated and how they travel.

Your body itself is electrically neutral. This means that the number of negative particles equals the number of positive particles (or almost equals, you know, give or take a particle or two…). The extracellular fluid (fluid outside cells) is electrically neutral, as is the intracellular fluid (the fluid inside cells). So if everything is neutral, where do these traveling electrical impulses come from? These electrical impulses are created by the movement of ions across the cell membrane of the neuron.

The cell membrane (also called the plasma membrane) of a neuron contains special proteins that extend through the full thickness on the membrane. The special proteins have channels or pores that allow passage of certain ions (charged particles) into and out of the neuron. These channels open and close in response to certain stimuli. For example, some channels open when in the presence of specific chemicals (perhaps a neurotransmitter, as we will see later). Other channels open in response to a change in voltage across the membrane itself.

When these channels open, ions move through them based on certain chemical principles. First, ions move from areas of higher concentration to areas of lower concentration. For example, let's say that in a certain situation there was a high number of potassium ions in the cytosol of the cell and a much lower number of potassium ions in the extracellular fluid around the cell. Then, a potassium ($K^+$) channel opens in the cell membrane that allows movement of $K^+$ ions. What do you think will happen? You are correct. $K^+$ from the inside of the cell will rush out of the cell toward the area of lower concentration. This is called a concentration gradient.

Second, ions move toward regions of opposite charge. Positively charged ions move towards regions of negative charge, and negatively charged ions move towards areas of positive charge. This is called an electrical gradient.

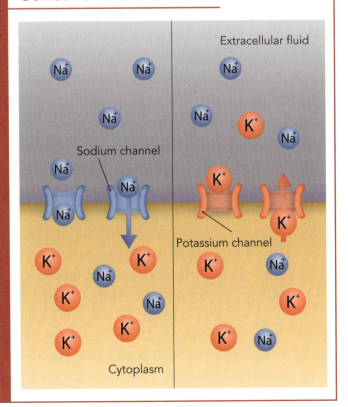

TAKING A CLOSER LOOK
**Concentration Gradient**

# The Resting Membrane Potential

As previously mentioned, the environments inside and outside the cell are each electrically neutral. That is, there are equal numbers of positive and negative charges. However, there does exist a small electrical difference across the cell membrane. This electrical difference is called the resting membrane potential.

If you measure the electrical difference between the area outside the cell membrane and the cytosol inside, you will find a difference of about minus 70 millivolts (- 70mV). The inside of the cell is negatively charged when compared with the outside of the cell. When a membrane potential, that is, a charge across a membrane, exists, the membrane is said to be polarized.

Membrane polarization is possible because of the different concentrations of ions in the intracellular and extracellular fluids and their abilities to cross the cell membrane. There is a much higher concentration of potassium ions ($K^+$) inside the cell than in the extracellular fluid surrounding it. And there is a higher concentration of sodium ions ($Na^+$) in the extracellular fluid than there is inside the cell. Inside the cell, there is electric neutrality because there are enough negative ions to balance the positive ones. And in the extracellular fluid outside the cell, there is electric neutrality for the same reason. But across the cell membrane, there is a measurable electrical difference, the resting membrane potential.

How is this possible? Remember we said $K^+$ and $Na^+$ ions are able to travel through the cell membrane. But the journey through the membrane is easier for some ions than others. Potassium and sodium ions differ in their ability to cross the cell membrane. At rest, the cell membrane is much more permeable to $K^+$ ions than to $Na^+$ ions. In other words, the protein channels that allow ions to pass are much more open to the movement of $K^+$ than to $Na^+$.

If there is a higher concentration of $K^+$ inside the cell than outside, in what direction do $K^+$ ions move? If you said, "Potassium ions move out of the cell," you would be right. By contrast, $Na^+$ is more concentrated outside the cell. Therefore, $Na^+$ "wants" to move into the cell. But since $K^+$ moves through the cell membrane much more easily than $Na^+$, lots of $K^+$ flows out, but much less $Na^+$ gets in. Therefore, on balance, more positive charges move out of the cell than move in. This leaves the inside of the cell a little bit negative compared to the outside. Positive ions build up outside the cell, leaving a relatively negative charge on the inside. This electrical difference is about -70mV.

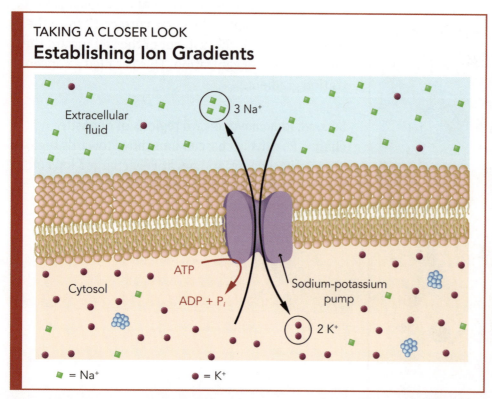

**TAKING A CLOSER LOOK**
**Establishing Ion Gradients**

# NERVE SIGNALS

## TAKING A CLOSER LOOK
### Nerve Impulse Action Potential

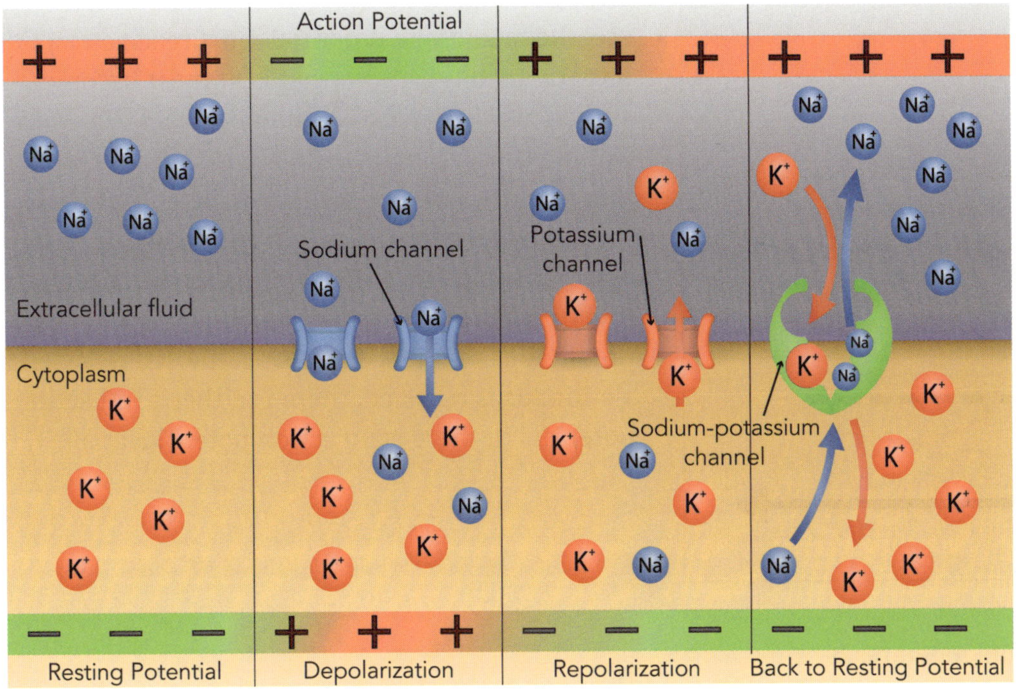

[Diagram showing four stages across a membrane: Resting Potential, Depolarization, Repolarization, and Back to Resting Potential. Labels include Extracellular fluid, Cytoplasm, Na+, K+, Sodium channel, Potassium channel, Sodium-potassium channel, and Action Potential.]

But wait a minute. If you think about it, eventually the concentration of K+ and Na+ would be the same on both sides of the membrane, right? After all, these ions move from higher concentration to lower concentration. Sooner or later they would reach a state where the concentrations are equal on both sides of the membrane. So would this not eliminate this electrical potential across the membrane? Yes, it would....but (isn't there always a "but"?).

There is a special type of pump in the cell membrane that helps keep the resting potential stable. This is called a sodium-potassium pump. The sodium-potassium pump transports Na+ out of the cell and K+ into the cell. This process maintains the concentration differences that create the resting membrane potential.

The sodium-potassium pump works hard, and it requires energy to do its job. Energy to fuel this sort of work is supplied by a special molecule called adenosine triphosphate (ATP). ATP stores chemical energy and is ready to supply it for all sorts of cellular work. ATP is the main energy currency of the cell.

The sodium-potassium pump is perfectly designed for its job. It is obviously meant to maintain the resting membrane potential because the number of Na+ and K+ it transports is not equal. For every two K+ brought into the cell, three Na+ are pumped out. This means that this pump maintains the -70mV inside the cell, because it consistently pumps more positive ions out of the cell than it brings in!

(You've probably noticed by now that we describe the membrane potential from the "point of view" of the inside of the cell. When there are more positive ions outside the cell than there are inside, the inside is negative compared to the outside.)

## The Action Potential

We've talked about how the neuron's cell membrane gets and stays polarized. The sodium-potassium pump pushes more positive charges to the outside than it brings in. A polarized membrane—an electrical difference between the inside and the outside—must exist before a neuron can send an electrical signal. So how exactly does a neuron send an electrical signal?

# THE NERVOUS SYSTEM

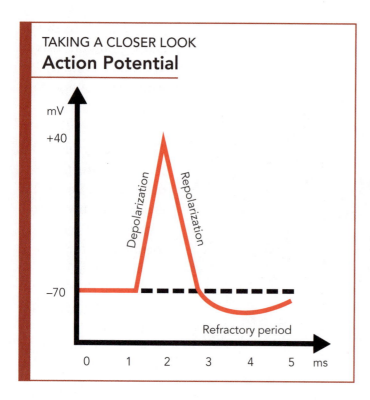

**TAKING A CLOSER LOOK**
**Action Potential**

Action potentials occur only in axons. When an axon is presented with an adequate stimulus, the nerve impulse is triggered. However, not every stimulus is strong enough to trigger the action potential. You see, the membrane potential must reach a certain level of depolarization (that is, it must become sufficiently less negative than its usual -70mV) to initiate the action potential. In a typical neuron, this level is -50mV or so. This is called threshold level. Any depolarization not reaching this level is sub-threshold and will not trigger a nerve impulse. A very important thing to remember about an action potential is that it is an all-or-none event. When a stimulus is received, there is either a full action potential or there is no action potential at all.

The electrical signal a neuron generates is called a nerve impulse. It is also called an action potential. An action potential is a change in the membrane potential from negative (more positive charges outside) to positive (more positive charges inside) and then back again.

The first thing that happens during an action potential is called depolarization. During depolarization, the membrane potential becomes less and less negative, and then positive. In other words, once a membrane is depolarized, there are more positive charges on the inside than there are outside. (Remember, these "negatives" and "positives" represent the point of view from the inside of the nerve cell.)

The next step is called repolarization. During repolarization, the membrane potential is reset, becoming negative once more. Once it is repolarized, there are more positive charges outside than inside. Having had its resting membrane potential reset, the membrane is prepared for the next action potential.

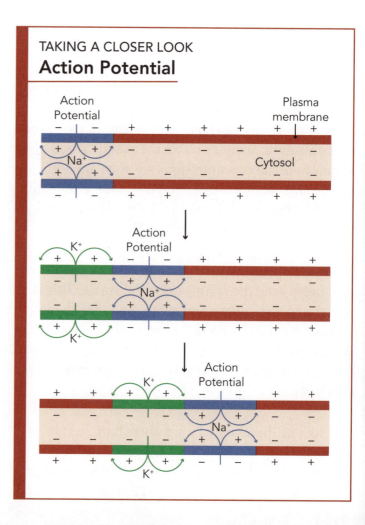

**TAKING A CLOSER LOOK**
**Action Potential**

# Depolarization

At rest the cell membrane is permeable to both K⁺ and Na⁺. That means both K⁺ and Na⁺ ions can pass through the membrane. This permeability is possible because of special proteins that create a channel to facilitate the movements of these ions through the cell membrane. There are, however, other special channel proteins involved with the movement of ions. These are called voltage-gated channels. These channels are opened when voltage across the membrane changes. As we will see in a moment, there are voltage-gated channels for Na⁺, and there are other voltage-gated channels for K⁺.

When a stimulus reaches the voltage-gated Na⁺ channels, the channels open and Na⁺ pours into the cell. This influx of positive ions makes the inside of the cell less negative. If the threshold level of -50mV is reached, many more voltage-gated Na⁺ channels open, and much more Na⁺ flows into the cell. At this point depolarization is in full swing, and the action potential moves on to a neighboring portion of the membrane, continuing down the entire length of the axon. At the end of the depolarization phase the membrane potential is around +30mV. You see, the inside of the cell is now positive with respect to the outside because a flood of Na⁺ ions has resulted in more positive charges inside the cell than outside.

This remarkably complex and precise series of events happens in less than a thousandth of a second.

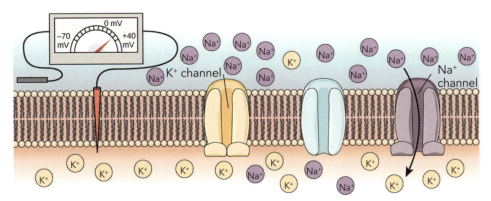

**Resting Potential:** At the resting potential, voltage-gated Na⁺ channels and voltage-gated K⁺ channels are closed. The Na⁺/K⁺ pumps moves K⁺ ions into the cell and Na⁺ ions out of the cell.

**Depolarization:** In response to a depolarization, some Na⁺ channels open, allowing Na⁺ ions to enter the cell. The membrane starts to depolarize (the charge across the membrane lessens). If the threshold of excitation is reached, all the Na⁺ channels open.

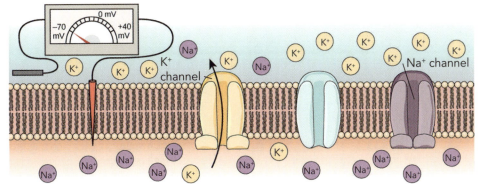

**Hyperpolarization:** At the peak action potential, Na⁺ channels close while K⁺ channels open. K⁺ leaves the cell, and the membrane eventually becomes repolarized.

# Repolarization

At the end of the depolarization phase, the membrane potential is around +30mV. Obviously this is far from the level of the resting membrane potential. In this state the neuron is not able to trigger another action potential. The neuron's negative resting membrane potential must be reset before another action potential can travel along that portion of the axon.

When the membrane potential reaches +30mV, the voltage-gated $Na^+$ channels begin to close. Therefore, movement of $Na^+$ into the cell stops almost completely. Then voltage-gated $K^+$ channels open.

$K^+$ rushes out of the cell through these channels as soon as they open. Remember that the resting concentration of $K^+$ is higher inside the cell than outside. When the "door" is opened, these ions rapidly move from an area of high concentration to an area of low concentration. Furthermore, the depolarization left the outside of the cell more negative than the inside. Positive charges are attracted to a more negatively charged area. Potassium ions therefore flood out of the cell through their channels.

As the membrane potential returns to negative—in other words, when there are more positive charges outside the cell than inside—these voltage-gated channels close, and the permeability of the membrane returns to its normal resting state. In some instances the outflow of $K^+$ is so rapid during repolarization that the membrane potential will overshoot to near -90mV. This is called hyper-polarization. When the voltage-gated $K^+$ channels are all closed, normal membrane potential is rapidly restored.

## TAKING A CLOSER LOOK
### Action Potential along Myelinated Nerve Fiber

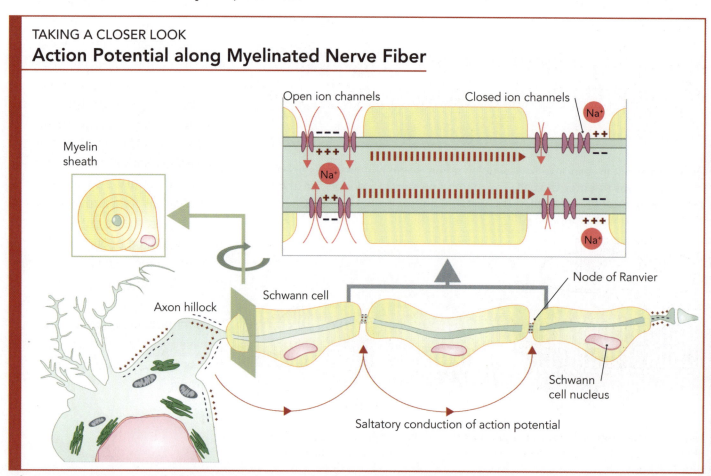

Repolarization, which prepares the axon for the next action potential, takes only a few thousandths of a second.

## Re-establishing Normal Resting Ion Concentrations

Even though a membrane is repolarized, or even hyperpolarized, after the potassium ions rush out of the cell, the sodium and potassium ions are not in their normal resting places. Normally, the $K^+$ concentration inside is greater than outside, and the $Na^+$ concentration outside is greater than inside. How do the ions return to this normal state?

Remember the sodium-potassium pump? The sodium-potassium pump gets busy. Using energy from ATP, the pump pushes out three $Na^+$ for every two $K^+$ that it brings into the cell. Soon the normal resting state is reestablished.

## Conduction of Action Potentials

An action potential moves down the length of the axon membrane like a series of dominoes each knocking the next down. A voltage change triggers voltage-gated channels to open in one area, producing the action potential there. This voltage change triggers the same response in the neighboring area of the membrane. Here one region directly triggers the next, and the next, and the next, and so on. This is known as continuous conduction. Continuous conduction occurs in unmyelinated axons. Because a bit of time is required to repolarize each section of the axon membrane before another action potential can be created there, continuous conduction is slow compared to the sort of conduction that occurs in myelinated axons.

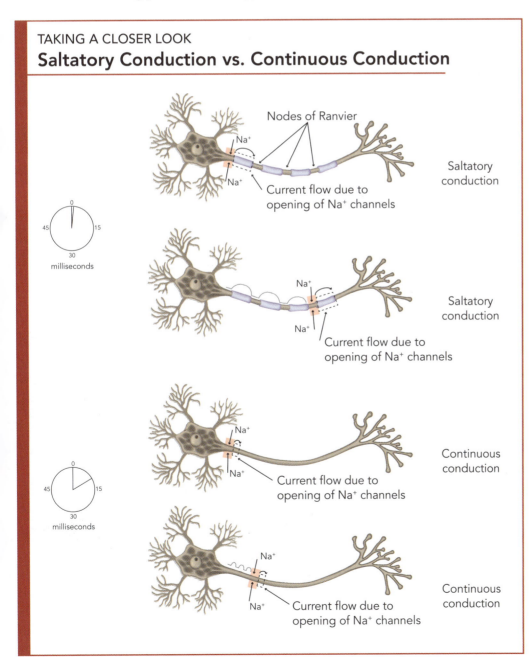

TAKING A CLOSER LOOK
**Saltatory Conduction vs. Continuous Conduction**

In myelinated axons, nerve impulse conduction is much more rapid. This is due to the myelin covering the axon itself. The myelin covering insulates the axon. Further, in myelinated axons, voltage-gated channels are concentrated near the gaps between the Schwann cells, the nodes of Ranvier. It is not necessary to repolarize then entire surface of the axon, just the portions at the nodes of Ranvier. By generating local currents around the myelin sheath, the action potential seems to "leap" from one gap to the next. The myelin sheath therefore speeds up conduction of the action potentials through the length of the axon. This is known as saltatory (from saltare, meaning "to leap or hop") conduction.

"How much faster?" you are probably wondering. Well, nerve impulse conduction speed varies based on the thickness of the axon and how much myelin covers it. But to get an idea of the sort of speeds we are talking about, figure a small unmyelinated axon might transmit nerve impulses at around 2 miles per hour. Depending on its diameter and degree of myelination, a myelinated axon might transmit nerve impulses at 30 to 300 miles per hour. When you realize how dramatically myelination speeds up nerve impulse conduction, you can understand how debilitating a disease that destroys myelin must be.

## Strong Versus Weak Signals

So we have seen that action potentials are all-or-nothing phenomena. When triggered the action potential fires completely. There is no such thing as a partial action potential. In a system as finely regulated as the nervous system, isn't this a rather clumsy arrangement? After all, if a stimulus triggers an action potential, and one action potential is pretty much like all others, how can such fine control be achieved?

We need to understand the difference between how a nerve responds to a weak stimulus compared to its response to a strong stimulus.

It is obvious that there are varying degrees of sensations we perceive every day, right? Water can be cold, warm, or hot. We can feel a gentle breeze on our face or strong winds on a stormy day. Walking on the beach, we can feel that the sand under our feet is warm and soothing or very hot and unpleasant. How do our neurons help us tell the difference?

Because all action potentials are the same, a "warm" stimulus triggers the same action potential as a "hot" one. However, a stronger stimulus triggers more action potentials in a given time than a mild one. A stimulus from sensory receptors sensing a strong wind will trigger more frequent action potentials by comparison to those triggered a mild breeze.

Also, as a general rule, more intense stimuli tend to activate more sensory receptors than mild stimuli do. Therefore, intense stimuli trigger a flurry of action potentials from more receptors than mild stimuli.

If you step barefoot onto hot sand, more action potentials are triggered in more temperature receptors than when you burrow your toes into soft, warm sand that is just right. Therefore, your CNS receives more input from a greater number of neurons from

your foot stepping on the hot sand than from your toes curling in the warm sand.

## Graded Potentials

Since action potentials occur only in axons, it is reasonable to ask how a nerve impulse is transmitted through the rest of the neuron. Nerve impulses travel through dendrites and neuron cell bodies using graded potentials.

When a dendrite receives a stimulus, there is a small change in the membrane potential in that area. However, an action potential is not triggered. Instead, a graded potential is generated. The stimulus causes a small number of channel proteins to open. This results in some ion movement across the cell membrane. However this small membrane response travels only a short distance and then fades away.

While an action potential is all-or-none, a graded potential varies in degree. Graded potentials vary with the strength of the stimulus. The greater the stimulus, the greater number of ion channels open. And the stronger the graded potential, the farther it travels. If this type of impulse travels far enough, reaching the initial part of the axon, it can ultimately lead to a full action potential.

## The Synapse

We have seen how the neuron transmits a signal along its length. A stimulus triggers graded potentials in the dendrites, and if a sufficiently strong impulse reaches the initial part of the axon, an action potential is triggered and passed down the axon's length. Thus we've now seen how a nerve impulse travels the whole length of a neuron.

That's all well and good you might say. But when a nerve impulse reaches the end of the axon, how does that impulse get to the next nerve, or to a muscle, or to anywhere for that matter?

A nerve signal is passed along via something called a synapse. A synapse is the place where a neuron communicates with another neuron or with a muscle cell.

At the far end of an axon are small branches called axon terminals. Axon terminals are positioned very close to the cell membrane of the next cell, the cell that is to receive the signal. The axon terminal, the membrane of the next cell, and the space between them (known as the synaptic cleft), are collectively known as a synapse.

The neuron that carries the impulse towards the next cell is called the presynaptic neuron ( pre- because this nerve is located "before" the synapse). The neuron that is to receive the nerve impulse is called the postsynaptic neuron (post or "after" the synapse). Neurons are usually not just pre- or post-synaptic. Most neurons are both. A given neuron is postsynaptic at its dendritic end where it receives impulses. The same neuron is presynaptic at its axon terminals where it sends impulses to the next cell. These terms are useful, however, when we describe the neurons around a particular synapse.

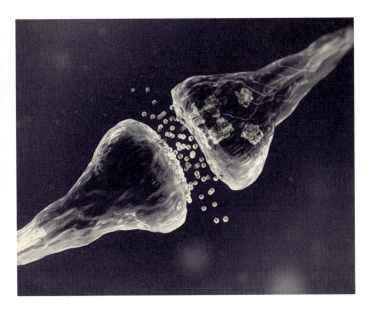

*Synapse and neuron cells sending chemical signals.*

There are two basic types of synapses: electrical and chemical.

At an electrical synapse, the action potential is transmitted directly to the next cell. This occurs by means of special connections called gap junctions. Gap junctions connect the cytosol of adjacent cells. Gap junctions allow the nerve impulse, like an electrical current, to pass directly to the next cell.

The most common type of synapse is the chemical synapse. In a chemical synapse, the nerve impulse does not travel directly from axon terminal to the post-synaptic neuron but must send a different sort of signal, a chemical signal, across the synaptic cleft. Let's take a closer look.

## Chemical Synapse

A chemical synapse is designed to transfer nerve signals by releasing special chemicals called neurotransmitters. This synapse is an incredible design.

The axon terminal of the presynaptic neuron is positioned adjacent to receptor region on the post-synaptic neuron. The space between the two cells is extremely small. It is usually between 20 and 40 nanometers. That is 20 to 40 billionths of an inch! (Not much wiggle room there.)

The axon terminal of the presynaptic neuron contains many small sacs, called synaptic vesicles. These vesicles contain the neurotransmitters. When an action potential reaches the axon terminal, voltage-gated $Na^+$ channels open, and $Na^+$ rushes into the cell. This is the same process we covered earlier. However, in the axon terminal there are also voltage-gated $Ca^{2+}$ (calcium ion) channels. The action potential triggers these also, and $Ca^{2+}$ pours into the cell. This increase in $Ca^{2+}$ inside the cell is a signal for the cell to begin to release the neurotransmitters.

Neurotransmitters are released from the synaptic vesicles into the synaptic cleft by exocytosis. This means that the vesicles containing the neurotransmitters merge with the cell membrane and release their chemical messengers into the synaptic cleft.

A single action potential will result in the release of only a small amount of neurotransmitter. If multiple, frequent action potentials are received, then more neurotransmitter is released. Thus a greater stimulus causes a larger of amount of neurotransmitter to be released into the synaptic cleft.

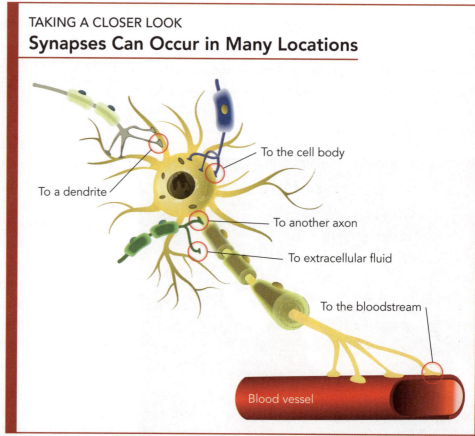

TAKING A CLOSER LOOK
### Synapses Can Occur in Many Locations

The neurotransmitter diffuses across the synaptic cleft and binds to special receptor proteins on the postsynaptic neuron. These receptor proteins then allow movement of ions across the cell membrane of the postsynaptic neuron. These receptor proteins are called chemically-gated channels. This means that they open and close not because of a change in voltage but rather due to the presence of certain chemicals, the neurotransmitters.

Now comes the really cool part. Neurotransmitters can either excite or inhibit the postsynaptic neuron. Some neurotransmitters cause the postsynaptic cell membrane to slightly depolarize. These promote excitation of the post-synaptic neuron. Other neurotransmitters cause the post-synaptic cell membrane to become slightly hyper-polarized, which tends to inhibit the production of another nerve impulse.

Remember that when the inside of the cell becomes less negative, and thus is brought closed to its threshold voltage, it is being depolarized. If enough excitatory neurotransmitter is released across the synaptic cleft, then a graded potential strong enough to trigger an action potential may result. Excitatory signals bring that postsynaptic neuron closer to firing an action potential.

The opposite is true if the neurotransmitter causes the postsynaptic neuron to become more negative inside. This takes the neuron farther away from the threshold. This is hyper-polarization.

Man, that's a lot to take in, so if you're feeling overwhelmed, you are probably normal. It can get quite complicated. (Remember this sentence when you are in medical school in a few years.) It really comes down to something simple. Let's have a look at the

### TAKING A CLOSER LOOK
### The Synapse

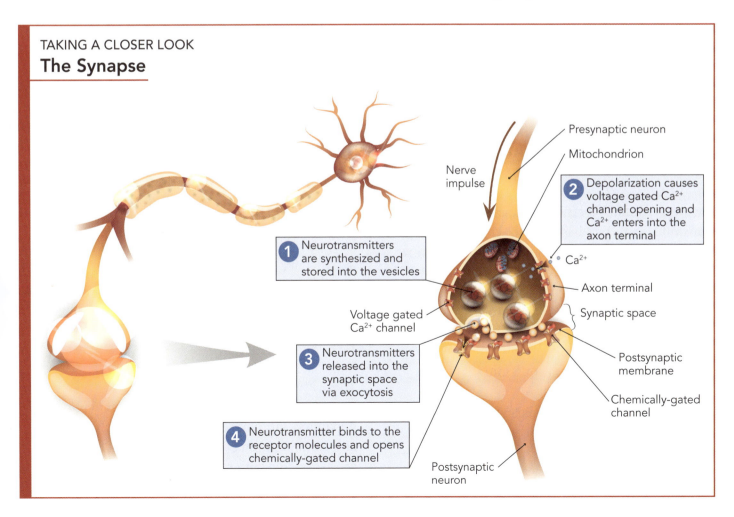

big picture, and I think this will make a lot of sense to you.

## The Role of the Synapse

The synapse is simply a mediator. It sort of "takes a vote." It allows many different stimuli or inputs to "have their say" in determining whether the postsynaptic neuron fires or not. See, easy.

Because the synapse works the way it does, it prevents any one single stimulus from triggering the postsynaptic neuron. It takes a significant stimulus to fire that neuron. The presence of both excitatory and inhibitory neurotransmitters makes it possible to achieve very fine control of neuron firing. This is crucial for the proper functioning of the nervous system. For example, if a postsynaptic membrane receives four excitatory stimuli and four inhibitory stimuli, the signals basically cancel each other out (4-4=0), so the postsynaptic neuron does not fire. If there are 10 excitatory stimuli and only two inhibitory stimuli, the postsynaptic membrane will be brought close to threshold, perhaps enough to trigger the action potential. You can imagine that if every stimulus, no matter how mild, were able to elicit a big response, your nervous system would be rather overloaded, and you would overreact to everything.

Therefore, unlike electrical synapses that directly transmit signals to adjacent cells, chemical synapses do not transmit electrical signals at all. They allow electrical signals in the presynaptic neuron to be converted to chemical signals. These chemical signals are then converted back into electrical signals in the postsynaptic neuron. While at first glance this might seem overly complex, it does allow very fine control of nerve signal transmission.

You really should not try to make it any harder than that. The synapse is a magnificent design from the Master Designer!

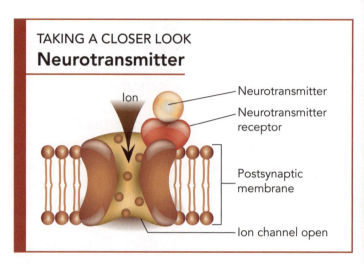

## Neurotransmitters

Neurotransmitters are the molecules that carry the signals across the synaptic cleft. They then interact with the channel proteins on the postsynaptic membrane, triggering ions channels to open. But what happens after that?

The actual work of neurotransmitter lasts only for a few milliseconds (thousandths of a second). Then it disengages from its receptor protein, ending its effect on the postsynaptic cell. But the neurotransmitter is still there, correct? It needs to be removed before the next batch of neurotransmitters is released, correct? Correct...both times.

There are several processes that are in place to remove the neurotransmitter after its work is done. In some cases the neurotransmitter simply drifts away from its receptor. Special enzymes can break down specific neurotransmitters to remove them from the synaptic cleft. And some neurotransmitters are pumped back into the axon terminal of the presynaptic neuron.

There are several dozen known neurotransmitters. This list includes acetylcholine, epinephrine, dopamine, glutamate, histamine, serotonin, and gamma-aminobutyric acid (GABA). Some are associated with specific areas of the brain or spinal cord, while some are found in the peripheral nervous system. Acetylcholine, for instance, is the neurotransmitter used by nerves synapsing with skeletal muscles but is also found in the brain. And norepinephrine is used in both the brain and the sympathetic nervous system. Some neurotransmitters send excitatory messages, while others are more inhibitory, and others can either excite or inhibit depending on the type of receptor. God used a wide variety of neurotransmitters in the design of the nervous system, fine tuning its many parts to communicate efficiently with each other and with the rest of the body.

## Homeostasis

As we have seen throughout our exploration of the human body, the body is constantly trying to maintain a "balance" among its many systems. This is known as homeostasis.

For the body to function properly and efficiently, its internal conditions must be kept within optimal ranges. There are numerous processes that the body must monitor and regulate from second to second, minute to minute.

The body's temperature must be kept within the correct limits, not too high, not too low. Blood pressure, heart rate, and respiration are constantly evaluated and assessed. Is the blood level of thyroid hormone appropriate? If it's too low, a person might feel sluggish. If it's too high then the body's metabolism can be accelerated. Is the level of oxygen in the blood high enough? It certainly needs to be. On the other hand, is carbon dioxide building up in the blood? If so, the respiratory system needs to be signaled to increase ventilation. And it goes on and on.

We will explore many of these processes during our studies. However, as we proceed you need to consider something. How could mechanisms so complex be an accident? The answer is that they are not. They could not possibly be the result of chemical reactions over millions of years.

The processes that aid in homeostasis bear the mark of amazing design. There is nothing random or accidental about them. They are the work of the Master Designer.

> For You formed my inward parts;
> You covered me in my mother's womb.
> I will praise You, for I am fearfully and wonderfully made;
> Marvelous are Your works,
> And that my soul knows very well.
> (Psalm 139:13–14)

# THE CENTRAL NERVOUS SYSTEM

Walking, talking, sitting, standing, learning, sleeping, laughing—just imagine all the things our nervous system has to process every second to make any of these things possible. It is quite literally beyond our imagination.

It seems like the more we understand about the nervous system, the more there is to learn. How does it all work, and how does it all work together?

The vast number of neurons, the even greater number of interconnections, the complex structures, and (probably the biggest thing) all those strange names given to all this stuff…it's all just too much!

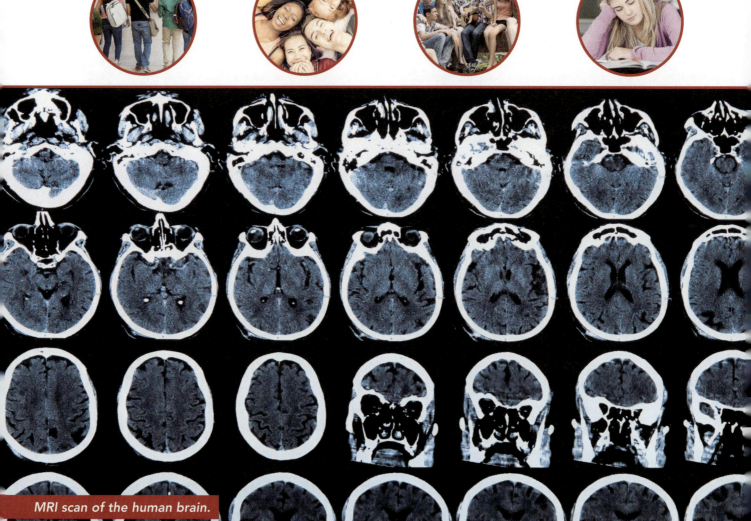

*MRI scan of the human brain.*

No, it really isn't. We will continue to take all this a step at a time.

Now that we've seen how nerve impulses travel, let's introduce the brain and spinal cord, the central nervous system.

# The Brain

The brain is the master control center of the human body.

Standing up without falling down? The brain controls this. Can you taste that ice cream? Yep, the brain is right on the job. Make an A on that math test? Good job, brain! Feeling sleepy? Your brain is telling you something.

The most amazing, incredible, complex thing in the universe is the brain. Even with all we have learned in recent decades, it is still the least understood organ in the human body.

The brain is so unimaginably complex and so astonishing in its function, it can only be the work of the Master Designer.

The human brain is quite recognizable to most people. It is an organ that is a sort of pinkish-gray in color and has many wrinkles or folds on its surface. An average brain weighs around 3 pounds. That's approximately 2 percent of the body weight of a typical adult. Even though its size is small, the brain consumes about 20 percent of the body's total energy.

How many neurons are packed into this 3-pound energy guzzler? The most often quoted estimate of the number is 100 billion! More recent research

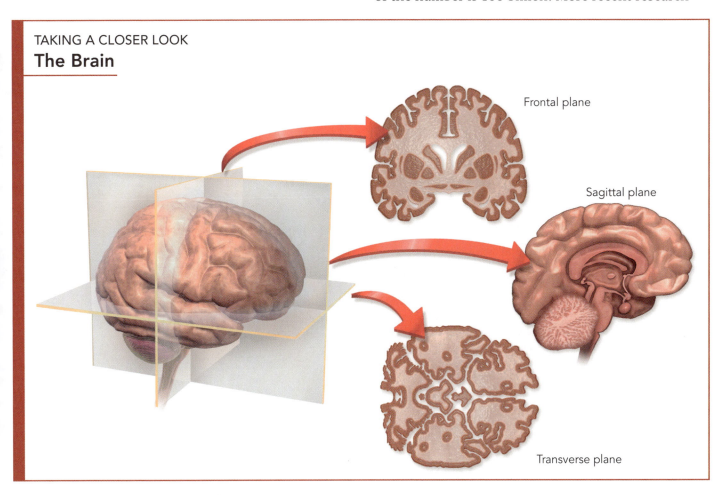

TAKING A CLOSER LOOK
**The Brain**

Frontal plane

Sagittal plane

Transverse plane

suggests that the actual number is closer to 86 billion, but that is still a lot of neurons. Each of these neurons communicates with many other neurons through synapses. The number of synapses in the brain, links between its neurons, has been estimated to be in the trillions! (Seriously, who counted them?)

## Brain - Its Protection

The brain is such a vital organ, and it makes sense that it is located where it can be well protected. The brain is located inside the cranial vault at the top of the skull. The skull consists of the cranial vault and the bones that make up your face and jaw. The cranial vault—also called the cranium—is the large open space in the skull. This bony container, shaped perfectly to house the human brain, provides significant protection from trauma.

Inside the cranium, the brain is cushioned by another layer of protection, the meninges. The meninges are really three layers of protection—three layers of connective tissue that cover the brain and spinal cord. The outermost layer is called the dura mater.

The dura (short for dura mater) is the thickest layer of the meninges and is itself composed of two layers. The outermost layer of the dura is attached to the inside of the cranium and is called the periosteal layer. (If you recall from Volume 1 of *Wonders of the Human Body* that the outer covering of bone is called the periosteum, it will help you remember that this periosteal layer is closest to the cranial bone.) The inner layer of the dura is called the meningeal layer. The two layers of the dura mater are fused together except in those areas where they separate to form

### Do We Only Use 10 Percent of Our Brains?

Is it true that we only use 10 percent of our brains? You've probably been told that your whole life by people who are sure it is true because they've heard it from folks who were sure it was so.

Well, no one was trying to pull the wool over your eyes, but it is not true. Not at all.

This "10 percent myth" is one of the most common myths in our modern culture. Somehow people have come to believe that we only use a small portion of our brain every day. I have so often heard, "Wouldn't it wonderful if we could unlock that unused part of our brains?" It would be. If it were true. But it isn't.

You see, with all the modern scanning and imaging techniques available to neuroscientists, no one has been able to discover exactly where this unused part of the brain is located. Plus, scanning and mapping studies of the brain have failed to find a part of the brain that is not active, even with basic activities such as walking or watching television.

So the 10 percent brain myth is 100 percent false.

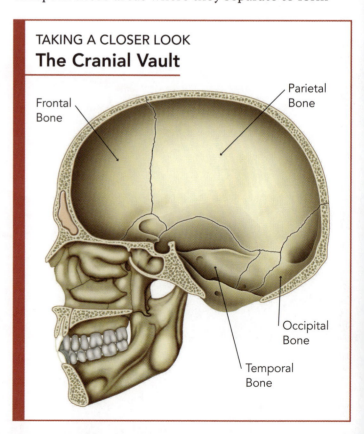

**TAKING A CLOSER LOOK**
**The Cranial Vault**

Frontal Bone
Parietal Bone
Occipital Bone
Temporal Bone

large veins called venous sinuses. These venous sinuses provide channels to drain venous blood from the brain on its way back to the heart.

The middle layer of the meninges is known as the arachnoid mater. This name might sound spidery to you, if you recall that spiders are called arachnids. Both words come from the Greek word for spider, *arachne*. From the arachnoid mater extend fine delicate fibers that resembles spider web. That's how it got its name! Like the dura mater, the arachnoid mater covers the brain and spinal cord. Neither of these layers dips down into the brain's many folds.

The third meningeal layer is the pia mater. This very thin layer of connective tissue lies next to the brain itself and is covered in very tiny blood vessels. The pia mater dips down into the folds and grooves in the brain.

The spaces between the meningeal layers also have names, naturally. The space between the dura and the arachnoid layer is called the subdural space. Sub means "under," so that's a pretty good name—subdural means "under the dura." The space below the arachnoid is called (logically enough) the subarachnoid space. And these spaces are a very important part of the cushion that protects the brain!

# Cerebrospinal Fluid

In the subarachnoid space is a liquid known as cerebrospinal fluid (CSF). This fluid flows around the brain and spinal cord, cushioning both. The CSF supports the brain by helping it float in the cranial vault. It's sort of a waterbed for your brain! This fluid layer also acts as a shock absorber to keep the brain from banging around inside the skull. Simply running up and down stairs could damage the brain and the surrounding structures if there were not a way to keep the brain from moving around and bumping into the bone that houses it. Another amazing design!

In addition to the subarachnoid space, CSF fills four chambers in the brain called ventricles. The two largest are the lateral ventricles, one in each side of the brain, inside the cerebral hemispheres. (The two cerebral hemispheres, as we will see in a moment, are

TAKING A CLOSER LOOK
## The Meninges

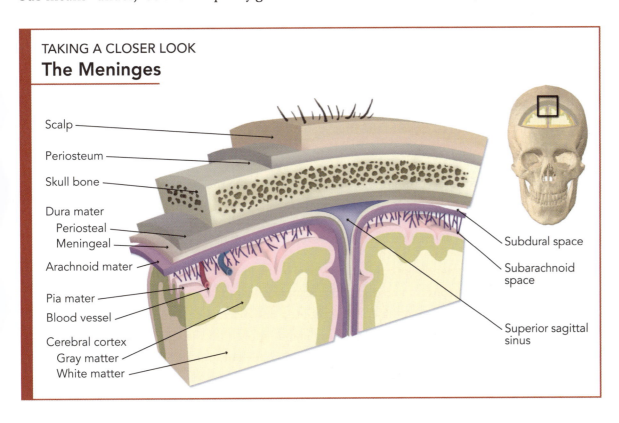

# THE NERVOUS SYSTEM

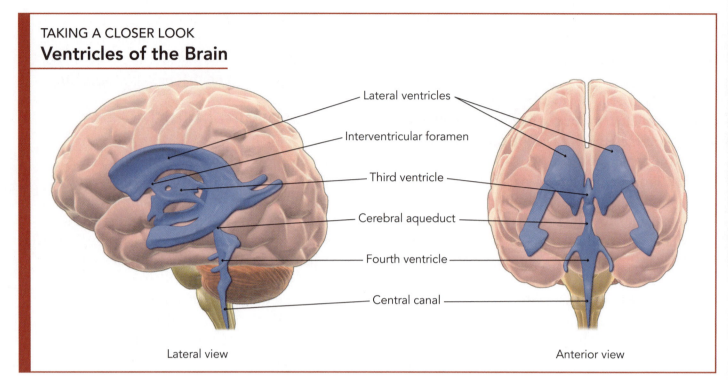

### TAKING A CLOSER LOOK
### Ventricles of the Brain

Lateral view — Anterior view

Labels: Lateral ventricles, Interventricular foramen, Third ventricle, Cerebral aqueduct, Fourth ventricle, Central canal

the two halves of the brain's upper part.) Between the cerebral hemispheres is the third ventricle. The lateral ventricles communicate with the third ventricle below them. Below the third ventricle is the fourth ventricle, and this ventricle connects to the central canal of the spinal cord. The interconnections between the four ventricles and the central canal allow the CSF to circulate around the brain and spinal cord.

The cerebrospinal fluid is clear and colorless. It is made from blood's clear liquid plasma, but it is completely separate from the blood. CSF is manufactured in the brain within the ventricles and suba-

## Meningitis

Meningitis is an inflammation of the meninges. It is most commonly caused by an infection, either by bacteria or viruses. Because any severe inflammation so close to the brain can have dire consequences, this illness is considered a medical emergency.

Bacterial meningitis can be fatal if untreated, so prompt intervention is crucial. On the other hand, viral meningitis has no specific treatment and generally resolves on its own. It is rarely fatal.

The most common symptoms of meningitis are severe headache, fever, and a stiff neck.

Meningitis is diagnosed by inserting a long needle into the spinal canal in the lumbar region of the spine. (This is below the spinal cord itself, so the spinal cord is in no danger.) A sample of cerebrospinal fluid is withdrawn and sent for immediate analysis. The cells present in the CSF are counted and examined. If bacterial meningitis is suspected, antibiotics are administered to the patient intravenously immediately after the CSF sample is taken, even before the test results are available, because timely treatment is so important. If the CSF sample does not show evidence of bacterial meningitis, no further antibiotics are given.

## Parts of the Brain

The human brain is made of four major parts. The largest part is called the cerebrum, which is made of two halves, the cerebral hemispheres. Next is the diencephalon. The diencephalon is made up of the structures near the third ventricle—the thalamus, hypothalamus, and the epithalamus. (We'll get to what those parts do later.) At the rear of the brain, tucked under the back of the cerebrum, is the cerebellum. And on the underside of the brain is the brain stem. The brain stem ultimately merges with the spinal cord.

Let's explore the structure and then the functions of each part in turn.

## Cerebrum — Gross Anatomy, the Parts

The most recognizable portion of the brain is the cerebrum. It is somewhat wrinkled in appearance, covered with many ridges and folds. The ridges are called gyri (singular is gyrus), and the folds are called

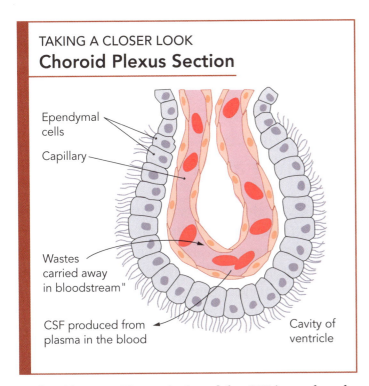

**TAKING A CLOSER LOOK**
**Choroid Plexus Section**

- Ependymal cells
- Capillary
- Wastes carried away in bloodstream
- CSF produced from plasma in the blood
- Cavity of ventricle

rachnoid space. The majority of the CSF is produced by the frond-like choroid plexus found in each ventricle. In these choroid plexuses, ependymal cells (remember they are a type of neuroglia cell) filter the plasma from the blood to produce the CSF.

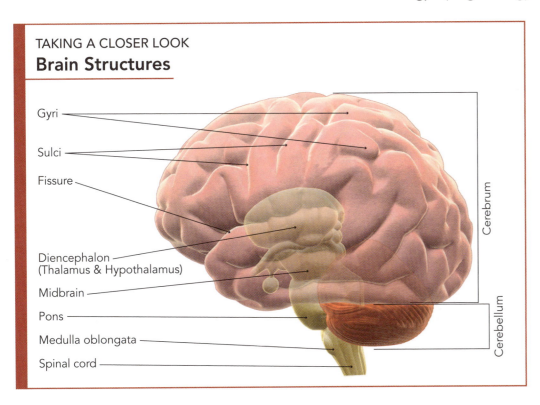

**TAKING A CLOSER LOOK**
**Brain Structures**

- Gyri
- Sulci
- Fissure
- Diencephalon (Thalamus & Hypothalamus)
- Midbrain
- Pons
- Medulla oblongata
- Spinal cord
- Cerebrum
- Cerebellum

sulci (singular is sulcus). In addition, there are a few deeper folds called fissures. The fissures separate major portions of the cerebrum.

If you examine the illustration provided, you can see that there are two halves to the cerebrum. These are called the cerebral hemispheres. These two hemispheres are separated by a large fissure known as the longitudinal fissure.

Several smaller fissures divide each hemisphere into smaller portions, called lobes. In the front of each hemisphere is the frontal lobe. Moving posteriorly (toward the back), you find the parietal lobe. The frontal and parietal lobes are separated by the central sulcus. Then, continuing posteriorly, separated from the parietal lobe by the parieto-occipital sulcus, is the occipital lobe. Below the parietal lobe is the temporal lobe. This lobe is bordered by the lateral sulcus. Make sure you can locate each of these lobes and the large sulci and fissures that demarcate them in the illustrations.

## Cerebrum — Gross Anatomy — the White and the Gray

Let's now examine a cross section of the cerebrum. The illustration below is a "frontal section," which is a cross section made by separating the front and back parts of a structure. Notice the cerebrum's gray outer layer. Have you ever heard of "gray matter"? Gray matter is made up of the cell bodies of neurons and neuroglia. And this gray matter is the cerebral cortex.

Deeper inside notice the cerebral white matter. White matter is made up of both myelinated and nonmyelinated axons.

We can get a glimpse deeper into the brain from a "sagittal section." The sagittal section on the next page is a cross-sectional view we get when we separate the right and left parts of a structure. The most important structure to note here is called the corpus

### TAKING A CLOSER LOOK
### Lobes of the Brain

- Frontal lobe
- Parietal lobe
- Occipital lobe
- Temporal lobe

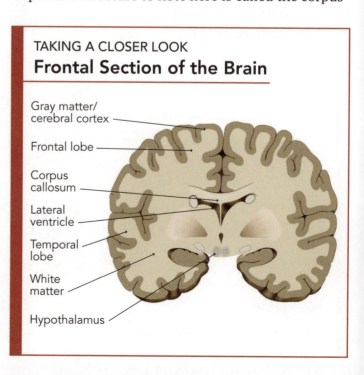

### TAKING A CLOSER LOOK
### Frontal Section of the Brain

- Gray matter/cerebral cortex
- Frontal lobe
- Corpus callosum
- Lateral ventricle
- Temporal lobe
- White matter
- Hypothalamus

callosum. The corpus callosum is a large band of white matter that connects the two cerebral hemispheres. Thanks to the axons that cross from one side of the cerebrum to the other through the corpus callosum, the two cerebral hemispheres communicate and work together.

That's enough for now about the gross anatomy of the cerebrum. We will mention a few other very important regions later, as we learn just what the cerebrum does.

## Cerebrum — Motor Functions

The cerebrum controls so many things in the body it is difficult to know where to begin. It controls voluntary movement, the movement you can consciously control. The cerebrum enables you to perceive the world by making you aware of the sensations you receive every second. The centers for speech and language are in the cerebrum. Signal processing in the cerebrum allows us to learn and understand. We can remember the things we learn and experience because the cerebrum helps store memories. The list could go on, but you get the idea.

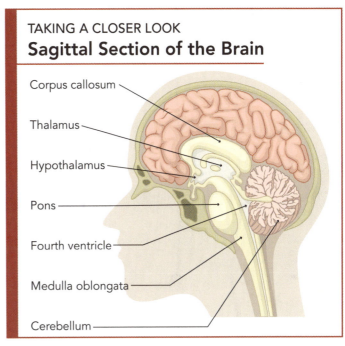

TAKING A CLOSER LOOK
### Sagittal Section of the Brain

The cerebrum consists of the two cerebral hemispheres. Each hemisphere is equally important, but they do not necessarily have all the same jobs. Each of the cerebral hemispheres controls one half of the body. Interestingly enough, they control the opposite side of the body. That is, each cerebral hemisphere controls the motor functions on the opposite side of the body. Each cerebral hemisphere also receives

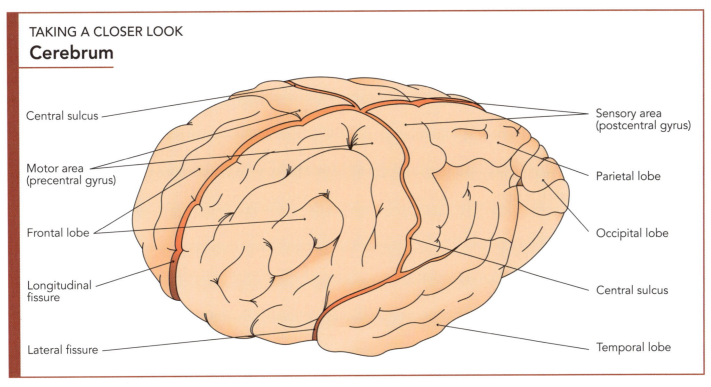

TAKING A CLOSER LOOK
### Cerebrum

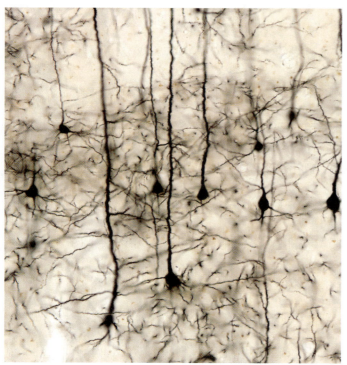
Pyramidal neurons of the cerebral cortex.

the sensory input from the opposite side of the body. Thus the right hemisphere controls and monitors the left side of the body, and the left hemisphere controls and monitors the right side of the body!

There are, however, some functions of the brain that are lateralized. That is, the responsibility for certain functions rests with one hemisphere or the other. We will see examples of these as we go on.

Finally, and perhaps most importantly, you need to understand that nothing that happens in the cerebrum, happens in isolation. The brain's neural pathways are very, very highly integrated—interconnected and coordinated. The signals into and out of the brain are constantly enhanced, inhibited, and modulated (changed) by input from all across the cerebrum. Thus, even while we describe the brain's structures and functions as simply as possible, we understand that most things about the brain are really more complicated than they appear on the surface.

Motor function is controlled by a region of the cerebrum called the primary motor cortex. The primary motor cortex is located on the precentral gyrus. This is the ridge just in front of the central sulcus. Pre- means "in front of," and gyrus means "ridge," so the name precentral gyrus makes sense: "the ridge in front of the central sulcus."

The primary motor cortex is made of a special type of neuron called pyramidal cells (or pyramidal neurons). Pyramidal neurons have very long axons that extend all the way into the spinal cord. Bundles of these axons are called pyramidal tracts. You probably recall that in the PNS bundles of axons were called nerves. Well we aren't in the peripheral nervous system now. In the central nervous system, bundles of axons are called tracts. The pyramidal tracts carry the message "move!" from the primary motor cortex to the spinal cord. (Watch for the pyramidal tracts to bulge into view again when we discuss the brainstem.) From the spinal cord, the "move" message is sent to the designated muscles to get the job done.

So how does the brain know to move my leg when I walk? How does it know how to send messages to the correct fingers when I type?

As it happens the primary motor cortex has specific areas that correspond to specific parts of the body. Over many years these areas have been mapped and studied by neuroscientists. You can see the various parts of the body mapped out in the illustration here. This type of visual representation of the motor regions is called a motor homunculus (meaning "little man"). The motor cortex maps in a similar way in each hemisphere. (But remember, left brain controls right body, and right brain controls left body. Right? No, left...seriously?)

Examining the homunculus, you see that it pictures the body upside down. The feet and legs are at the top, and the face is at the bottom. This doesn't mean the brain considers the feet more important than the face. Far from it. As you look more closely, you will find that some body regions take up more of the motor area than others. In fact, the hand and face take up the majority of the motor cortex. Completely logical, right? The hands require very precise motor control, as do the muscles of the face and jaw. Writing, typing, playing an instrument, smiling, talking, chewing...all require lots of muscles acting in just the right way. We aren't surprised then that God designed the brain with a lot of territory devoted to muscles in the hands and face.

But, as you might have guessed, muscular control is more complex than the homunculus illustration suggests. This map of the "little man" gives you a general idea of where the major control areas are, but voluntary muscle movements are fine-tuned by a vast number of neural inputs from areas throughout the cerebral cortex.

When studying the control of motor function, there is another very important region to examine. This is the area just anterior to (in front of) the primary motor cortex. Remember pre- means "in front of." Therefore this important part of cerebral cortex—"in front of the motor cortex"—is called the premotor cortex. Here all the planning happens before any movement begins! You see, something like reaching for a glass or picking up a pencil is not as easy as it might seem. The coordination of many muscles is necessary to perform even relatively simple tasks. The preparation for these movements occurs in the premotor cortex, even though you don't often realize you are making plans for how to move. Which muscles must move, and in what order they move are all laid out in the premotor cortex. The premotor

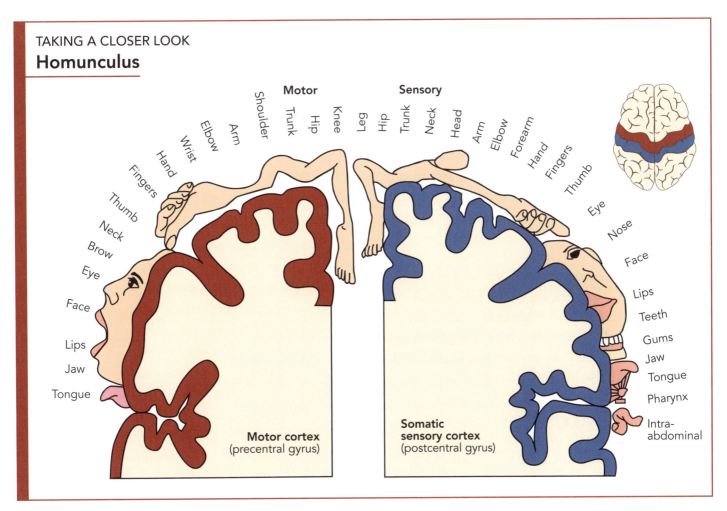

TAKING A CLOSER LOOK
## Homunculus

cortex then sends appropriate signals to the primary motor cortex to get voluntary muscle movement (or movements) underway.

In addition to the areas we have covered thus far, there is one more very interesting region to examine. It is known as Broca's area. In most people, Broca's area is located in the left cerebral hemisphere, though in some folks Broca's area is on the right. Broca's area is thus lateralized. It is found in only one of the hemispheres.

Broca's area controls muscles involved with speech. Muscles in the mouth and larynx are coordinated by Broca's area, enabling us to speak. Neurons here interact with both the premotor cortex and the primary motor cortex to manage this complex process of communicating using spoken language.

## The Cerebrum — Sensory Functions

The cerebrum is also involved with processing sensory information from throughout the body. Sensory information is processed in the primary somatosensory cortex. This enormous name makes sense, if we break it down. Primary means is the area first in importance for this function. Somato- means "body." And sensory, well that means having to do with input to the brain, such as information from your sensory organs.

The primary somatosensory cortex is located in the parietal lobe of the cerebrum on the postcentral gyrus. Does that name tell you where it is located? If you said, "behind the central gyrus," (because post- means "behind"), then you are catching on to the pattern behind these big names.

The cerebrum's primary somatosensory cortex receives sensory inputs from all over the body. Touching a piano key, sitting in a chair, drinking warm cocoa, stubbing your toe...all these actions

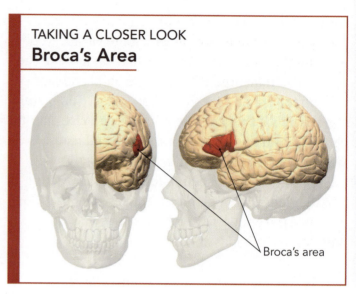

**TAKING A CLOSER LOOK**
**Broca's Area**

Broca's area

produce sensory inputs that are ultimately processed in the primary somatosensory cortex.

AND...just as the primary motor cortex has specific areas that send signals out to specific areas of the body, the primary somatosensory cortex has specific areas that receive input from specific areas of the body. Just as with the primary motor cortex, neuroscientists have mapped the primary somatosensory cortex. This mapping can be represented by a somatosensory homunculus, another "little man" map. Once again, the senses on the hands and lips map onto a greater territory than touch sensations from other parts of the body. The primary somatosensory cortex maps in a similar way in each hemisphere.

## Expressive Aphasia

As we have seen, Broca's area controls the muscle activity in involved with speech. When people suffer damage to Broca's area, perhaps after a stroke, they often lose the ability to speak. In this very unfortunate situation, people can comprehend and process everything that is said to them. And they know what they want to say, but they are not able to make the appropriate muscle movements to form words. This inability to speak, to "express themselves," is called expressive aphasia.

Remember that motor functions start by being planned in a special area (the premotor cortex) before instructions to move something are issued by the primary motor cortex. We see the same pattern on the sensory side of things.

Just behind the primary somatosensory cortex is the somatosensory association cortex. This area is a processing area for sensory inputs. Here many sensory inputs are received and integrated. This signal processing allows the cerebrum to take many individual inputs and understand their greater meaning. Because the somatosensory association cortex combines, evaluates, and interprets a variety of signals and signal types, our brain can, for instance, distinguish whether a round object in your hand is a marble or a grape, even without looking. Sensory integration is incredibly complex, but we need to praise God that our brains have this ability. We could not make sense of our world without it.

## Cerebrum — Association Areas

Other important areas of the cerebrum receive and process information from many sources. These are known as association areas. There are many of these, but here we only consider a few:

The frontal association area is located in front part of the frontal lobe. This extraordinarily complex area controls many of our intellectual functions. It is involved with learning, reasoning, planning, and abstract reasoning (you know, important questions like, "Why did the chicken cross the road?"). Many aspects of our personality are determined by the frontal association area.

The visual association area is in the occipital lobe. This area processes visual information to allow us to understand what we are looking at. Are we staring at a bee or a butterfly? When you are looking for a snack, be thankful that your visual association area helps you distinguish between an apple and an onion.

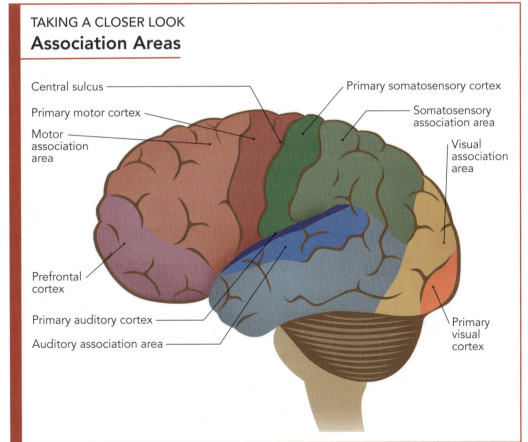

TAKING A CLOSER LOOK
**Association Areas**

Wernicke's area is found in the temporal and parietal lobes of the left cerebral hemisphere. It is in Wernicke's area that the language we hear is processed, allowing us to understand speech. Wernicke's area, like Broca's area, is a lateralized part of the brain associated with language.

The auditory association area is in the temporal lobe. This area helps us distinguish between types of sounds. Are we hearing

# Right-Brained? Left-Brained? Scatter-Brained?

In recent years much has been written about being "right-brained" or "left-brained." This idea is the result of many popular misconceptions about how the brain works.

It has been said that the right hemisphere of the brain is devoted more to emotions, music, pattern recognition, and artistic endeavors. At the same time, it said that the left hemisphere is the seat of logic, reasoning, and planning. To a certain degree, these things are true, as certain functions in the brain are lateralized.

This discovery has caused people to claim that artistic and thoughtful people are "right-brained." At the same time, people who are analytical and logical are called "left-brained." In the final analysis, this is a gross oversimplification. No one is truly that "dominant" on either side of the brain. In fact, we all do much better when we use "all" of our brain at the same time (Some people don't. . . . We call them "scatter-brained.")

To further dispel this myth, let me use myself as an example.

Over the years, I have felt myself to be a reasonable (though not gifted by any means) guitar player. Thus, I am a musician. Also, my primary hobby for over 30 years has been photography. Therefore I am an artist. That would mean I am "right-brained," right?

On the other hand, for over 20 years, I worked in an intensive care unit taking care of very sick people. In that setting logic, reason, and critical thinking are first and foremost. Since I took good care of my patients, I must be "left-brained," right?

Well both can't be true. Apparently this left-and right-brained stuff isn't so straightforward as popular wisdom would suggest.

Yeah, I know what you are thinking. Scatter-brained, right?

music or laughter? Are we hearing thunder or a firecracker? Like other association areas, this one helps us interpret and understand our world.

## Which Is the Important Side?

So which side of the brain is the most important? That's easy. The answer is BOTH!

There really is no dominant side of the brain. One cerebral hemisphere is not more important than the other. Remember that the left cerebral hemisphere gets sensory signals from the right side of the body. It also controls the voluntary motor function of the right side of the body. The right cerebral hemisphere gets sensory input from the left side of the body and controls voluntary muscle movement on the left side. So asking which side if the brain is more important is like asking which side of the body is more important.

It is true that certain functions are performed by either one hemisphere or the other, Broca's area, for example. However, just because some specific things are lateralized in the brain does not mean that side of the brain is more important. The white matter bands in the corpus callosum connect the right and left hemispheres to allow communication between the hemispheres. This communication is very, very important. With essentially every process, the two hemispheres are in constant communication, assisting and assessing each other.

Speaking and understanding language are lateralized functions of the brain. And the precise hand control seen in a person's dominant hand is achieved, you remember, from control from the opposite side of the brain. Does this mean that one side of the brain is more important for those uniquely human characteristics and abilities? Let's explore that issue.

In the vast majority of people, the left hemisphere is predominant when it comes to speech and language processing. Speaking of course requires precise control of all the muscles used to produce sound. Since dominant handedness also indicates superior motor control of one hand, some claim there is a relationship between language and handedness. They claim that being right handed indicates a dominant left hemisphere. If that were always true, then logically in a left-handed person, the right hemisphere would be dominant. In fact, your author, who is left handed, often makes the statement that he is "left handed, but in my right mind." (Everyone agrees but my wife, sad to say.)

In the final analysis, it is just is not that simple. One study has shown that 95 percent of right-handed people, whose dominant hand is controlled by their left hemisphere, have language dominance in the left hemisphere. The same study indicates that only 18 percent of left-handed people, whose dominant hand is controlled by their right hemisphere, have language dominance in the right hemisphere. So handedness, as with everything else about the brain's function, does not lend itself to simple explanations. And claims that one side of the brain is more important than the other collapse when the facts are examined.

## Diencephalon

The diencephalon is the portion of the brain between the cerebrum and the brainstem. It surrounds one of the fluid filled chambers we mentioned earlier, the third ventricle. The two main parts of the diencephalon are the thalamus and the hypothalamus.

The thalamus makes up 80 percent of the diencephalon. The thalamus consists of two egg-shaped masses (called thalamic nuclei) connected by stalk. The thalamus plays several major roles in the brain. First of all, the thalamus relays sensory input from the spinal cord to the primary somatosensory cortex. Next, it facilitates the transmission of signals from the cerebellum to the motor cortex. Further, it plays a role in regulating consciousness. It many ways it acts as a gatekeeper for the sensory and motor cortices.

The hypothalamus is located below the thalamus. The hypothalamus controls many body functions that seem fairly automatic to us. It is one of the main regulators of homeostasis. If you remember, homeostasis is the body's tendency to maintain internal balance. Diverse mechanisms work together to regulate conditions that must remain fairly steady in a healthy living body. These include temperature, blood pressure, balanced levels of many hormones, and the concentrations of various substances in the blood.

## TAKING A CLOSER LOOK
## The Diencephalon and Brain Stem

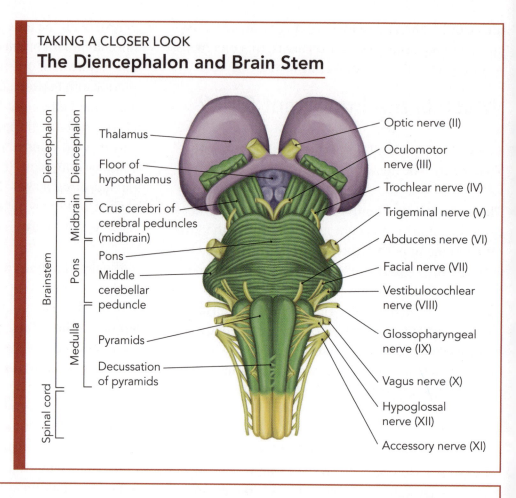

## TAKING A CLOSER LOOK
## Hypothalamus and Its Function

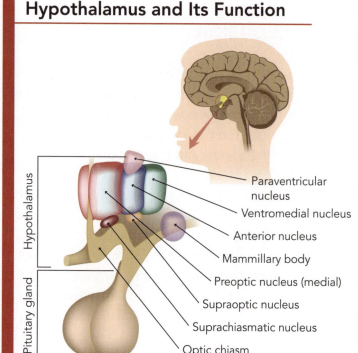

| HYPOTHALAMIC NUCLEI | FUNCTION |
|---|---|
| Paraventricular nucleus | Thyrotropin-releasing hormone release, corticotropin-releasing hormone release, oxytocin release, somatostatin release |
| Suprachiasmatic nucleus | Circadian rhythms |
| Ventromedial nucleus | Neuroendocrine control |
| Anterior hypothalamic nucleus | Thermoregulation, sweating |
| Mammillary body | Memory |
| Supraoptic nucleus | Vasopressin release, oxytocin release |
| Posterior nucleus | Increase blood pressure, pupillary dilation, shivering |
| Tuberomammillary nucleus | Attention, wakefulness, memory, sleep |
| Arcuate nucleus | Growth hormone-releasing hormone (GHRH), feeding |

# THE CENTRAL NERVOUS SYSTEM

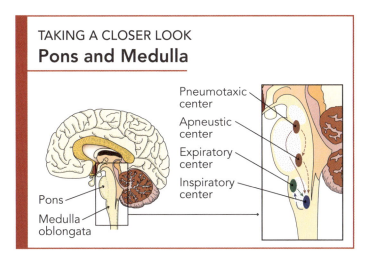

### TAKING A CLOSER LOOK
#### Pons and Medulla

### TAKING A CLOSER LOOK
#### Decussation

The hypothalamus receives sensory input from many places throughout the body. It regulates your body's temperature. It tells you when you are thirsty. It helps sense when you are hungry and when you are full. Your attentiveness is affected by the hypothalamus, as are your sleep cycles. Heart rate, blood pressure, and even the movement of food through your digestive tract are subject to control by the hypothalamus.

## Brain Stem

Proceeding downward from the diencephalon, we come to the brain stem. The brain stem consists of three parts: the midbrain, the pons, and the medulla oblongata. The brain stem provides a path for fibers extending into the spinal cord. Also, many vital body functions are controlled or regulated in the brain stem.

The midbrain is located between the diencephalon and the pons. It is roughly one inch long. It consists of two bulges in the front called cerebral peduncles. The peduncles are bundles of axons. In the rear are four nuclei known as colliculi. (In the CNS, a collection of neuron cell bodies is called a nucleus; the plural is nuclei) The colliculi are involved in both hearing and vision.

The pons (meaning "bridge") is positioned between the midbrain and the medulla oblongata. It name—bridge—reveals its function. The pons serves as a bridge linking various parts of the brain together. Through the pons run fibers from the spinal cord up into the brain and also from the cerebellum to the motor cortex. In addition there are areas in the pons that assist in the control of breathing and balance.

The medulla oblongata is below the pons, where it connects the brain to the spinal cord. Through the medulla ascend all the sensory tracts going to the brain. These are called ascending tracts. (Remember, bundles of axons are not called nerves in the CNS;

they are called tracts.) Also descending through the medulla are all the motor tracts going down into the spinal cord. These are called descending tracts. The descending tracts form bulges on the anterior part of the medulla. Remember the pyramidal cells in the primary motor cortex? These bulges in the medulla contain the pyramidal tracts, the axons from those pyramidal cells. The bulges themselves are called pyramids.

Remember that each half of the cerebrum controls the muscles on the opposite side of the body. Therefore, just before reaching the spinal cord, most fibers in the pyramidal tracts decussate (cross over) to the opposite side. After this they descend into the spinal cord. This is how the right side of the brain controls the left side of the body, and the left side of the brain controls the right side of the body. It is because of the crossing over that occurs in the medulla.

## Headache

Headache is one of the most common complaints in modern society. Some studies have estimated that nearly 50 percent of adults have at least one headache a year. There are many different types of headaches, but here we will concern ourselves with just two: tension headaches and migraine headaches.

Tension headaches are the most common type of headaches. The pain of a tension headache can be mild to very severe. The discomfort can involve the sides and back of the head, the neck, and eyes. Very often the pain occurs on both sides of the head at the same time.

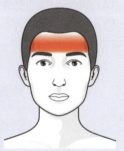
Tension

The most commonly cited cause of tension headaches is stress. Lack of sleep, eyestrain, and hunger can also play a role. Muscle tension in the head or in the neck often contribute.

Treatment of tension headaches is usually straightforward. In most cases a mild analgesic (pain medication) is all that is required. In the case of patients with more frequent or more severe tension headaches, a combination of medications may be required.

Migraine headaches are generally (but not always) more severe than tension headaches. Migraine pain usually affects one side of the head. It is often described as a throbbing, pounding pain. Migraines are frequently associated with nausea, vomiting, and extreme sensitivity to light and sound. It is not uncommon for someone with a migraine to retreat to a dark, quiet room to try and ease the pain. Some migraines are preceded by a sensation of seeing flickering lights that partly obscure normal vision. These are called scintillating scotoma.

The cause of migraines in unknown, although there are several schools of thought. Some researchers feel that migraines are linked to dilation of blood vessels in the brain. Others feel that dilation of arteries outside the cranium are the root cause. Some scientists feel that over-excitable neurons in the brain play a role in migraine headaches. Most likely all of these play a role. At present research is ongoing.

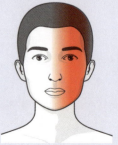
Migraine

Treatment of migraines is more complex than treatment of tension headaches. Milder analgesics are not as effective with migraines but are still the first line of treatment. Due to the complex nature of migraines, very often a combination of medications is used.

Also, the medulla contains several nuclei (collections of neuron cell bodies). These nuclei help control and regulate many vital body functions. The vasomotor center changes the diameter of blood vessels to adjust blood pressure. The cardiac center regulates the heart rate. Other nuclei in the medulla regulate such things as hiccupping, vomiting, and swallowing.

Before ending our review of the brain stem, there is one other thing I should alert you to—the reticular formation. The word reticular means "net." The reticular formation is a net-like collection of interconnected nuclei that runs through the midbrain, pons, and medulla. A part of the reticular formation is made of sensory axons that run up to the cerebral cortex. This part of the reticular formation is called the reticular activating system (RAS). The RAS stimulates the cerebral cortex during the transition from sleep to wakefulness, helping you to become alert.

## Cerebellum

The final major portion of the brain is the cerebellum. The cerebellum is located posterior to (behind) the medulla. The cerebellum and brainstem are connected by three bundles of white matter called peduncles.

The cerebellum itself consists of a central area called the vermis. Lateral to the vermis are two cerebellar hemispheres. Each hemisphere is composed of three lobes: the anterior lobe, the inferior lobe, and the flocculonodular lobe.

The cerebellum is responsible for making you aware of your position. That is, where the body is, how it is positioned, how fast it is moving. Position sense is called proprioception. The cerebellum receives sensory input from muscles and tendons throughout the body, and processes all these inputs so you will know where all your body parts are located. The cerebellum also helps maintain muscle tone.

Furthermore, the cerebellum helps us maintain our balance. Just think about all the muscle activity that is involved in getting up from a chair. This seemingly simple action takes the coordinated effort of dozens of muscles exerting just the proper amount of force at just the right time. Thank your cerebellum the next time you stand up. Or walk outside on a really windy day. Get the idea?

## Blood Supply to the Brain

The brain is the master control center of the body. As such it requires constant high levels of oxygen and nutrients to function correctly and efficiently. Even though the brain makes up only about 2 percent of the body's weight, it requires about 20 percent of the body's oxygen and glucose. Since the brain's metabolic needs are so high, it is not surprising that an interruption of its oxygen supply can have very serious consequences. A loss of oxygen for as little as four minutes can result in permanent brain damage. Let's see how the blood flow to the brain works.

The blood supply to the brain primarily comes from the internal carotid arteries. If you press on your

### TAKING A CLOSER LOOK
**Cerebellum**

- Anterior lobe
- White matter
- Gray matter
- Pons
- Medulla oblongata
- Central canal of spinal cord
- Posterior lobe

neck gently on either side of your Adam's apple, the pulse you feel is from your internal carotid artery. (Do not check the pulse on both sides at once. Putting pressure on both arteries at once may cause you to lose consciousness.)

After entering the cranium, the right and left internal carotid arteries each branch to form the anterior cerebral arteries and the middle cerebral arteries. As you might guess, the anterior cerebral arteries supply the front part of the brain, and the middle cerebral arteries supply the middle parts of the brain. To help ensure the brain's uninterrupted blood supply, the right and left anterior cerebral arteries are connected by the anterior communicating artery.

Near the base of the neck arteries branches from both the right and left subclavian arteries. These branches are the vertebral arteries. They ultimately enter the posterior portion of the cranium. After entering the cranium, the two vertebral arteries join to form the basilar artery. After proceeding under the base of the brain, the basilar artery divides into the two posterior cerebral arteries. It will be no surprise that the

### TAKING A CLOSER LOOK
### Blood Supply to the Brain

- Right anterior cerebral artery
- Anterior communicating artery
- Right internal carotid artery
- Right posterior communicating artery
- Right posterior cerebral artery
- Left middle cerebral artery
- Basilar artery
- Left vertebral artery

## Subclavian Steal Syndrome

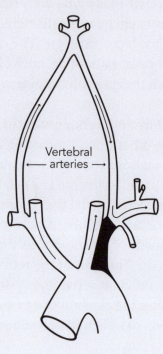

Recall that the subclavian arteries, which are located under the collarbones, supply blood to the arms. You've just learned that the subclavian arteries also supply blood to the brain via the vertebral arteries that branch from them. And since those vertebral arteries merge to form the basilar artery, they are connected to the circle of Willis, giving the vertebral arteries—and hence the subclavian arteries from which they branch—a connection to the brain's blood supply.

Suppose a subclavian artery develops a problem close to its point of origin, becoming so narrow that it cannot keep the arm well supplied with blood. Some of the blood flowing though the circle of Willis can be re-directed away from the brain and flow back down the vertebral artery to the place where the subclavian artery is still open above the blockage. This blood, having bypassed the blockage, helps keep the arm supplied with oxygenated blood.

This condition is called subclavian steal syndrome because blood is "stolen" from the circle of Willis by this subclavian connection. Fortunately, the blood flow to the brain is usually so good that most people with subclavian steal syndrome have no symptoms, though some people experience fainting.

posterior cerebral arteries supply the posterior parts of the brain. Again, to ensure an uninterrupted blood supply to the brain, each posterior cerebral artery is connected, via a posterior communicating artery, to one of the middle cerebral arteries.

Tucked underneath the brain is an amazing structure called the circle of Willis. It is named after Thomas Willis, the 17th-century physician who discovered it. The circle of Willis is made up of the internal carotid arteries, the anterior cerebral arteries and the anterior communicating artery that connects them, the basilar artery, and its branches, the posterior cerebral arteries, as well as the posterior communicating arteries that connect them to the middle cerebral arteries. This incredible structure provides much protection for the brain. If a vertebral or carotid artery becomes blocked, circulation to the brain can be maintained (to a certain degree, at least), by blood

## Stroke

A stroke occurs when cells in the brain are killed by loss of blood flow. Strokes can result in permanent neurologic damage or even death.

The most common type of stroke is called an ischemic stroke. In this case, blood flow to the brain is interrupted by a blockage in one or more arteries supplying the brain. With no blood flow past the blockage, the nervous tissue supplied by that particular artery is damaged due to lack of oxygen and nutrients. If only a small amount of circulation is interrupted, the resulting damage may be small. On the other hand, if a blockage interrupts a larger circulatory pathway, the damage can be extensive. Blockages can result from the buildup of atherosclerotic plaques in an artery (imagine a clog slowly building up and eventually clogging the kitchen sink) or a blood clot that blocks blood flow.

Another type of stroke is called a hemorrhagic stroke. In this case there is bleeding directly into the space around the brain. Sometimes the bleeding is the result of the rupture of an aneurysm. An aneurysm is a dilated area on the wall of an artery, an area of weakness resembling a thin, weak section of a balloon. At other times an artery in the brain simply ruptures.

Both types of stroke are very dangerous.

The symptoms of a stroke include sudden loss of function of a part of the body (face, hand, arm, leg, etc.), sudden inability to speak, dizziness, loss of sensation, change in sense of smell or taste, or a sudden change in vision. These are only a few potential symptoms. At times the diagnosis of a stroke can be quite challenging. However, the sooner a stroke victim gets medical attention, the better. Time is of the essence with a stroke.

Rick factors for stroke include high blood pressure, smoking (NEVER, ever, ever, EVER start smoking!!), high cholesterol, diabetes, lack of exercise, and poor nutrition.

flow coming from the other arteries making up the circle of Willis.

This arrangement found in the circle of Willis—somewhat circular connections between arteries from both sides—is called collateral circulation. Collateral circulation helps ensure a good supply blood to this important organ.

## Blood Brain Barrier

We noted earlier that the brain requires a high percentage of the body's oxygen and glucose to function properly. The brain is also quite sensitive to many substances. To prevent potentially harmful things from coming into contact with brain tissue, there is a blood brain barrier.

Recall that an astrocyte is a type of neuroglia cell. The processes of astrocytes surround the capillaries in the CNS. In doing so, the astrocytes stimulate the endothelial cell lining of the capillary to form tight junctions. Most capillaries in the body have fairly permeable linings that permit many substances to exit the bloodstream and enter the surrounding tissues. Not so in the brain! These tight junctions make the capillaries in the CNS much less permeable

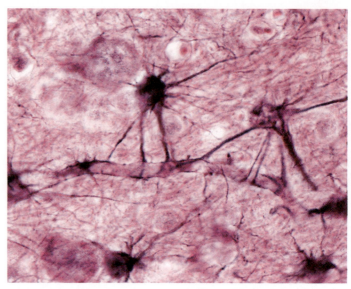

*Astrocyte processes in contact with the walls of capillaries.*

than capillaries found elsewhere in the body. This loss of permeability prevents many substances from coming into contact with tissues in the CNS.

From what sort of substances does the brain need to be protected? Various metabolic waste products, proteins, some toxins like the sort that causes botulism food poisoning, and many types of drugs are stopped by this altered capillary membrane, thus protecting the CNS from their effects. However, oxygen, glucose, and carbon dioxide can easily pass

### TAKING A CLOSER LOOK
### Electroencephalogram (EEG)

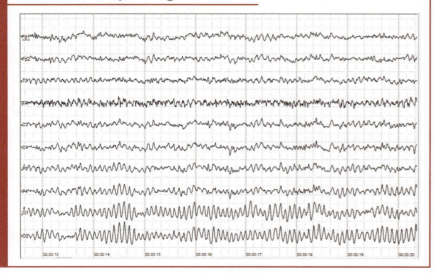

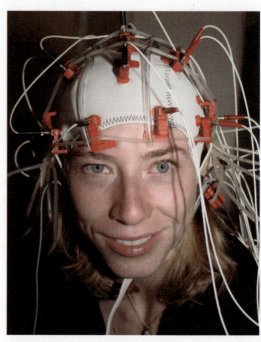

across the blood brain barrier. Fortunately, anesthetics can also pass across the blood brain barrier, making it possible to put people to sleep during surgery (which is fortunate if you ever need your appendix or your wisdom teeth removed).

# Brain Waves

The electrical activity that makes your heart beat can be measured on the body's surface with an electrocardiogram (ECG). Similarly, when the brain works, its electrical activity can be measured by means of an electroencephalogram (or EEG, for short). The electrical activity being measured by an EEG is not from action potentials racing down the length of axons. An EEG records the synaptic activity of neurons close to the surface of the brain, those located in the cerebral cortex. The various patterns of electrical activity seen are called brain waves.

There are four basic brain wave patterns, and they can be distinguished by their frequency—how fast they oscillate. Frequency is measured by the number of peaks per second in a waveform. 1 Hz (1 Hertz) means one peak each second. You can see easily distinguish the higher and lower frequencies by examine the waveforms in the illustration.

From lower to highest frequency, the basic brain wave types are:

Delta waves — These are usually 1–4 Hz. This pattern is the lowest frequency and occurs in deep sleep. Delta waves in an awake adult usually indicate some type of brain damage.

Theta waves — These are usually 5–8 Hz. In adults, theta waves occur during meditation or drowsiness.

Alpha waves — These are usually 8–13Hz. Alpha waves occur during wakeful but relaxed times.

Beta waves — These are usually 15–40 Hz. Beta

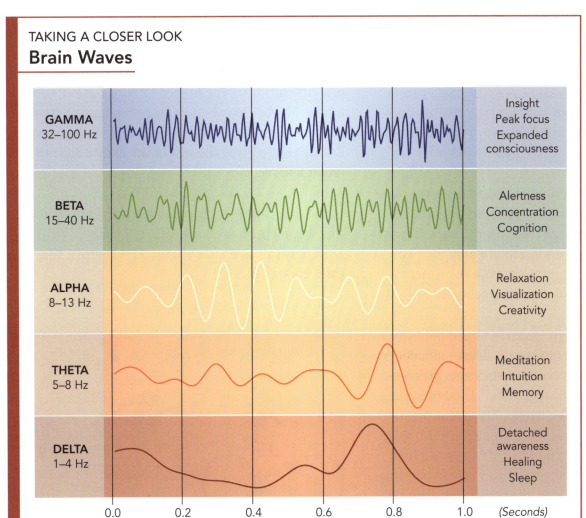

TAKING A CLOSER LOOK
**Brain Waves**

waves occur when the mind is active, like when you are concentrating or trying to communicate.

Gamma waves — These are 32–100Hz. This wave type is the least understood of the brain waves. It may be related to consciousness and awareness.

There are, of course, brain wave patterns that are grossly abnormal. The brain wave patterns seen during a seizure would be an example. EEG studies are of primary importance when treating and caring for patients with seizure disorders.

Because the brain is active round the clock even during deep sleep, some degree of brain activity is always detectable by EEG. In the tragic circumstance where there is no detectable brain activity on the EEG, the patient is said to be "brain dead."

## Sleep

Everybody sleeps. While it is true that some people sleep better than others, the vast majority of people get at least a few hours of sleep every day. If sleep is something everybody does, exactly what is it?

Sleep is a state in which an individual achieves a degree of unconsciousness from which he or she can be aroused. Although there are several different stages of sleep, here we will explore only the two major types. These are non-rapid eye movement sleep (NREM) and rapid eye movement sleep (REM).

A person passes through stages of non-rapid eye movement sleep as he or she falls deeper and deeper into sleep. The brain wave frequency gets progressively slower with each stage. After 60-90 minutes, the pattern on the EEG changes. This is the beginning of a period of rapid eye movement (REM) sleep. The brain wave pattern during REM sleep is a high frequency pattern, reflecting an increase in neuronal activity. During REM sleep, the body is very relaxed but the eyes move rapidly underneath the eyelids, which of course is how rapid eye movement sleep got its name.

Periods of REM sleep last anywhere from 10 to 60 minutes and recur about every 90 minutes. In the average adult, REM sleep takes up about 25 percent of the total time asleep. Most dreaming occurs during REM sleep. People who do not get enough REM sleep tend to be moody and irritable.

So why do we sleep? What is its ultimate purpose? When it comes right down to it, nobody really knows!

There are many theories about why we sleep. Perhaps sleep is a time that the brain can "take stock of itself." This suggests that the brain needs this time to process all the information from the day that needs to be tucked away and stored or discarded. Others feel that sleep is perhaps a time in which neurons in the brain reset and prepare for the next wake period. Perhaps they do this by removing residual waste products left from the higher activity state of being awake. Maybe sleep helps conserve energy, for we are certainly less active while asleep that while awake. Perhaps sleep gives the rest of the body a time to repair and recover.

In the final analysis, there are a lot of theories about why we sleep, but nobody can say for sure. When I get 8 hours of sleep each night, I feel better. That's good enough for me!

# Learning and Memory

Life sure would be boring if we did not have the ability to learn new things. No one would survive very long if the information in our brain the day we were born was all the information we would ever have. We could not even know how to care for ourselves.

Just consider all the new things you have learned over the past 5 years. Perhaps you have begun to play an instrument or ride a bike. Have you started learning a new language? Or perhaps you have developed an interest in computer programming? You see, learning is not only necessary, it can be fun too!

To learn, we must be exposed to new information. Sometimes this happens because we are being taught something new, like all this incredible stuff you are learning from the *Wonders of the Human Body*! Other times you can learn just by being exposed to situations or circumstances often enough that you eventually learn how something is done. Ride with your dad to the grocery store enough times, and you eventually learn where the store is and how to get there. Check into a cabin with your family on vacation, and you will not only figure out where the kitchen is, but you will soon easily remember where you found it. If your room is upstairs, you very quickly learn that it is easier to go down the stairs than it is to go up the stairs. You do not need to have someone teach you these things. You figure them out by experience.

A vital part of learning is memory. After all, anything you learn is useless to you if you cannot recall it at the proper time. (If you recall your last history test, you will get the idea...or maybe it was all the stuff you could not recall that was the problem.)

The problem is that we do not remember everything we see or experience. Most often we must be exposed to information multiple times before we learn it. Fortunately, God designed our brains to be able to accommodate to new information. When presented with new information, our brains can actually change. Many scientists feel that our ability to remember is the result of synaptic plasticity (now there's your phrase for the day!). This is the ability of synapses to change their strength. This change then results in the encoding of memories.

## Amnesia

Amnesia is a loss of memory. Amnesia can result from trauma or severe illness. Certain drugs are also known to cause episodes of amnesia.

There are two primary types of amnesia. The first is called retrograde amnesia. It this case, the person is unable to remember things that occurred before a given event. For instance, after a blow to the head, a severe accident or illness, or a major surgery, a patient may lose to ability to recall some memories from before the event.

The other type of amnesia is anterograde amnesia. Here the person is unable to make new memories after a particular event. In cases like this, it is not unusual for a person to be unable to remember things that happened only minutes before.

# Consciousness and the Mind

There are lots of people who believe that the human body is nothing more than a cosmic accident. They would have everyone accept that millions of years ago all the matter in the world simply appeared... out of nothing...from nowhere. Then this matter that appeared from nowhere began to interact with other matter that also appeared from nowhere. As a result of matter interacting with matter, life sprang into being...from lifeless matter.

Then over millions of years the first simple life forms became more and more complex until humans were produced. This process is called evolution. And it is a cosmic fairy tale. You see, if evolution is the correct explanation for our existence, then the human body is nothing more than a chemical accident. Just the result of atoms bumping together over those millions of years.

So...how then do evolutionists explain things like consciousness and the human mind? In their view these things must be the result of chemicals interacting with other chemicals. They really have no other explanation. But those people need to understand something: *Professing to be wise, they became fools* (Romans 1:22).

### TAKING A CLOSER LOOK
### Spinal Nerves

- Cervical Nerves
- Thoracic Nerves
- Lumbar Nerves
- Sacral / Coccygeal Nerves

- Base of skull
- Cervical enlargement
- Lumbar enlargement
- Conus medullaris
- Cauda equina

C1, C2, C3, C4, C5, C6, C7, C8
T1–T12
L1–L5
S1–S5
Coccygeal nerve

### SO SIMPLE YET SO COMPLEX — Designed by the Master

### Body and Soul

The human body was designed by the Creator God, and it is more than just a collection of chemicals. Certainly we have a physical body, and it is truly amazing, as you are learning. However, there is much more to being human than that. We not only have a physical body. We also have a spirit and a soul. Humans were created as rational, thinking beings. We are self-aware, thinking creatures. We are able to think abstract thoughts and use language to express them and to communicate them to others.

We have a vast ability to interpret, perceive, and interact with the world around us. We are aware of ourselves and the world around us. This is called consciousness.

Our ability to think and understand is more than just the reactions between neurons and chemicals in the brain. Much more… We are made to think logically, make moral judgements, and use reason to make decisions. A mind is so much more than just chemical reactions and the movement of ions. The mind is a gift from God.

*For God has not given us a spirit of fear, but of power and of love and of a sound mind* (2 Timothy 1:7).

*Who has put wisdom in the mind?*
*Or who has given understanding to the heart?*
(Job 38:36)

# The Spinal Cord

The spinal cord is the other major part of the central nervous system. It is a very necessary component, for without the spinal cord the brain would be unable to receive sensory information from your body or tell your body what to do. The spinal cord and spinal nerves provide a pathway for sensory information to reach the brain. Equally important, they provide the means for motor output for the brain to reach the body.

The spinal cord starts at the medulla oblongata and extends to roughly the level of the second lumbar vertebral bone in the lower back. In the average adult the spinal cord is about 18 inches in length.

Like the brain, the spinal cord is well protected. It is covered by the same three meningeal layers as the brain, the dura mater, the arachnoid mater, and the pia mater. The meninges, along with the cerebrospinal fluid in the subarachnoid space, cushion the delicate nervous tissue of the spinal cord. Further, the spinal cord is housed in the vertebral canal. The vertebral canal is a long cylinder created by openings in the stacked vertebral bones. The vertebrae surrounding the spinal cord protect it in the same way the cranium provides protection to the brain, even though the vertebral bones move in ways the fused bones of your cranium do not. Thanks to the way the vertebrae are designed, you can bend and twist without damaging your spinal cord or the nerves emerging from it.

## Spinal Cord — Gross Anatomy

The spinal cord is roughly cylindrical structure about thickness of a garden hose. As previously described, it extends from the medulla oblongata to the level of the second lumbar vertebra. At its distal end, the spinal cord tapers into a conical-shaped structure known as the conus medullaris.

Emerging sequentially all along the length of the spinal cord are 31 pairs of spinal nerves. The two nerves in each pair exit the spinal cord at the same level, one on the right and one on the left. As nerves

### TAKING A CLOSER LOOK
### Spinal Cord and Vertebra

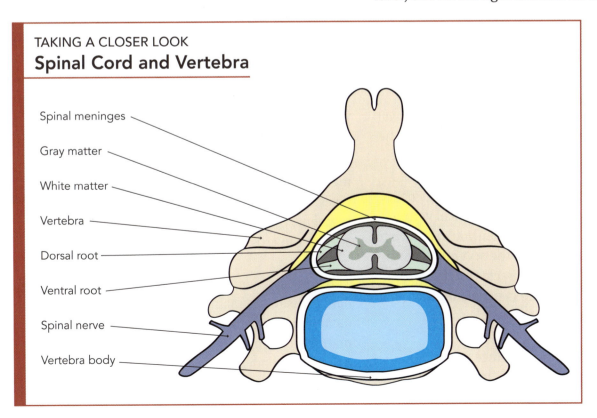

- Spinal meninges
- Gray matter
- White matter
- Vertebra
- Dorsal root
- Ventral root
- Spinal nerve
- Vertebra body

branch from the spinal cord, each leaves the vertebral canal by passing through a window called an intervertebral foramen on their way out to the body.

The name of these windows—intervertebral foramina (plural)—tells us a lot about how they are constructed. Inter- means "between," and the windows are between the vertebrae. Each vertebra in the stacked vertebral column has notches in it, top and bottom. Where a notch on the bottom of one vertebra aligns with a notch on the top of the next one, a window (foramen) is created.

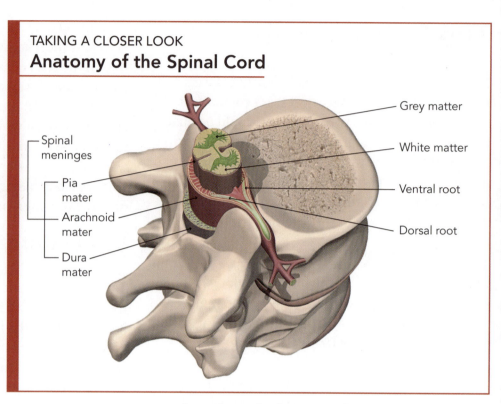

TAKING A CLOSER LOOK
### Anatomy of the Spinal Cord

There are eight pairs of cervical nerves (C1-C8), twelve pairs of thoracic nerves (T1-T12), five pairs of lumbar nerves (L1-L5), five pairs of sacral nerves (S1-S5), and one pair of coccygeal nerves (Co1). You might notice something a little odd here. The spinal cord ends at about the second lumbar vertebra, so why are there spinal nerves named for the levels below this?

The naming of the spinal nerves is based on the level of the vertebral column near which they exit, not the level of the spinal cord at which they originate. For cervical nerves one though seven, names are based on the vertebra that is below where they exit. For example, C2 exits above the second cervical vertebra. Then this relationship changes when you get down to C8. Spinal nerve C8 exits below the seventh cervical vertebra and above the first thoracic vertebra. From T1 on down, each spinal nerve is named for the vertebra above its exit.

Here is another puzzler for you. Since the spinal cord ends at about the level of the second lumbar vertebra, where do the nerves from L3 to Co1 come from? That is simple. These remaining spinal nerves emerge from the distal portion of the spinal cord and then travel downward until they exit. This group of spinal nerves extending down from the end of the spinal cord looks like the fibrous tail of a horse. Therefore it is called the cauda equina, which is Latin for "horse's tail"!

## Spinal Cord — A Closer Look at the Horns of the Matter

Now that we have seen the big picture, let's take a much closer look at the spinal cord. Each area that gives rise to a spinal nerve is called a spinal cord segment. We will examine a typical spinal cord segment in cross section.

The first thing you likely will notice are the light and dark areas in the main portion of the spinal cord. The dark area is gray matter, and the light area is white matter. Remember, gray matter consists mostly of neuron cell bodies, and white matter consists mainly of axons.

The gray matter projects out in several directions. These projection are called horns, and in cross section they look like a butterfly. The projections directed toward the front are the right and left ventral horns, also known as the anterior horns. Here are found cell bodies of the motor neurons that send signals to "move!" to the skeletal muscles.

The projections to the rear are the right and left dorsal horns (also called the posterior horns). In addition to the cell bodies of interneurons, the dorsal horns contain the axons of sensory neurons. These sensory neurons bring sensory information from the body to the CNS. The cell bodies of these sensory neurons are actually located just outside the spinal cord in a series of dorsal root ganglia. (A ganglion is a collection of nerve cell bodies located outside the CNS.) From each dorsal root ganglion, the sensory nerve axons enter the nearby dorsal horn.

Interneurons are just what their name sounds like they should be: neurons between neurons. As we discussed a while back, interneurons are found in both the brain and spinal cord, forming connections between sensory and motor neurons. Signals from sensory neurons are delivered to interneurons. Interneurons pass those impulses on to the appropriate motor neurons.

In the thoracic and lumbar regions, there are also small horns between the dorsal and ventral horns. These are called lateral horns. Lateral horns contain cell bodies of sympathetic neurons in the autonomic nervous system. These are the neurons that help get the body's organs ready for an emergency.

## Spinal Cord —- A Closer Look at White Columns and Roots

Just as the spinal cord's gray matter is divided into horns, so the spinal cord's white matter is divided into columns. If you think of the horns in the cross-sectional view of the spinal cord like the wings of a butterfly, the white matter columns are the in front of, behind, and beside the butterfly's wings. In the rear is the posterior white column. In front is the anterior white column. To each side is a lateral white column.

Each column is made up of tracts. If you recall, bundles of axons in the CNS are called tracts. Each tract is a group of axons headed to the same place (convenient, huh?).

### TAKING A CLOSER LOOK
### Spinal Cord: Internal Anatomy

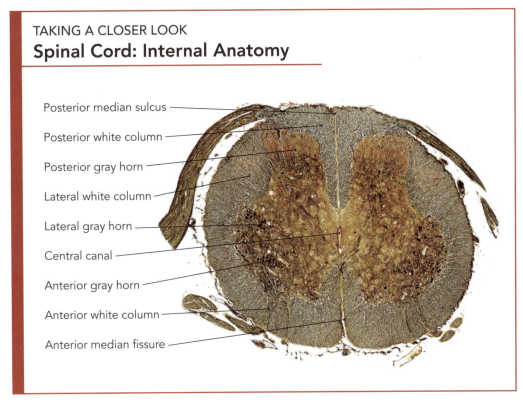

- Posterior median sulcus
- Posterior white column
- Posterior gray horn
- Lateral white column
- Lateral gray horn
- Central canal
- Anterior gray horn
- Anterior white column
- Anterior median fissure

In the spinal cord, the sensory tracts that carry signals to the brain are called ascending tracts. Logical name, right? Sensory signals are sent to the spinal cord and then up (ascending) to the brain. Then the brain integrates all the information. (Remember the three functions of the nervous system?) The brain sends the appropriate signal—the motor output—out to the body. So what do you think the motor tracts are called that take motor output signals down the spinal cord? Yep, they are called descending tracts.

We've seen that the spinal nerve segments have horns and columns, and that the columns consist of lots of tracts. Now let's seen how the spinal cord connects to the spinal nerves that emerge from it. The spinal nerves connect to the spinal cord by means of two bundles of axons, called roots. On each side (right and left) there is a ventral root and a dorsal root. The ventral root contains axons of motor neurons carrying nerve signals from the CNS out to muscle and glands. The dorsal root contains only sensory axons bringing input from sensory

## Amyotrophic Lateral Sclerosis

Amyotrophic Lateral Sclerosis (ALS) is a degenerative disease of the nervous system. It is a progressive illness that attacks motor neurons in the ventral horn and the pyramidal tracts. Accompanying this loss of motor neuron function is loss of control of voluntary muscles.

Patients typically present with *fasciculations* (muscle twitching), *atrophy* (loss of muscle mass), and weakness. Over time they often develop difficulty speaking, swallowing, and walking. In later stages of the illness, the patient may be unable to use his arms or legs and lose the ability to swallow or speak. Increasing weakness in the diaphragm and intercostal muscles can result in the patient finally being unable to breathe.

In approximately 90 percent of cases, the cause of ALS is not understood. This is called sporadic ALS. There is another form of ALS, known as familial ALS. This accounts for around 10 percent of cases. This form of ALS is inherited, so there must be a genetic component to the illness in these situations.

At present there is no cure for ALS. Although numerous medications have been tried to aid in slowing the progression of the disease, treatment still consists primarily of supportive care. Most patients die within 3 to 5 years.

Amyotrophic lateral sclerosis also known as Lou Gehrig's disease, so-called after a legendary New York Yankees baseball player who died of the disease in 1941.

*Lou Gehrig June 1923*

receptors throughout the body. Remember the dorsal root ganglia mentioned above? On each dorsal root is a dorsal root ganglion. The dorsal root ganglion contains the cell bodies of sensory nerves. Sensory input from the body reaches the cell bodies in the dorsal root ganglia, and that input is relayed onward through the dorsal roots to the dorsal horns in the spinal cord. The ventral and dorsal roots merge to form the 31 pairs of spinal nerves.

## Tracts in the Spinal Cord

As you might expect, the spinal cord is far more complex than a simple set of two-way streets in which some tracts take signals to the brain and other tracts bring signals from the brain. While the basic concepts remain the same, different types of signals are carried on a wide variety of tracts and processed in different ways.

Thus, the spinal cord is not just a pair of anterior white columns, a pair of lateral white columns, and a pair of posterior white columns. It's more. Much more. Just look at the graphic illustration below. There are many smaller tracts within the bigger areas we defined earlier. This level of complexity is necessary to efficiently integrate all the information the nervous system encounters every second.

Before you feel too overwhelmed with all the fancy names, there is a simple way to work your way through many of them. (And if you decide to go to medical school someday, you will want to remember this simple "trick"!) If you break down the tract names, very often you will know where it goes. For example, the corticospinal tract begins in the cerebral cortex and ends in the spinal cord. Get it? It goes from "cortico-" to "spinal." (The corticospinal tract is one of the pyramidal tracts we talked about earlier.) The "spinothalamic tract" begins in the spinal cord and ends in the thalamus. From "spino-" to "thalamic." So, can you figure out the spinocerebellar tract? Sure you can!

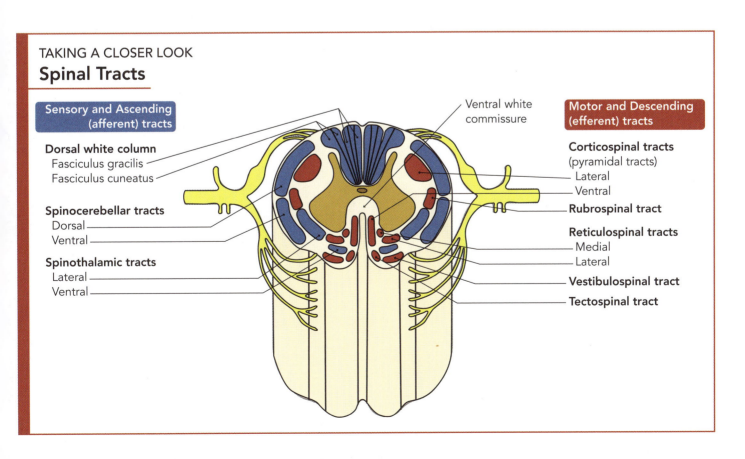

### TAKING A CLOSER LOOK
### Spinal Tracts

**Sensory and Ascending (afferent) tracts**

**Dorsal white column**
Fasciculus gracilis
Fasciculus cuneatus

**Spinocerebellar tracts**
Dorsal
Ventral

**Spinothalamic tracts**
Lateral
Ventral

Ventral white commissure

**Motor and Descending (efferent) tracts**

**Corticospinal tracts**
(pyramidal tracts)
Lateral
Ventral

**Rubrospinal tract**

**Reticulospinal tracts**
Medial
Lateral

**Vestibulospinal tract**

**Tectospinal tract**

# THE PERIPHERAL NERVOUS SYSTEM

The peripheral nervous system is the portion of the nervous system outside the brain and spinal cord. It is basically everything else in the nervous system that we have not yet covered. That includes the cranial nerves and the spinal nerves. (We've seen how the spinal nerves begin—we've looked at their roots!—but we haven't seen how they get where they are going.) Cranial nerves emerge from the brain, and spinal nerves emerge from the spinal cord.

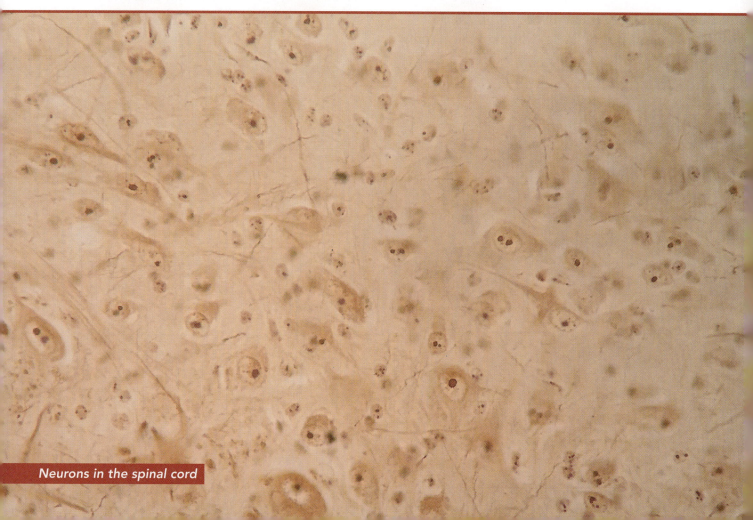

Neurons in the spinal cord

The peripheral nervous system brings sensory signals from the body to the central nervous system, and it takes motor signals out to the various part of the body after the CNS integrates the information. Without the peripheral nervous system, we would not be aware of the world around us. Let's begin by looking at the cranial nerves.

# Cranial Nerves

There are 12 pairs of cranial nerves. These nerves emerge directly from the brain and pass through holes (called foramina) in the cranium. Remember, a foramen is a "window," and foramina are "windows." Your cranium has many little windows through which your brain connects to the world using cranial nerves. Even though cranial nerves connect directly to the brain, they are considered a part of the peripheral nervous system.

These nerves are numbered and named, as seen below on the Cranial Nerves image. First, each pair of cranial nerves is numbered based on where it arises from the brain, from front to back. Then, the name of each cranial nerve reflects its function or its path. For instance, the first pair of cranial nerves, I, are associated with smell, so they are is called the "olfactory (I) nerves." The second pair to emerge from the brain, cranial nerves II, are for sight, so they are called the "optic (II) nerves."

We will now present all the cranial nerves in order. Please refer back to the image below to see the nerve locations.

### TAKING A CLOSER LOOK
### Cranial Nerves

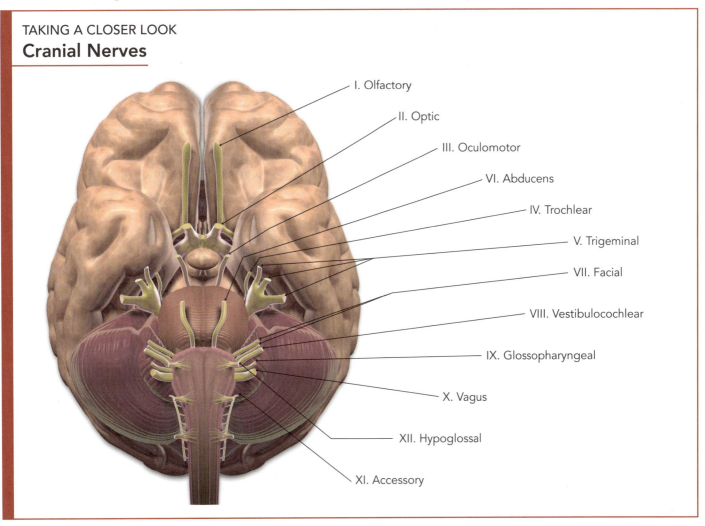

- I. Olfactory
- II. Optic
- III. Oculomotor
- VI. Abducens
- IV. Trochlear
- V. Trigeminal
- VII. Facial
- VIII. Vestibulocochlear
- IX. Glossopharyngeal
- X. Vagus
- XII. Hypoglossal
- XI. Accessory

## TAKING A CLOSER LOOK
### The Olfactory (I) Nerve

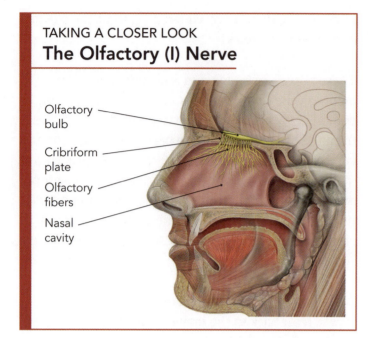

Cranial Nerve I: The olfactory nerve is a sensory nerve. It transmits impulses about smells, or olfaction. Specialized receptors that are tailor made to detect many sorts of molecules in the air are embedded in the lining at the top of the nasal cavity. Axons run from these receptors through a thin bony plate full of holes — the cribriform plate — to the brain's olfactory bulb just above it, and eventually to the olfactory cortex. Cranial nerve I is purely sensory: it detects smells, but it doesn't send your nose any instructions.

Cranial Nerve II: The optic nerve carries nerve impulses for vision. Fibers from each eye's retina coalesce to form an optic nerve, one for the right eye and one for the left. Each optic nerve runs through an opening in the eye's orbital socket to reach the brain.

The optic nerve—cranial nerve II—is purely sensory: it carries information about what you see to your brain, but it does not bring back any instructions for your eyes. Those instructions—instructions telling you which way to point your eyeballs—arrive via three other pairs of cranial nerves—III, IV, and VI. Do you get the idea that what you see and where you look are pretty important? Four of the twelve pairs of cranial nerves are devoted to your eyes!

## Trigeminal Neuralgia

Trigeminal neuralgia is a pain syndrome affecting the trigeminal (V) nerve. It is sometimes called tic douloureux.

Trigeminal neuralgia is characterized by episodes of pain across the face in areas supplied by the trigeminal nerve. (Neuralgia means "nerve pain.") These episodes can last from a few seconds to a few minutes, and they can recur over and over for hours. The pain is usually moderate to severe, but can be so intense that the patient is incapacitated.

The pain feels like a stabbing or burning sensation. It can affect the scalp, forehead, cheek, nose, teeth and gums. At times the simplest activity or movement precedes an attack. Brushing the teeth, blowing the nose, or even shaving may trigger an episode.

Trigeminal neuralgia may be caused by loss of the myelin sheath surrounding the trigeminal nerve or by pressure from a blood vessel compressing the trigeminal nerve.

Treatment of trigeminal neuralgia is challenging. Controlling the episodes may require very potent medications. Even if they successfully control the neuralgia, such medications can have significant side effects. Pain medications, even very potent ones, are often ineffective in controlling the intense pain. In the most severe cases, surgery may be needed. The goal of the surgery is often to decompress the nerve by putting a cushion between the nerve and local blood vessels. At present several different surgical techniques to control trigeminal neuralgia are being investigated.

Cranial Nerve III: The oculomotor nerve carries motor fibers to four of the six extrinsic eye muscles and the muscles of the upper eyelid. Each eyeball is surrounded by six muscles that enable it to look up, down, left, right, and up and down at angles. These are called extrinsic muscles because they are located outside the eyeball. The oculomotor nerve controls most of the movements of the eyeball.

Each eye also has some intrinsic muscles—muscles located inside the eye itself—that adjust the size of the pupil and focus the lens. Autonomic fibers that cause the pupil to constrict are also present in the oculomotor nerve. Cranial nerve III is exclusively a motor nerve; it has no sensory functions.

Cranial Nerve IV: The trochlear nerve controls the movement of the superior oblique muscle of the eye, another of the eye's extrinsic muscles. Cranial nerve IV is a motor nerve.

Cranial Nerve V: The trigeminal nerve provides sensory input from the face. It has three divisions—the ophthalmic, the maxillary, and the mandibular—which supply bring sensory information from the upper, middle, and lower thirds of the face, respectively. The cells bodies from all three divisions are found in the trigeminal ganglion. (Remember, a ganglion is a collection of cell bodies in the peripheral nervous system.)

In addition to its important sensory functions, the trigeminal nerve's mandibular division also has motor fibers that control chewing. Since cranial nerve V has both sensory and motor fibers, it is called mixed nerve.

Cranial Nerve VI: The abducens nerve controls the lateral rectus muscle of the eye, another of the eye's extrinsic muscles. Thanks to this nerve, you can look out to the side. It enables your right eye to look to the right, and your left eye to look to the left. Cranial nerve VI is a motor nerve.

Cranial Nerve VII: The facial nerve supplies motor function to the facial muscles. Thanks to your facial nerves, you can smile, grimace, wrinkle your forehead, and squeeze your eyes tightly shut. The facial

## Bell's Palsy

Bell's palsy is a one-sided paralysis of the muscles in the face. It results from damage to the facial (VII) nerve. Patients with Bell's palsy typically come to the doctor complaining of an inability to smile or frown on one side of the face. Sometimes they cannot close an eyelid, or they may lose taste sensation on the front part of the tongue. Some people with Bell's palsy cannot wrinkle the forehead. The exact symptoms depend on just where the facial nerve is damaged.

The first priority when someone presents with paralysis like this is to be certain the person has Bell's palsy and has not had a cerebrovascular accident (stroke). Usually, a careful examination makes this distinction quickly.

So how does a facial nerve get damaged? Bell's palsy is caused inflammation of the facial nerve, usually due to a viral infection. This inflammation interrupts motor signals to the facial muscles. Most cases of Bell's palsy resolve spontaneously within 3-4 weeks.

nerve also carries taste sensation from the front two-thirds of the tongue. Cranial nerve VII is a mixed nerve.

Cranial Nerve VIII: The vestibulocochlear nerve consists of a cochlear branch for hearing and a vestibular branch for balance. You will learn more about this nerve later when we study the ear. Cranial nerve VIII is a purely sensory nerve.

Cranial Nerve IX: The glossopharyngeal nerve contains motor fibers to control swallowing and sensory fibers carrying taste sensation from the rear third of the tongue. It also controls how much you salivate! Cranial nerve IX is obviously a mixed nerve.

Cranial Nerve X: The vagus nerve is the only cranial nerve that extends beyond the head and neck. The vagus nerve provides parasympathetic motor input to the heart, lungs, and abdominal organs. It carries sensory information from the aortic arch near the heart as well as the carotid bodies—collections of sensory receptors near the fork of the carotid arteries in the neck. The brain uses information from these vital avenues of arterial blood flow to help regulate blood pressure and respiration. Cranial nerve X is a mixed nerve.

Cranial Nerve XI: The accessory nerve delivers motor impulses to the sternocleidomastoid and trapezius muscles. These muscles enable you to move your neck and shoulders. Cranial nerve XI is a motor nerve.

Cranial Nerve XII: The hypoglossal nerve provides almost all the motor input to the tongue. Control of the tongue's movement is essential not only for speech but also swallowing. Cranial nerve XII is a motor nerve.

Knowing that an entire pair of cranial nerves is devoted to control of the tongue is a good reminder of the scriptural truth that an uncontrolled tongue can cause a lot of damage.

*Even so the tongue is a little member and boasts great things. See how great a forest a little fire kindles! And the tongue is a fire, a world of iniquity...*

(James 3:5-6)

# Spinal Nerves and Their Distribution

Spinal nerves carry sensory input to the CNS and motor output away from the CNS. These 31 pairs of nerves run the entire length of the spinal cord, exiting at each level. But where do they go after that? There must be some pattern to this. After all, the nervous system controls the entire body, so the nerves must somehow make their way everywhere, right? Yep, right again.

So do the spinal nerves just go directly out to the body structures they supply? Well, yes and no. The spinal nerves in the thoracic (chest) region, T1 to T12, mainly just go to their target areas. Pretty straightforward. However, in the cervical (neck) and

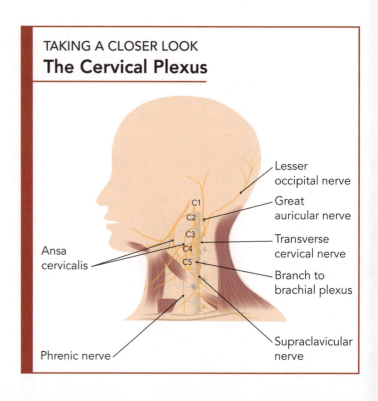

TAKING A CLOSER LOOK
## The Cervical Plexus

## TAKING A CLOSER LOOK
### The Brachial Plexus

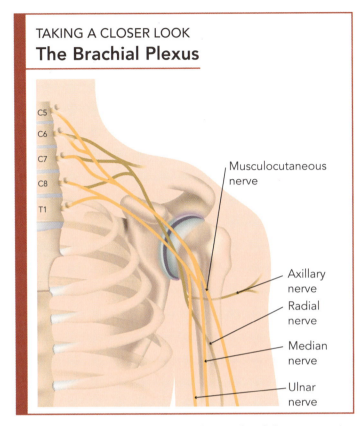

The brachial plexus is composed of fibers from the C5 to T1 spinal nerves. This plexus is located in the neck and axilla (armpit). The brachial plexus provides the nerve supply to the shoulder and arm. It is incredibly complex, combining and recombining into an array of trunks and cords and branches. The major nerves that extend from the brachial plexus are the axillary nerve, the radial nerve, the median nerve, and the ulnar nerve.

The axillary nerve supplies the deltoid and teres minor muscles. The radial nerve provides motor input to the triceps and the extensor muscles of the forearm. The median nerve supplies muscles of both the thumb's side of your forearm and hand. The ulnar nerve supplies muscles on the medial side (your pinky's side) of the forearm and most of the muscles in the hand.

lumbar (lower back) regions the path of the nerves is not so direct.

In these areas when the spinal nerves exit they branch to form complex networks called plexuses. Within a plexus, nerve fibers cross over one another, intermingle, and regroup. Therefore, each branch that ultimately leaves a plexus contains fibers from more than one spinal nerve. The four major plexuses are the cervical, brachial, lumbar, and sacral. The branches that arise from each plexus serve that particular region of the body. We will briefly examine each in turn.

The cervical plexus is formed by branches from the first five cervical nerves (C1 to C5). Through this plexus is carried sensory information from the back of the head, neck, and shoulder. Motor branches supply muscles in the neck. The most important motor nerve coming from the cervical plexus is the phrenic nerve. The phrenic nerve provides motor input to the diaphragm, regularly instructing you to breathe whether you are thinking about it or not.

## TAKING A CLOSER LOOK
### Upper Limb Dermatome

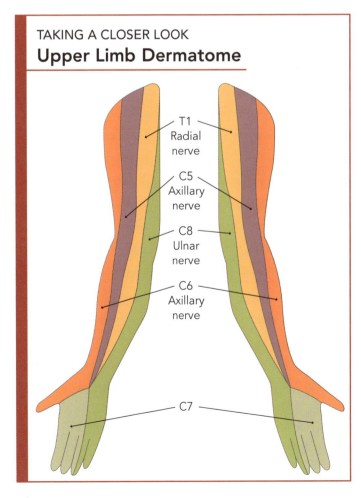

Just an FYI. Have you ever bumped your elbow and hit your "funny bone"? Wasn't really funny, was it? What happened was that something struck your ulnar nerve at a point where it was very close to the surface. You can actually feel your ulnar nerve. There is a small notch (or groove) on the back of the elbow on the pinky side of the arm. If you press too hard in this area, you can feel it tingle. That's your ulnar nerve.

Moving on we come to the lumbar plexus. Thankfully, this plexus is not as complex as the brachial plexus. The lumbar plexus arises from lumbar segments 1 though 4 (L1 to L4). The two large nerves that come from the lumbar plexus are the femoral nerve and the obturator nerve. The femoral nerve is the motor supply to muscles that flex the hip and extend the knee. Among other things, the obturator nerve supplies the adductor muscles of the thigh, the muscles you use to pull your thigh inward.

Finally we come to the sacral plexus. It is composed of fibers from lumbar and sacral spinal nerves (L4 to S4). The nerves from the sacral plexus primarily innervate the buttocks and lower limbs. The primary nerve coming from the sacral plexus is the sciatic nerve. The sciatic nerve supplies the muscles on the back of the thigh, the lower leg, and the foot. It also carries sensory information from the foot and leg.

## Phrenic Nerve Injury

As the phrenic nerves control the movement of the diaphragm, injury to these nerves is a very severe situation. If one phrenic nerve is damaged somewhere along its course, perhaps by trauma or a tumor, the diaphragm on that side can be paralyzed. Even more seriously, if the spinal cord is damaged, let's say by a broken neck, above the C3 to C5 level, the diaphragm can be completely paralyzed and the patient will be unable to breathe.

A spinal injury below the C5 level may result in paralysis of the limbs, but in this case the patient would still be able to breathe on his own. This is because the phrenic nerve would be spared in a lower spinal injury.

### TAKING A CLOSER LOOK
### The Lumbar Plexus

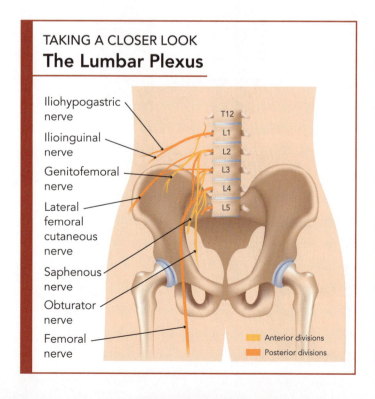

### TAKING A CLOSER LOOK
### The Sacral Plexus

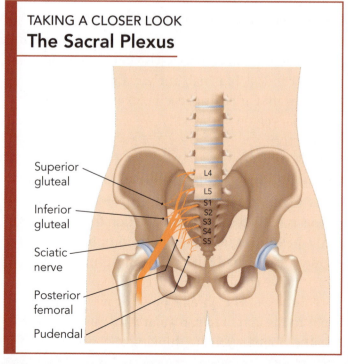

# Carpal Tunnel Syndrome

Carpal tunnel syndrome (CTS) can develop when the median nerve is compressed as it runs though the wrist. Carpal tunnel syndrome usually presents as pain and numbness in the thumb, index, and ring fingers—the fingers whose sensation is supplied by the median nerve. The pain can be mild to very severe, even debilitating. As CTS progresses, the pain may even extend up the arm.

The median nerve passes along the palmar surface of the wrist through an area known as the carpal tunnel. You've probably already guessed (correctly) that carpal means "wrist." This "tunnel" is bordered by wrist bones on one side and a strong, flat, fibrous band (the flexor retinaculum) on the other. Compression of the nerve between the bones and the flexor retinaculum had been cited as the cause of CTS, although there are other theories as well.

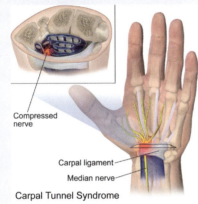
Carpal Tunnel Syndrome

Many factors increase the likelihood of developing CTS. These include obesity, arthritis, and diabetes. Although still controversial, the most cited risk factor for CTS is repetitive movement. People who perform repetitive tasks such as typing, using a computer mouse, hammering, or any other activity that puts pressure on the palm side of the wrist are thought to therefore be at higher risk for CTS. This debate will likely continue for some time.

Treatment of CTS includes anti-inflammatory medications, steroids, and the use of wrist splints (particularly during sleep to prevent excessive wrist flexing). In more severe cases, surgery to "release" the flexor retinaculum can be attempted.

# Spinal Segments and Dermatomes

If we map the sensory input from your skin to the spinal cord, we see a very interesting pattern. You see, there are sensory receptors in the skin all over the body. These cutaneous receptors send their input to sensory nerves, which transmit these signals to the brain. The regions from which these cutaneous sensory inputs come match the spinal segments. This pattern is very similar in every human body.

## Shingles

Shingles, also known as herpes zoster, is a viral disease characterized by painful blisters in localized areas of the body, typically within a particular dermatome. The virus causing this is the varicella zoster virus. The same virus causes chickenpox.

When a bout of chickenpox is over, the varicella zoster virus becomes dormant. It can reside in the dorsal root ganglia of the spinal nerves. If the virus reactivates, it can travel down the sensory nerve fiber and produce blisters on the skin. This is why a characteristic of shingles is patches of painful blisters that follow the distribution of a dermatome on one side of the body.

Treatment of shingles includes both pain medication and antiviral medications. The pain and rash typically resolve in 3-4 weeks. However, some patients have persistent pain in the region affected by the shingles. This can continue long after the rash has resolved. This pain is called postherpetic neuralgia.

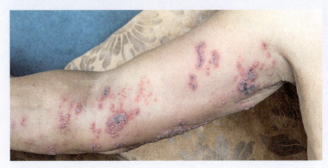

# 270
## THE NERVOUS SYSTEM

The region of the body that provides sensory input to a particular spinal nerve (or segment) is called a dermatome. Scientists have been able to divide the body into dermatome segments. However, the dermatome segments are not absolute. There is some degree of overlap. For example, the segment mapped to the L2 spinal segment may well send some sensory input to the L1 and L3 segments also. This overlap varies from person to person. Nonetheless, doctors are able to test sensations in the various dermatomes when assessing patients with possible spinal cord damage.

## Reflexes

Everyone know what a reflex is, or at least what it does. These are things that the body does automatically, or so it seems anyway. Reflexes are fascinating. They allow the body to perform certain functions or activities without us having to think about them.

A reflex is an automatic motor response triggered by a stimulus. This motor response happens before

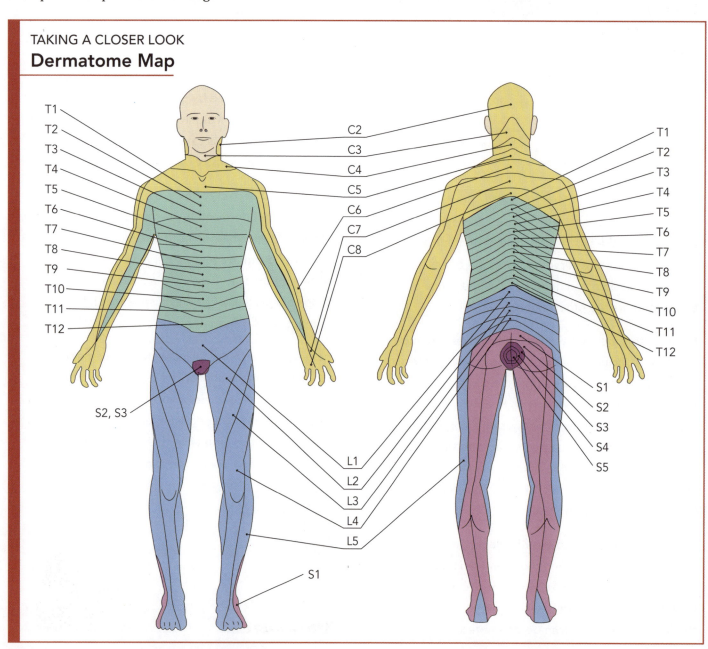

TAKING A CLOSER LOOK
**Dermatome Map**

you realize it. Your brain is not involved in the motor output the produces a reflex.

The two primary types of reflexes. One is the somatic reflex. This results in contraction of skeletal muscles. The other is the autonomic reflex. This triggers a response in smooth muscle or glands. With the somatic reflex, you will ultimately be aware of what happened. With an autonomic reflex, you remain unaware of what happened.

Here is the classic example of a somatic reflex. You are walking through the kitchen. You casually place you hand on the top of the stove, not knowing it is very hot. Practically as soon as your hand hits the stove top, it pulls back to get away from the heat. By the time you yell out, you realize that your hand is in the air. Before your brain told you, "Get your hand off the stove; it's really hot!" your hand was off the stove.

Let's examine this typical somatic reflex in detail.

When you placed your hand on the hot stove, sensory receptors in the skin of the hand sensed the heat. These receptors produced a signal that was then transmitted to a sensory neuron. This neuron then carried the impulse through the dorsal root of the spinal nerve, ending in the dorsal horn of the gray matter in the spinal cord.

In the dorsal horn, the sensory neuron can synapse in either of two ways. It may synapse directly with a motor neuron. This involves a single synapse, and it is known as a monosynaptic reflex. (Note that in this case, the integration function of the nervous system

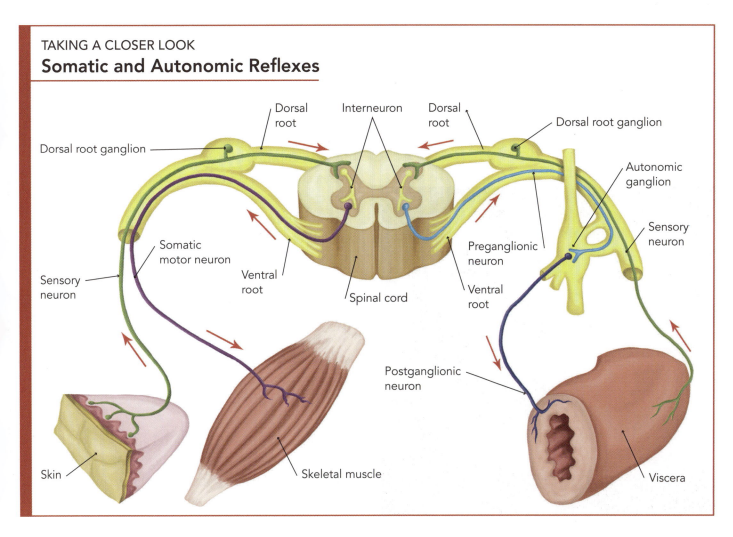

TAKING A CLOSER LOOK
**Somatic and Autonomic Reflexes**

is this single synapse.) In other cases, the sensory neuron synapses with one or more interneurons. This is a polysynaptic reflex.

A nerve signal is then sent out to skeletal muscles by way of motor fibers in the ventral root. The signal is received by the appropriate muscles, prompting you to take you hand off the hot stove! All before you brain even knew.

Of course your brain does find out. Other pathways and connections inform your brain that your hand is now red and burning. This information is processed in the brain and hopefully stored in your long-term memory. Then the next time you are walking though the kitchen, you will ask yourself, "I wonder if that stove is hot?" before you put your hand on it. Learning is a wonderful thing.

Broken down into its basic components, this reflex arc consists of: a sensory receptor, a sensory nerve, an integration center in the spinal cord, a motor nerve, and an effector (the skeletal muscle that withdraws your hand).

## Sensory Receptors

When the wind blows on your face, you touch that hot stove top (remember?), you detect a pebble in your shoe, or you smell brownies baking in the oven, you instantly recognize and identify what you are feeling. How does your body detect and process these things? Your sensory receptors must first detect the stimulus—the wind, the heat, the pressure, the smell.

Sensory receptors in the nervous system allow us to be aware of changes in our environment. These changes are called stimuli, and they occur all around us, to one degree or another, practically all the time. A sensory receptor "receives" a stimulus. This is turn triggers a nerve impulse that can be delivered to the central nervous system.

There are many ways to classify sensory receptors. Each classification scheme has its strengths and weaknesses. We will group sensory receptors according to the type of stimulus they detect.

A thermoreceptor is triggered by changes in temperature. There are—not surprisingly—two types of thermoreceptors. One senses warm, and the other senses cold. These receptors are located near the skin surface. They are activated by moderate degrees of warm and cold. Extremes of warm and cold activate special receptors called nociceptors, which respond to more painful stimuli.

An osmoreceptor senses changes in osmotic pressure. This type of receptor is found mainly in the hypothalamus. Osmoreceptors in the hypothalamus detect changes in the concentration of substances dissolved in the blood because of the pressure created on them when concentrations change.

When such changes in the blood are detected, the hypothalamus can send a message to the posterior pituitary gland nearby, instructing it to increase or decrease its secretion of a hormone called vasopressin. Vasopressin, in turn, lets the kidneys know how much water to hold onto or eliminate from the body. You can see that this is an important homeostatic mechanism without which the concentrations of dissolved substances in the blood could soon change more than is good for us.

A photoreceptor responds to changes in light. Receptors like these are found in the retina of the eye.

Chemoreceptors are triggered by exposure of certain chemicals. Olfactory receptors are a good example. This receptor can detect chemicals in the air. Different chemicals are interpreted by the nervous system as different smells. Taste buds also work by means of chemoreceptors.

Mechanoreceptors sense mechanical force such as stretch, touch, pressure, or vibration. Tactile sensations like touch, vibration, itching, and tickling are mediated by mechanoreceptors.

Proprioceptors sense position in relation to other parts of the body. Muscle spindles within your skeletal muscles are proprioceptors.

A nociceptor are the receptors for painful and potentially damaging stimuli. Things that might damage tissue can trigger nociceptors. Extremes of heat or cold, excessive pressure, and even tissue irritation due to chemical exposure are sensed by nociceptors.

Sensory receptors can be found in skeletal muscle, skin, joints, and visceral organs. (Visceral organs are organs in the chest or abdomen.) By means of these sensory receptors, the CNS can keep up with everything that is happening in and to the body.

# The Autonomic Nervous System

Up to this point, our study has focused on the somatic nervous system. Here sensory inputs that we are conscious of such as touch, temperature, pain, taste, and sight are received by the CNS. Then motor output signals are sent to skeletal muscle. When the skeletal muscle is stimulated, its membrane is excited, and as a result it contracts. Therefore, the motor output from the somatic nervous system can only stimulate. We will see that the autonomic nervous system, in contrast, can either "dial up" (stimulate) or "dial down" (inhibit) the targets it affects.

The autonomic nervous system (ANS) receives sensory input from sensory receptors in visceral organs—like the heart, stomach, and intestines—and blood vessels. Baroreceptors, chemoreceptors, and mechanoreceptors monitor blood pressure, heart rate, and respiration. We are not generally aware of these sensory inputs. Our autonomic nervous system monitors all these things day and night. When these inputs are integrated, the motor output from the autonomic nervous system goes to smooth muscle, cardiac muscle, and glands. Again these targets—the effectors—that autonomic motor outputs affect are not under our conscious control.

Remember we said that the autonomic nervous system can either stimulate or inhibit? Well there are two divisions to the autonomic nervous system, and they tend to have opposite effects on any given target. These divisions of the ANS are called sympathetic and parasympathetic. Although many organs receive input from both divisions of the ANS, there are exceptions. There are several effectors that receive only sympathetic innervation.

Sympathetic and parasympathetic signals have opposite effects on a given organ. In any given situation, one will stimulate and the other inhibit. One division will excite; the other will depress. Therefore, in contrast to the somatic nervous system which only excites effectors (namely, skeletal muscles, right?), the ANS has the capability to both excite and inhibit.

The ANS plays a vital role in maintaining homeostasis. You recall that homeostasis is the body's ability to use many interacting mechanisms to maintain balance or "equilibrium" among its many systems. The two divisions of the ANS make it possible to adjust many vitally important variables upward or downward to keep them in the narrow range that is safe for us.

Being able to detect changes in the body's internal environment and to process that information is one thing. But to then be able to immediately either excite or inhibit the effector organs to maintain internal balance, or homeostasis, is quite another thing entirely. Quite a brilliant system.... One that we need to explore further.

## Anatomy of the Autonomic Nervous System

You recall that the autonomic nervous system is part of the peripheral nervous system. And you recall that a ganglion (plural, ganglia) is a collection of neuron cell bodies in the peripheral nervous system. Well, ganglia are an important part of the ANS.

### TAKING A CLOSER LOOK
### Sympathetic Nerves in the ANS

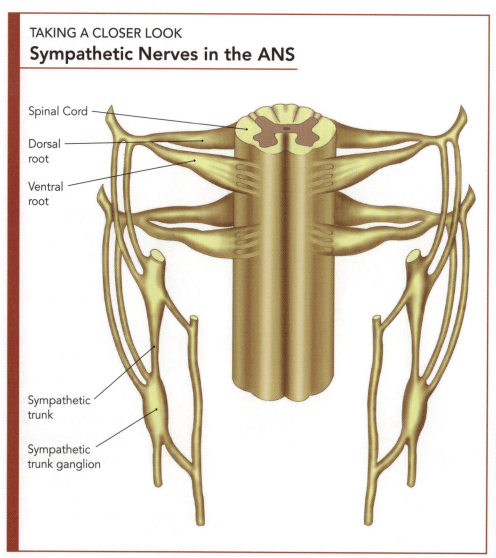

Each motor pathway in the ANS is composed of two neurons, the preganglionic neuron and the postganglionic neuron. The cell body of the preganglionic neuron is in the brain or the gray matter of the spinal cord. Its axon exits the CNS by way of a spinal nerve or a cranial nerve. That axon goes to a ganglion, where it synapses with the dendrites of a postganglionic neuron.

The dendrites and cell bodies of postganglionic neurons reside in ganglia. In the ganglion, the postganglionic cell can synapse with many preganglionic nerve cells.

In the sympathetic division of the ANS, preganglionic neuron cell bodies are located in the lateral horns of the 12 thoracic segments and the first two lumbar segments. These neurons are relatively short, sending their axons to ganglia located very close to the vertebral column in the sympathetic trunk. A chain of these ganglia run parallel to the vertebral column, one on each side. Another chain of sympathetic ganglia are found in front of the vertebral column and are called the prevertebral ganglia. Preganglionic neurons synapse with postganglionic neurons in the sympathetic trunk and in prevertebral ganglia. Because their preganglionic cell bodies are located from T1 to L2, the sympathetic division is also called the thoracolumbar division.

In the parasympathetic division, cell bodies are found in brain in the nuclei of four cranial nerves (III, VII, IX, and X) and in the lateral horns of the gray matter in sacral segments two through four (S2 to S4). With preganglionic neuronal cell bodies located in the brain or sacral spinal segments, the parasympathetic division is also called the craniosacral division. The axons of preganglionic neurons in the parasympathetic division tend to be much longer than those in the sympathetic division. They synapse with their postgangionic neurons in ganglia (called terminal ganglia) away from the spinal cord and nearer the visceral organs.

### TAKING A CLOSER LOOK
### Synapse in Trunk Ganglion at the Same Level

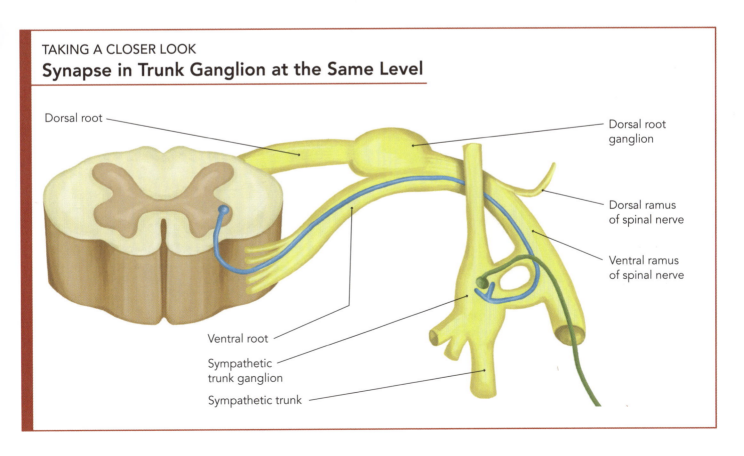

## Autonomic Plexuses

Neurons of the somatic motor system formed networks of fibers called plexuses. The same cam be said for ANS. In the thorax, abdomen, and pelvis, there are multiple autonomic plexuses composed of axons from both sympathetic and parasympathetic neurons. These frequently are found near larger arteries. Each plexus is located near the effector organ or region it serves.

In the thorax the cardiac plexus supplies the heart with autonomic input. The pulmonary plexus serves the lungs and bronchial tree.

In the abdomen is the celiac plexus, which innervates the major abdominal organs (stomach, spleen, pancreas, liver, adrenal glands). The two mesenteric plexuses supply the small and large intestines with autonomic nerves. The renal plexus supplies the kidneys and ureters.

## Function of the Sympathetic Nervous System

The operation of the sympathetic nervous system is most evident when we are frightened or excited. Sympathetic output supports processes that are required for vigorous physical activity. Therefore, the sympathetic nervous system is called the "fight or flight" system.

"Fight or flight" responses to stress include: an increase in heart rate and blood pressure, an increase in blood flow to skeletal muscles, dilation of bronchial tubes to allows more oxygen intake, break down of glycogen stored in the liver to obtain glucose for energy, and dilation of the pupils. These are actions are all stimulated by the sympathetic nervous system. But the sympathetic nervous system, at the same time, slows down some physiologic functions that are unnecessary drains of valuable energy when

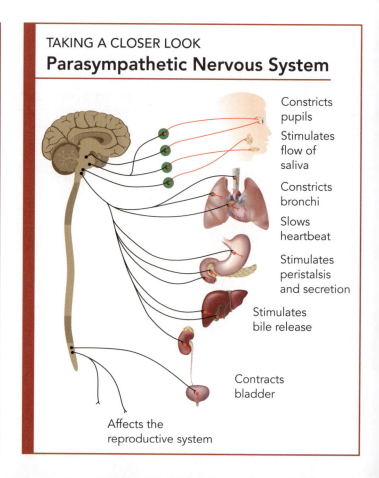

TAKING A CLOSER LOOK
**Sympathetic Nervous System**

- Dilates pupils
- Inhibits salivation
- Relaxes bronchi
- Accelerates heartbeat
- Inhibits peristalsis and secretion
- Stimulates glucose production and release
- Secretion of adrenaline and nonadrenaline
- Inhibits bladder contraction
- Affects the reproductive system

TAKING A CLOSER LOOK
**Parasympathetic Nervous System**

- Constricts pupils
- Stimulates flow of saliva
- Constricts bronchi
- Slows heartbeat
- Stimulates peristalsis and secretion
- Stimulates bile release
- Contracts bladder
- Affects the reproductive system

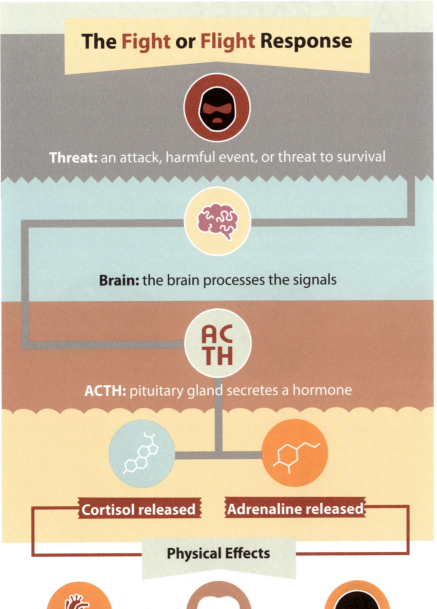

## Function of the Parasympathetic Nervous System

fight or flight is necessary. Thus the sympathetic nervous system can decrease in blood flow to the digestive tract and kidneys, systems not essential to urgent physical activity.

In contrast to the sympathetic nervous system, the parasympathetic nervous system could be called the "rest and digest" system. This system is geared to support the rest and recuperation activities of the body. For the parasympathetic nervous system, the support of functions that result in conservation and storage of energy is the goal.

During times of rest, parasympathetic signals to the gastrointestinal tract are often active. These signals promote digestion of food and elimination of waste products. All this aids in preparing the body for any upcoming episode of activity.

# SPECIAL SENSES

How many "senses" do we have?

The usual answer is five. So often we hear about using our "five senses." These five senses are taste, smell, touch, sight, and hearing. And it's true. We do smell things. We can taste stuff. We can feel things. We can see objects. We hear sounds.

Some people suggest that we have many more than just five senses. This is true also. It just depends on how you define the concepts of senses.

Remember all the different sensory receptors we listed earlier? That was quite a list! There were probably even a few words—a few "senses"—that you'd never heard of. Mechanoreceptors sense mechanical stress, such as pressure or stretch. Chemoreceptors respond to chemical changes, such as changes in pH (acidity) or the presence of various molecules or ions. Photoreceptors sense light. Thermoreceptors respond to temperature changes. Proprioceptors sense position change. And this list could go on.

Aren't all these things "senses"? Yes, they are. But we don't have to let those senses complicate things by crowding into the classical list of senses.

Let's think instead in terms of what is called a "special sense." These special senses involve sensory receptors contained in specialized organs or structures in the body. Sight, hearing, and taste are examples of special senses.

Specialized photoreceptors are contained in the retina of the eye. Specialized chemoreceptors that allow us to taste are present in the tongue. These are very special types of receptors, and they are confined in organs specifically designed for them. The special senses are taste, sight, smell, hearing, and balance (or equilibrium).

The other sense modalities are better thought of as "general senses." The general senses utilize sensory receptors scattered throughout the body.

But what about touch? Isn't it a sense? Yes, it is, but touch requires the input of many receptors that are not localized in one area or in one organ. Touch is certainly a real sense, but it is better to consider it more of a general sense.

*The hearing ear and the seeing eye, The Lord has made them both.*

(Proverbs 20:12)

# Smell

We are surrounded by smells. Cookies baking in the oven, the smell of freshly cut grass, the pungent aroma of your gym shoes—lots of things have smells. Some things don't, but that's another story. We are going to sniff out the facts about our sense of smell. (See what I did there?)

The sense of smell is called olfaction. It is mediated by special cells, cleverly enough called olfactory sensory neurons. These neurons are embedded in the olfactory epithelium in the roof of the nasal cavity. Air entering the nose passes by the olfactory epithelium on its way to the airways that take it to the lungs.

The olfactory epithelium covers only one-and-a-half square inches but contains millions of olfactory sensory neurons. As you can see in the illustration, the olfactory epithelium rests on the cribriform plate of the ethmoid bone. There are columnar (column-shaped) cells in this layer. These cells act as a support structure. Between the columnar cells are the olfactory sensory neurons. These olfactory neurons are unusual in that they are bipolar neurons. That it, each has only one dendrite and one axon.

The dendrite of an olfactory neuron ends in several cilia. These cilia lie on the surface of the olfactory epithelium. They are covered and protected by a thin layer of mucous.

The bundles of axons of the olfactory neurons extend through holes in the cribriform plate. These bundles of axons make up the olfactory (I) nerve, the first cranial nerve. After passing through the cribriform plate, they synapse with neurons in the olfactory bulb.

At the base of the olfactory epithelium is one more kind of cell. This is called a basal cell. (Clever name, right?) The basal cells of an epithelium are the cells at its base, in the bottom layer. You might want to have a look back at Volume 1 of the *Wonders of the Human Body* for a refresher on epithelium. The basal cells here are actually stem cells that divide to produce replacement olfactory sensory neurons. The olfactory neurons live for only a few weeks. A mechanism for replacing them regularly is vital.

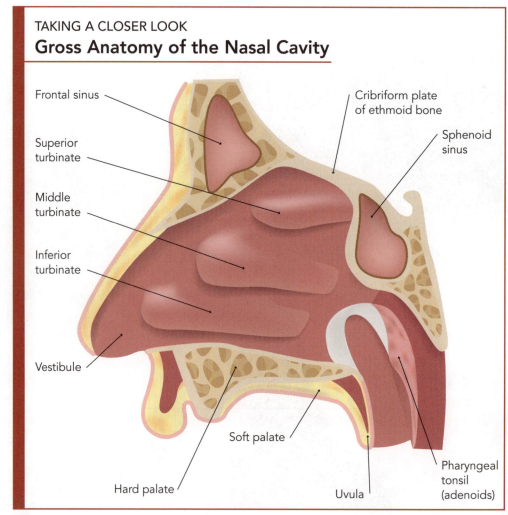

TAKING A CLOSER LOOK
**Gross Anatomy of the Nasal Cavity**

## How Does Smelling Work?

Now that we know the anatomy, how does smelling work?

Substances that can trigger smell are called odorants. In order for an olfactory sensory neuron to be triggered, an odorant molecule must reach its receptor. This occurs after the odorant molecule is breathed in. As this molecule passes the olfactory epithelium, it is trapped by the mucous layer covering the epithelium.

In the mucous layer the odorant contacts the olfactory cilia and binds to a receptor site. This binding triggers the opening of ion channels nearby. If enough odorant molecules trigger the receptor cell, an impulse is sent down the entire length of the neuron.

Once a full action potential is triggered, the nerve impulse reaches the olfactory bulb. Then the signal moves down the olfactory tract to the olfactory area of the cerebral cortex. We then perceive a smell!

We do know that our threshold for smell is low. That means it only takes a few molecules of some odorants to trigger smell. Certain odors we can perceive even at extremely low concentrations (like 1 molecule in 50 billion!) while other odorants require a higher concentration for perception.

### Odor Number

For many years the traditional thinking was that humans could detect about 10,000 different odors. A recent study has challenged that number. This new study claims that humans may be able to detect as many as one trillion smells (again, who counted them?)! Research is obviously ongoing.

*And walk in love, as Christ also has loved us and given Himself for us, an offering and a sacrifice to God for a sweet-smelling aroma* (Ephesians 5:2).

### TAKING A CLOSER LOOK
### How Smelling Works

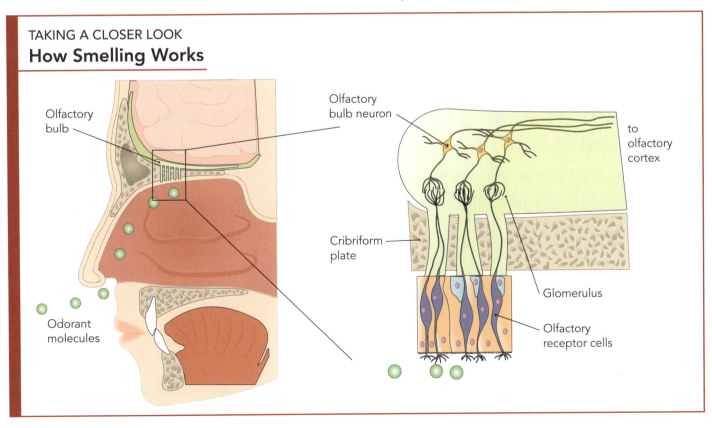

# Taste

*And the house of Israel called its name Manna.
And it was like white coriander seed,
and the taste of it was like wafers made with honey.*

(Exodus 16:31)

The sense of taste, like the sense of smell, is a chemical sense. Our ability to taste depends on our ability to detect, and then react to, certain chemicals in our environment. We can do this by means of our taste buds.

If you look at your tongue in the mirror, you will see that it has a rather rough-looking surface. This appearance is due to the presence of many protuberances called papillae on its surface. Filiform (thread-like) papillae cover the major portion of the front two-thirds of the tongue. Unlike the tongue's other papillae, these lack taste buds. Fungiform papillae are mushroom-shaped and are scattered over the entire surface of the tongue. Taste buds are found on the top of fungiform papillae. Foliate papillae are leaf-like folded ridges on the sides of the tongue near the rear. Nodular-appearing circumvallate papillae form a row across the back of the tongue. In these two types of papillae, Taste buds are found in the side walls of the foliate and circumvallate papillae.

In the olfactory epithelium, the dendrite of the olfactory neuron was the sensory receptor, but taste buds are different. In a taste bud, the neuron is not the sensory receptor. The actual sensory cell is the gustatory epithelial cell. (Gustatory means "taste." Bet you guessed that.) Each of these cells has microvilli that extend through a taste pore on the epithelial surface. These microvilli are the sites that trigger a reaction in the gustatory cells. Also in contact with the gustatory epithelial cell are sensory neurons. These are the neurons that start a nerve signal on its way to the brain.

Because of their location, gustatory epithelial cells are subjected to lots of wear and tear. They are thus easily damaged and have a very short life span, usually just a week or so. Just as the basal cells in the olfactory epithelium divide to make replacement cells, so the basal epithelial cells in the taste buds divide and produce replacement gustatory epithelial cells regularly.

## Physiology of Taste

As mentioned, the sense of taste is a chemical sense. When exposed to a certain chemical, a specialized receptor reacts to the stimulus and ultimately a nerve signal is produced. Even though smell and taste are both chemical senses, the process works differently for taste.

Remember the actual receptor cell for taste is the gustatory epithelial cell. This cell is not a nerve cell.

### TAKING A CLOSER LOOK
**Tongue**

- Epiglottis
- Palatine tonsil
- Lingual tonsil
- Vallate papillae
- Fungiform papillae
- Foliate papillae
- Filiform papillae

## SPECIAL SENSES

### TAKING A CLOSER LOOK
### Taste Buds

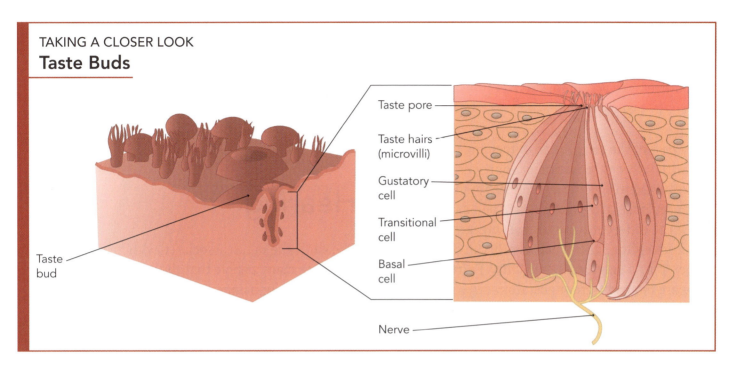

That's a big difference right there. So what gives? Well, it's really easy.

A gustatory epithelial cell is exposed to a stimulatory chemical called a tastant. This happens at the microvillus at the taste pore. When the tastant binds to the chemical receptor on the microvillus, no nerve signal is generated. The gustatory cell is not a nerve cell. What happens is that the gustatory epithelial cell releases a neurotransmitter that then stimulates receptors in the dendrites of the sensory neurons that are in the taste bud. This then triggers the action potential in the neurons.

The afferent (sensory) fibers that carry taste signals to the brain are mainly in two of the cranial nerves. The facial nerve (VII) carries impulses from the anterior two-thirds of the tongue, and the glossopharyngeal nerve (IX) carries impulses for the rear third of the tongue. Fibers from these nerves synapse in the solitary nucleus of the medulla oblongata. Then the signals are taken to the thalamus and ultimately on to the gustatory cortex.

## Types of Tastes

At present, taste sensations have been categorized into five basic groups:

1. Bitter taste is perceived as unpleasant. Bitter taste can result from bases or alkaloids (like the medication quinine, for example). This is the most sensitive of the taste modalities.

2. Sweet taste is generally pleasant. Sugars are an obvious source for this taste. Other sources include some alcohols and some amino acids.

3. Sour taste detects acidity. Citrus fruits tends to be naturally acidic.

4. Salty taste detects inorganic salts. Sodium chloride (table salt) is the best example.

5. Umami is the taste sensation produced by the amino acids, glutamate and aspartate. Some describe this as the "savory" taste. Beef and cheese can elicit this taste.

There is one other "taste" that may be added to the list in the future. This is, at present, called oleogustus. This is the taste for fats. So far this has not been universally accepted, but over time it may be the "sixth" taste.

Every part of the tongue can detect any of the taste modalities. Sweet sensation or sour sensation are not localized to specific area of the tongue as has been previously taught. Any taste bud can detect any of the different tastes, However, each individual gustatory epithelial cell apparently can only be triggered by a single taste modality. Fortunately, there are many different gustatory cells in every taste bud!

Another factor greatly influencing how we perceive a taste is our sense of smell! If our olfactory sense is triggered when we eat, it enhances the pleasure our food gives us. On the other hand, if you have ever had a really bad cold or an allergy attack, you understand that the opposite of this is also true. When your sinuses are congested, you cannot smell things. Funny how during those times, food just doesn't taste as good!

# Hearing

The ear is the organ associated with hearing. It surprises many people that the ear is also the organ responsible for balance (or equilibrium).

These senses are so amazing. We can hear the booming fireworks on the Fourth of July and yet are

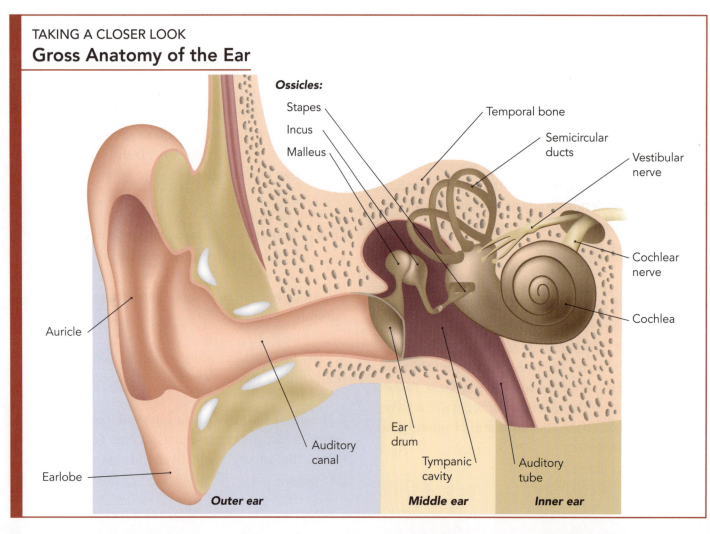

TAKING A CLOSER LOOK
**Gross Anatomy of the Ear**

# SPECIAL SENSES

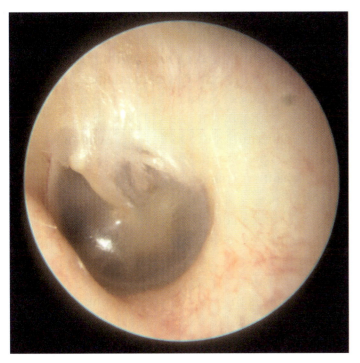

The eardrum

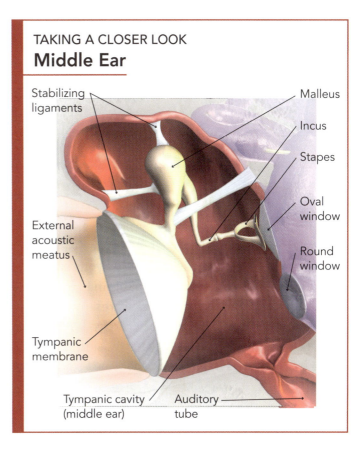

### TAKING A CLOSER LOOK
### Middle Ear

able to heard the gentle buzzing of the bee in our backyard. We can remain standing, balanced, with our eyes closed. We can run the bases and climb the stairs without giving it a thought. Let's see how they work.

## Anatomy of the Ear

The ear is composed of three main regions: the external ear, the middle ear, and the inner ear.

The external ear is what most people refer to when they talk about the ear. The external ear is the part that is visible to the world. The shell-shaped protrusion from the side of your head is called the auricle. It is made of elastic cartilage covered by skin. The external auditory canal is a tube through which sound waves move toward the tympanic membrane (eardrum). This auditory canal is about an inch long.

The external auditory canal is lined with skin, a few hairs, and special glands called ceruminous glands.

These glands produce cerumen, or earwax. Although earwax can be a problem if it builds up and blocks the ear canal, it is there to keep foreign objects from reaching the delicate tympanic membrane.

At the end of the external auditory canal is the tympanic membrane, or eardrum. It is round and coned slightly inward. As we will see soon, the tympanic membrane vibrates when sound strikes it. The tympanic membrane marks the boundary between the external ear and the middle ear.

The middle ear is a small cavity in the temporal bone of the skull. The tympanic membrane separates it from the external ear. The middle ear is separated from the inner ear by a bony wall containing two openings, the oval window near the top, and the round window below. Both of these windows are covered by membranes. Also opening into the middle ear is the Eustachian tube. This tube connects to the nasopharynx (high in the back of the throat). Each

Size of the stapes.

The inner ear has two parts, one inside the other. The outermost is the bony labyrinth, which is a series of cavities in the temporal bone. Within this is the membranous labyrinth. The membranous labyrinth is a series of tubes and sacs that lies within the bony labyrinth. Both labyrinths are filled with fluid. Inside the bony labyrinth, surrounding the membranous labyrinth, is perilymph. The fluid inside the membranous labyrinth is called endolymph.

There are three major parts of the bony labyrinth. They are the vestibule, the semicircular canals, and the cochlea.

The vestibule is the bony labyrinth's central chamber. It is located medial to the middle ear. It is separated from the middle ear by the oval window, to which the stapes attaches. The membranous labyrinth

end of the Eustachian tube is open to air, the one end to the air in your throat, which is pretty much at the same pressure as the air outside your ears in the room, and the other end to the air in the middle ear. Therefore, it helps equalize the pressure on both sides of the tympanic membrane. This equal pressure allows the eardrum to vibrate freely as sound waves strike it.

Strung across the middle ear is a chain of three small bones. In fact, these are the three smallest bones in the body. These bones are named based on their appearance. The first is the malleus (Latin for "hammer"). On one side this hammer attaches to the inner surface of the tympanic membrane, and on the other side it connects to the middle bone in the series, the incus. The incus (Latin for "anvil") then connects to the third bone, the stapes. The base of the stapes (Latin for "stirrup") then fits into the oval window. So you see, the vibrations that start on the eardrum are going to be transmitted to the hammer, then to the anvil, then to the stirrup, and on to the oval window.

### TAKING A CLOSER LOOK
### Inner Ear

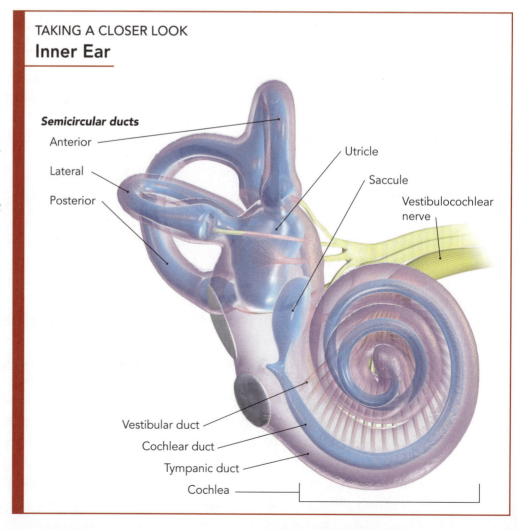

components in the vestibule consist of two sacs, the utricle and the saccule. Both of these are important in controlling equilibrium, as we will see later.

Posterior to the vestibule are the semicircular canals. There are three semicircular bony tubes oriented are right angles to one another. Through each of these canals runs a duct called a semicircular duct. At the end of each duct is an enlarged area called the ampulla. These ducts—parts of the membranous labyrinth—open into the utricle.

Anterior to the vestibule is the cochlea. This is a spiral chamber made of bone. It is shaped pretty much like a snail's shell. Running through the middle of the cochlea is the cochlear duct. Inside the cochlear duct is the spiral organ (also called the organ of Corti) which is the hearing receptor. By its placement in the cochlea, the cochlear duct effectively divides the cochlea into three chambers. These chambers are called scala. There is the scala vestibuli, which begins at the oval window. The scala media is the cochlear duct itself. Finally, there is the scala tympani, which ends at the round window.

At the distal portion of the scala vestibule is a small opening, called the helicotrema. This small opening connects the fluid in the scala vestibule and the scala tympani.

A closer look at a cross-section of the cochlea shows these three chambers. The cochlear duct is separated from the scala vestibuli by the vestibular membrane. The floor of the cochlear duct, separating it from the scala tympani, is the basilar membrane.

Resting on the basilar membrane is the spiral organ (or organ of Corti). This organ is composed of supporting cells and cochlear hair cells. The hair cells are arranged in rows and are covered by the tectorial membrane. At the base of the hair cells are fibers of the cochlear branch of the vestibulocochlear (VIII) nerve.

Yep, that's an awful lot of anatomy, but we will make sense of all this very soon. Hang in there, and keep a close eye on the illustrations as we go through the path of sound and the way it gets transmitted to your brain.

## Sound

When we hear something, we are sensing sound waves from the environment. Understanding what is happening when we hear requires some knowledge about sound itself.

Sound is a series of vibrations. It can be illustrated as a series of waves as shown. These vibrations are pressure waves, and they must travel through a medium, such as air or water. Because they can vibrate, even the bones in the skull are capable of transmitting sounds waves. You can compare sound waves to the ripples you produce when you throw a rock into a quiet pond. Sound cannot travel through a vacuum, such as in space.

Different sounds sound different, right? Otherwise, there would be only one sound. Sounds can vary in pitch or in amplitude. Pitch is the frequency of the sound. The more waves per second, the higher the pitch. The sound from a bass drum has a much lower frequency than the sound from a policeman's whistle. The whistle has a much higher frequency.

Sounds also different in their loudness, or intensity. A bass drum can make a soft sound or a very loud

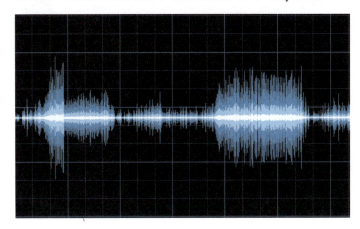

one. The frequency of the two sounds will be very close to the same, but the amplitude of the sounds will differ. In the illustration, the greater the size of the wave, the greater the amplitude. The greater the amplitude, the louder the sound. Take a train whistle as an example. If a train is one half mile away and blows its whistle, you will hear it and most likely recognize it as a train whistle. It will be distinct, but not comfortably loud. If that same train blows its whistle when it is only fifty yards away, the sound will be very loud. When it is louder, it has a higher amplitude.

Both of these concepts, pitch and amplitude, are important as we explore our sense of hearing.

> "I have heard of You by the hearing of the ear,
> But now my eye sees You.
>
> (Job 42:5)

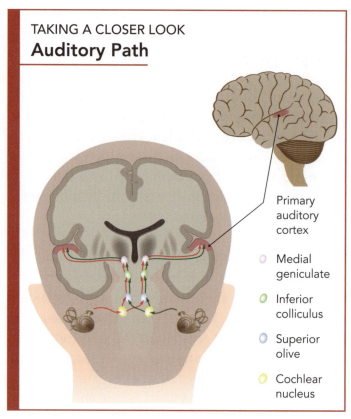

**TAKING A CLOSER LOOK**
**Auditory Path**

- Primary auditory cortex
- Medial geniculate
- Inferior colliculus
- Superior olive
- Cochlear nucleus

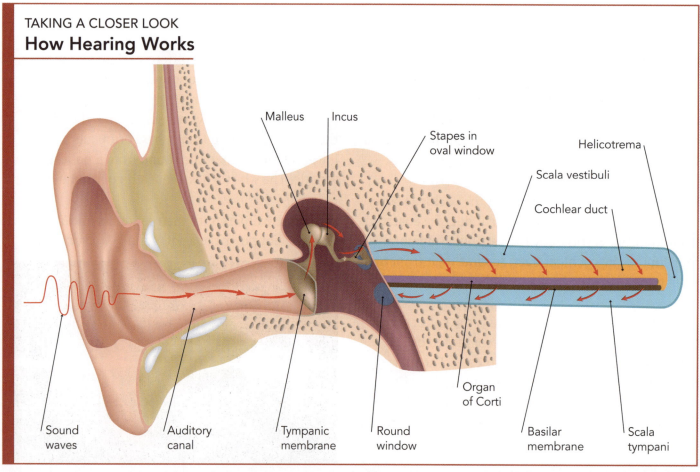

**TAKING A CLOSER LOOK**
**How Hearing Works**

Labels: Sound waves, Auditory canal, Tympanic membrane, Malleus, Incus, Stapes in oval window, Round window, Organ of Corti, Basilar membrane, Scala tympani, Helicotrema, Scala vestibuli, Cochlear duct

# How We Hear

Hearing is our ability to convert the pressure waves (sound waves) in our environment to action potentials that can be transmitted to the brain. In the brain, these nerve signals are processed and perceived as sound.

We will follow a sound wave, step by step, though the ear.

1. The external ear (the auricle and external auditory canal) captures sound waves (pressure waves) and directs them toward the tympanic membrane.

2. When it reaches the tympanic membrane, a sound wave strikes the membrane and causes it to vibrate at the same frequency as the sound wave. This matching of the frequency transmits the pitch

### TAKING A CLOSER LOOK
### Organ of Corti

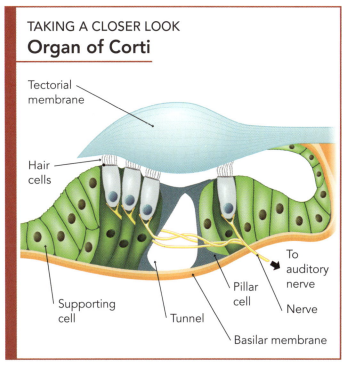

of the sound. More than that, the amplitude of the sound wave is transmitted also. The louder the sound reaching the tympanic membrane, the farther the membrane is pushed as it vibrates. So it is not only how fast the membrane vibrates, but it is also important how far the membrane moves with each vibration, that passes the pitch and amplitude of the sound wave accurately to the middle ear.

3. When the tympanic membrane vibrates, it causes movement in the three bones of the middle ear—first the malleus, then the incus, and finally the stapes. Both the frequency of the vibration and its intensity are transmitted, via these bones, to the oval window.

### TAKING A CLOSER LOOK
### Anatomy of the Cochlea

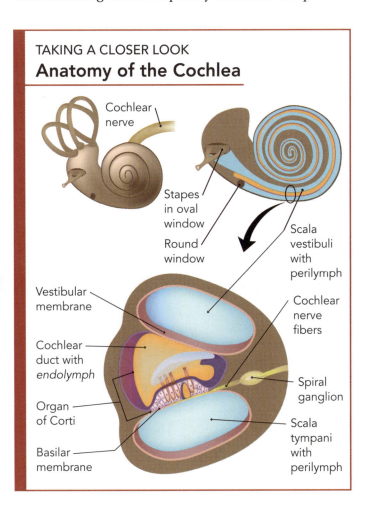

4. The base of the stapes rests against the oval window. The vibrations transmitted from the tympanic membrane are thus transferred to the oval window, and it begins to vibrate accordingly. These vibrations are now transmitted onward, but now the transmission is through fluid, not air (as in the external ear) or bones (as in the middle ear). The pressure waves from the oval window are now carried by the perilymph. These waves travel down the scala vestibuli, through the helicotrema, and back down

the scala tympani. When the waves reach the end of the scala tympani, they reach the round window. The pressure waves cause the round windows to flex and bulge. This bulging effectively damps out and ends the pressure waves.

5. As the perilymph carries the sound wave throughout the scala, it causes the basilar membrane to vibrate. (Keep your eye on those illustrations as you trace the path of sound's vibrations!) When the basilar membrane vibrates, the cochlear hair cells move against the tectorial membrane. The movement of the hair cells generates receptor potentials.

6. The receptor potentials produced by the cochlear hair cells trigger action potentials in the neurons associated with them. These nerve impulses are then carried by the cochlear branch of the vestibulocochlear (VIII) nerve.

Another fascinating aspect of the sound transmission process is how the ear is able to process different frequencies efficiently. It has to do with the structure of the basilar membrane. The basilar membrane is "tuned" to respond to different frequencies along its length. On the end nearer the oval window, the membrane responds better to higher frequencies. On its more distal end, the membrane responds better to lower frequencies.

## Balance

It is incredible that we don't fall down all the time. It takes an enormous amount of information processing to keep all the right muscles contracting at just the right time with just the right amount of force…to keep us from falling down. All this cannot happen if we don't know where we are and what

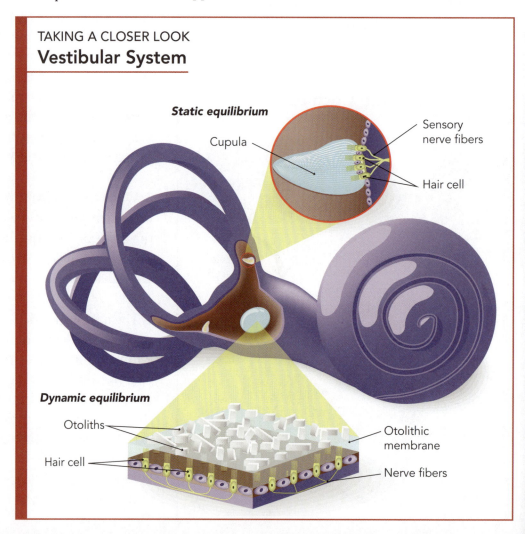

TAKING A CLOSER LOOK
**Vestibular System**

*Static equilibrium*
- Cupula
- Sensory nerve fibers
- Hair cell

*Dynamic equilibrium*
- Otoliths
- Hair cell
- Otolithic membrane
- Nerve fibers

direction we are moving in right now. Is our head up or down? Are we looking to the left or to the right? There must be something that tells the brain all this. That something is called the vestibular apparatus.

The vestibular apparatus consists of the utricle, the saccule, and the semicircular canals. From these structures signals are sent to the brain to keep it informed of the head's position in space.

In the walls of both the utricle and the saccule is found a structure called a macula. The utricle and the saccule are positioned perpendicular to each other (at a 90-degree angle to one another). Therefore they are able to monitor movement in two different planes.

A macula is a layer of two different types of cells: supporting cells and hair cells. The supporting cells do exactly what their name suggests; they provide support. The hair cells have microvilli and are the sensory receptor cells. The hair cells are covered by a jelly-like layer called the otolithic membrane. In the otolithic membrane are many small stones (actually calcium carbonate crystals) called otoliths. The otoliths add mass to the otolithic membrane. When the head moves, the otolithic membrane, weighted down by these little rocks in your head, is pulled by gravity. As the membrane moves, the hair cells underneath are moved. This movement triggers impulses in the nerve cells near the base of the hair cells. These are fibers of the vestibular portion of the vestibulocochlear (VIII) nerve.

But there's more! Balance is pretty important. These semicircular canals—another vital part of the vestibular apparatus—are also very important in maintaining balance.

If you recall, the three semicircular canals are oriented at right angles to one another. Each canal opens into the utricle. In the ampulla of each canal is found a receptor called the crista ampullaris (or crista, for short). The structure of the crista is very similar to the structure of the macula we just studied. There are supporting cells and hair cells. Again, the support cells support the hair cells, and the hair cells are the sensory cells. Covering the crista is another jelly-like mass. This is called the cupola.

As the head rotates, the endolymph in the semicircular canals moves also. As the endolymph moves, it moves the cupola, which in turn moves the hair cells. When the hair cells move, receptor potentials are generated, which then trigger action potentials in their associated neurons. Again, the nerve impulses are carried by the vestibular branch of the vestibulocochlear (VIII) nerve.

Quite an elaborate system for keeping our balance, wouldn't you say?

## Vertigo

Vertigo is the sensation a person experiences when he feels like he is moving but he isn't. Vertigo can be spinning sensation or a falling sensation. Very often this dizziness is accompanied by nausea, breaking out in a sweat, or feeling faint. Movement of the head to either side usually makes the symptoms worse.

The most common disease that results in vertigo is benign paroxysmal positional vertigo. This is probably due to loose otoliths moving into a semicircular canal. These loose otoliths are felt to cause abnormal movement of the endolymph, triggering the vertigo.

Another cause of vertigo is labyrinthitis. Labyrinthitis in an inflammation of the inner ear. It may be caused by a virus.

Other possible causes of vertigo include Ménière's disease, migraine, stroke, multiple sclerosis, and Parkinsonism.

# Sight

I don't have to tell you where you would be without your eyes. In the dark, that's where. We use our eyes for things we take for granted every day. Reading a book, watching a movie, seeing your mom's smiling face, seeing your dad's frowning face when you forgot to do your homework (well, you can probably do without that one…)—without eyes you could not really picture what you'd be missing. From the time you wake up in the morning until the time you fall asleep at night, your eyes are involved in almost everything you do.

There is an enormous amount of visual information that we process each day. This is how it all works.

## Corneal Abrasions

A corneal abrasion is a scratch on the surface of the cornea. As the cornea has many nerve endings, a corneal abrasion can be quite uncomfortable, causing pain, redness, and sensitivity to light.

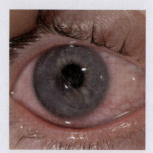

Corneal abrasions are very common and are usually caused by a finger poked in the eye. Abrasions are also caused by windblown objects or even contact lenses.

There is no specific treatment, although some people have some degree of pain relief if the eye is held shut with a patch. Most patients recover fully in 3-4 days.

*Now therefore, stand and see this great thing which the Lord will do before your eyes.*

(1 Samuel 12:16)

### TAKING A CLOSER LOOK
### Gross Anatomy of the Human Eye

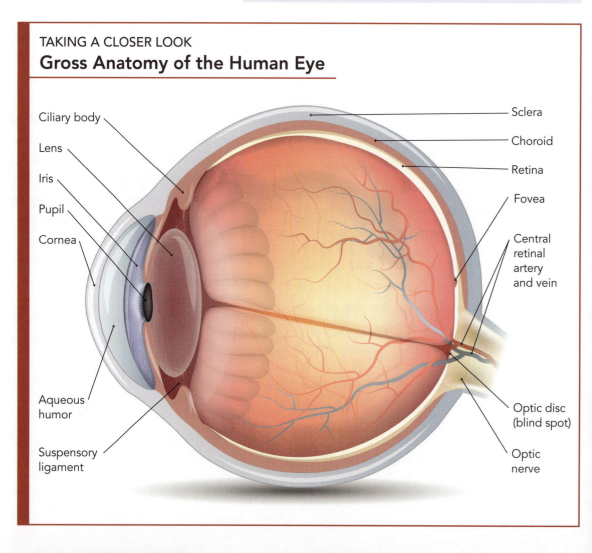

# Anatomy of the Eye

The eye is a fascinating thing. It is small compared to other some organs (usually about 1 inch in diameter), but it is quite complex in both its structure and its function.

The outside layer of the eye consists of two parts. The sclera is a layer of dense connective tissue. It is quite durable, and it helps the eye maintain its round shape. When you see the "white" of someone's eye, you are seeing the sclera. The other part of the outer layer of the eye is the cornea. This is the clear portion of the eye in the front. It is a good thing that this portion is clear. This is where the light enters the eye! The cornea is curved in the front. This curvature bends light as it enters the eye, beginning the process of focusing it.

The cornea has lots of nerve endings, so it is very sensitive to being touched or scratched. On the other hand, it has no blood vessels, so it is completely clear.

The middle layer of the eye is called the vascular tunic. The posterior 5/6 of this layer is the choroid. The choroid contains many blood vessels and provides oxygen and nutrients to the retina. It is dark

# Glaucoma

The fluid in the anterior chamber of the eye, the aqueous humor, is produced by the ciliary processes. As aqueous humor is produced, it is also drained from the anterior chamber. If for some reason, the drainage system is blocked, the fluid pressure inside the eye can increase. This increase in pressure can eventually damage the retina. This increased pressure in the eye is called glaucoma.

Glaucoma is a painless condition. This is one of the reasons it is so insidious. Many people have permanent damage from glaucoma even before they know they have it. Often the visual loss in the affected eye goes unnoticed because it is compensated for by the other eye, so the patent remains unaware there is a problem until there is permanent damage.

Glaucoma can be detected by a simple measurement of intraocular pressure. This is usually done during a routine eye exam. The screening test is painless.

At present, there is no cure for glaucoma. There are treatments to help slow the progression of the disease. These include medications to lower the pressure inside the eye and interventions using a laser. In some patients there are surgical options available.

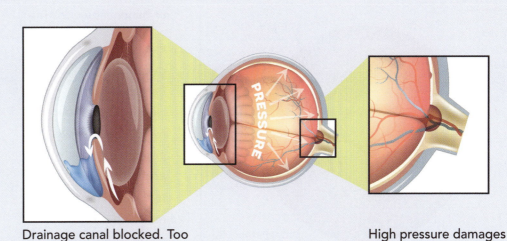

Drainage canal blocked. Too much fluid stays in the eye; this increases pressure

High pressure damages optic nerve.

brown in color, due to the presence of a large amount of melanin, a dark pigment. The melanin helps the choroid absorb story light rays. This helps the image on the retina to be sharp.

The front portion of the choroid becomes the ciliary body. It is made up ciliary muscles and the ciliary processes. Ciliary muscles help change the shape of the lens to improve its ability to focus. The ciliary processes secrete the fluid that fills the anterior chamber of the eye, in front of the lens.

The iris is a round, flat layer of smooth muscle with an opening in the middle. The opening is called the pupil. The iris is the colored portion of the eye. The function of the iris is to regulate the amount of light entering the eye by changing the size of the pupil.

## The Retina

The inner layer of the eye is the retina. Its outer surface can be seen by looking through the pupil with an ophthalmoscope, a medical instrument that has a lens and a light source to provide a magnified view into the eye.

The surface of the retina has several landmarks. Nearest the nose is a lighter area known as the optic disc. This is the area where the optic (II) nerve exits the eye. The optic disc is also knowns as the "blind spot." There are no photoreceptors in this area. You will also see several retinal blood vessels through an ophthalmoscope, and they tend to converge at the optic disc.

Lateral to the optic disc is a small area, essentially in the center of the retina. This is the macula lutea. In the center of the macula is the fovea centralis. The fovea is the area of highest visual resolution.

Taking a microscopic look the retina, you find an array of complicated structures. We will take them one at a time, starting from the choroid and moving inward.

### TAKING A CLOSER LOOK
**Retina**

- Fovea
- Macula
- Optic disc
- Central retinal vein
- Central retinal artery
- Retinal venules
- Retinal arterioles

## Cataracts

A cataract is a clouding of the lens of the eye. As this clouding progresses, there is an increasing loss of vision. People with cataracts also have problems reading and seeing at night and often complain of having difficulty with bright lights.

Cataracts are a major cause of blindness worldwide.

Cataracts are primarily due to aging. Fortunately, the cloudy lens can be removed surgically and replaced with an artificial lens. Lens replacement usually results in substantial restoration of visual function and an associated improvement in the patient's quality of life.

# SPECIAL SENSES

The first thing we see is the pigmented layer of the retina. This layer is made of melanin-containing cells that, like the choroid, help absorb stray light.

The next layer contains the photoreceptor cells, the rods and cones of the eye. This is where all the work is done, so to speak. Then comes a layer of bipolar cells. Then, closest to the inner surface of the retina, is a layer of ganglion cells and their axons. (The retina's two sorts of neurons are called bipolar cells and ganglion cells.)

So when you think about it, light has to go all the way through the retina—past blood vessels and past two layers of neurons—to stimulate the photoreceptor cells so that a nerve signal can make its way back out of it? Umm, well, yes.

When light enters the eye, it travels through the cornea, through the lens, through the blood vessels on the surface of the retina, through the ganglion axons, through the ganglion cell bodies, and through the bipolar cells, just to get to the photoreceptor cells. When this light stimulates the photoreceptors cells, the process of generating a neural impulse begins. That signal then begins its journey back out to the surface of the retina. It has to go to the bipolar cells, then on to the ganglion cells, then through the axons of the ganglion cells. All this to get to the optic nerve and get to the brain.

But (there's that "but" again), it's the very best way the retina could have been designed.

## Is the Film in Backwards?

There are so many people in the world who mock God, so many who deny Him as Creator. So many people suppress the truth in unrighteousness.

Those who deny God as Creator must then have a way to explain all the marvelous things we see. Those people generally believe that blind evolution, millions of years of chemicals just randomly bumping together, somehow assembled all the incredibly complex things in our world. Including the human body. Including the eye.

### Designed by the Master — So Simple Yet So Complex

**Wonderful Eyes**

A very common complaint against God is the design of the human eye. Many learned people have said the design of the eye proves that there is no God, or at least, if God exists He is a bad designer! After all, no engineer would design the eye in such a fashion. God put the retina in backwards, they say. The light receptors are on the wrong side. The light must go all the way to the back to be detected. A good engineer would have built the retina with the photoreceptors on the surface of the retina. So, in their view, God did it wrong. Not only did God get it right. He got it really right! Maybe no human engineer would have designed the retina as it is, but be very glad God did.

You see, the photoreceptor cells use an enormous amount of energy in the process of converting light into nerve impulses. These cells use lots of different chemicals, need lots of nutrients, and generate lots of waste products. This high level of metabolism also generates a lot of heat that must be carried away. The design of the retina is perfect to accomplish these tasks.

The amount of blood flowing in the choroid layer is very high. It is able to provide oxygen and nutrients needed by the photoreceptors. Further, this blood flow helps dissipate the heat that is generated, keeping the photoreceptors functioning efficiently without overheating. The pigment layer also plays role in that it has mechanisms to break down harmful molecules generated by the action of light on the photoreceptors.

So, if God did put the photoreceptors in front, like so many people say He should have, there would be a huge problem. For the photoreceptors to then work properly, the pigmented layer and the choroid would have to rest on top of the photoreceptors on the INNER surface of the retina. Then how could any light get through at all?

The virtues of this great design don't end there. Scattered in these retinal layers are another kind of cell, Müller cells. Müller cells act like fiberoptic cables, efficiently transmitting the light that strikes the surface of the retina to the photoreceptor cells. In so doing, they clean up the light that has entered the eye, sharpening the images it transmits by removing any distorting light reflections and ensuring that all the colors are well-focused when they reach the photoreceptors.

I'll take God's engineering anytime!

*He who planted the ear, shall He not hear?*
*He who formed the eye, shall He not see?*

(Psalm 94:9)

## Rods and Cones

Photoreceptors are the cells that convert light into nerve impulses. There are two main types of photoreceptors, rods and cones.

There are far more rods in the retina than cones. Rods are the receptors that are the most sensitive to light. Without them, we could not see in dim light or at night. Rods do not provide color vision, nor do they produce sharp images. These are found mostly in the peripheral retina.

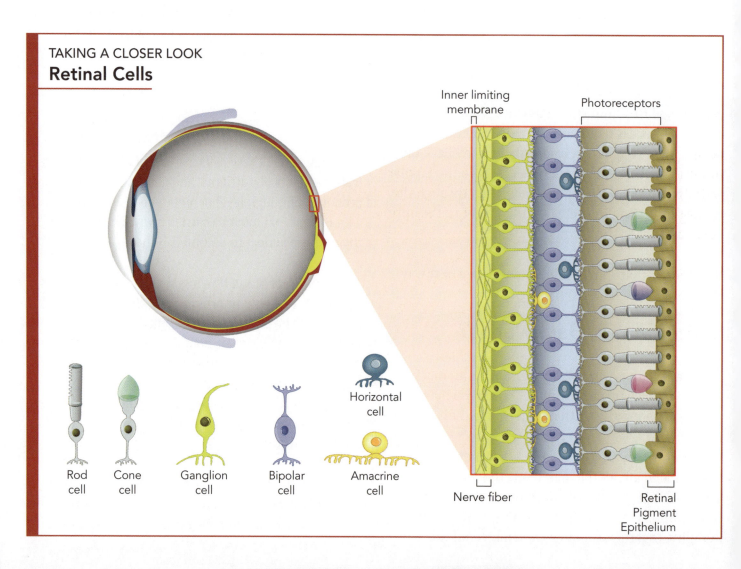

TAKING A CLOSER LOOK
**Retinal Cells**

Cones are far fewer number in the retina. These photoreceptors function much better in bright light. These are more concentrated in the central retina.

Cones are also responsible for our color vision. There are three types of cones: blue cones, which are obviously sensitive to blue light, red cones, which sense red light, and green cones, which detect green light. All other colors that we perceive are mixtures of these.

## How the Eye Focuses

In order for us to see things clearly, the light that enters the eye needs to be focused sharply on the retina. The focusing, or bending, of the light is done by both the cornea and the lens of the eye. The majority of the light bending, about 75 percent or so, is done by the cornea.

When light passes through a transparent object, it can be bent, or refracted. The more curved the transparent object is, the more the light is bent. Therefore, when light passes through the cornea, the curved cornea bends it. Ideally, this bending of the light places the focus precisely on the surface of the retina.

In a normal-shaped eye, with the lens in its most flattened shape (and thus having the least effect on bending the light), light from any object more than 20 feet away should focus on the retina. This point on the retina is called "the far point" of vision. Then as we progressively focus on things closer than 20 feet, the lens shape changes. It becomes more rounded and is thus able to bend the light more. This additional refraction keeps the focus point on the retina as our eye accommodates to see closer and closer objects. This is basically how the eye focuses.

But not all eyes are perfectly round.

In people who are nearsighted (like your author), the far focus point is in front of the retina, rather than on it. Their eyes are myopic. Nearsighted people can see close objects well, but have difficulty seeing more distant objects. The distant objects appear blurry or fuzzy because the light focuses too soon and then spreads out again by the time it reaches the retina.

Farsighted people have the opposite problem. In their eyes the far focus point falls behind the retina. Their eyes are hyperopic. These people can see distant objects well, but closer things appear blurry and fuzzy.

## Color Blindness

Color blindness is the inability to see color or distinguish between colors. It is the result of a deficiency in one of the three types of cones. It is far more common in males. Some estimates state that as many as 10 percent of males have some degree of color blindness.

The most common form of color blindness is red-green, which signifies a lack of either red or green cones. Those with red-green color blindness see reds and greens as the same color. The second most common form of color blindness is blue-yellow.

There is no cure for color blindness. Can you tell a difference between the two squares in this illustration? If they look the same to you, you may be red-green color blind.

And then there is the problem that your grandparents may well experience. As people grow older, their lenses often become stiffer. Stiffer lenses cannot adjust to become rounder when looking at things that are close. Therefore, it may become difficult to read without holding a book at arm's length. This is called presbyopia.

Within certain limits, nearsightedness, farsightedness, and presbyopia can be corrected with glasses.

We have come to the end of our exploration of the nervous system. I am certain that because of your new knowledge about the nervous system, you now want to become a neuroscientist, or a neurologist, or a neurosurgeon! With all we know, much remains to be learned about the nervous system, so there are plenty of things left to be discovered.

At the very least, you have been introduced to the amazing complexity of the nervous system and have some understanding of the love and care used by God in its design.

### TAKING A CLOSER LOOK
### Vision Disorders

Normal vision

Myopia

Hyperopia

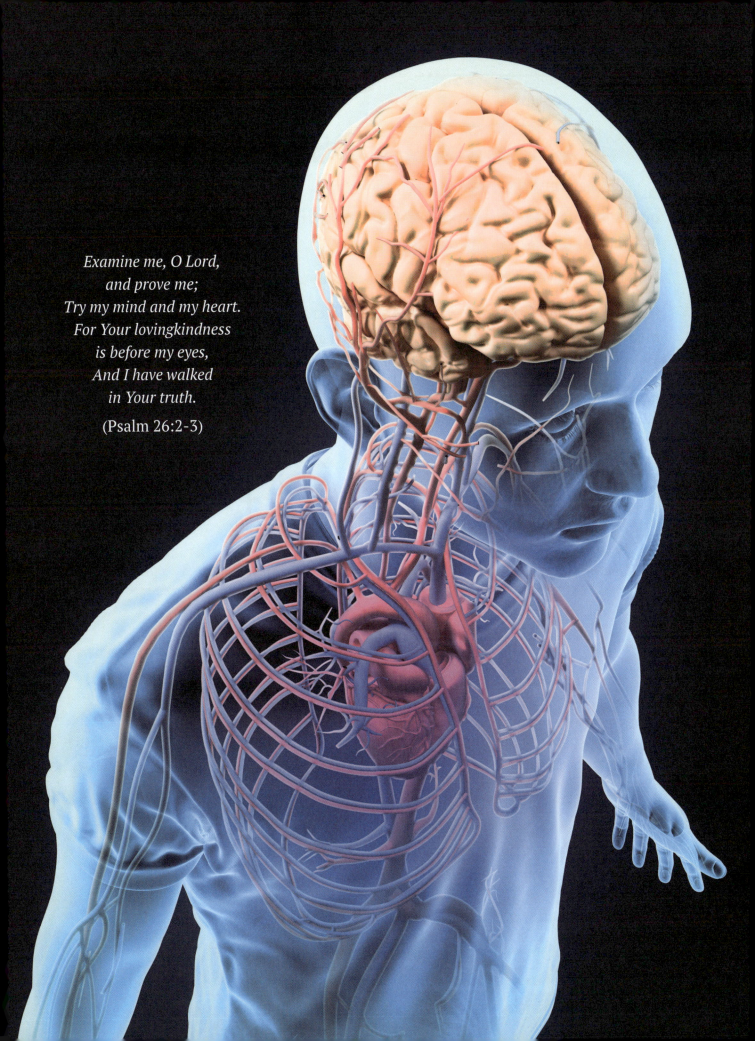

# THE GOSPEL

As incredible as our bodies are, they do not last forever. Eventually, they wear out or become damaged from disease or injury. But why is that exactly? If the body is so amazing, why do we die?

God's Word gives us the answer to this question.

In the beginning God created everything. He created everything is six days. He made the first man, Adam, and the first woman, Eve. They were created in the image of God.

*So God created man in His own image; in the image of God He created him; male and female He created them (Genesis 1:27).*

Creation

Sin and death

Redemption

Resurrection

After He finished creating, He looked on all He had made and called it "very good." It was a perfect world where there was no death. And in the perfect world, God gave man a choice, obey Me or disobey Me.

*And the LORD God commanded the man, saying, "Of every tree of the garden you may freely eat; but of the tree of the knowledge of good and evil you shall not eat, for in the day that you eat of it you shall surely die"* (Genesis 2:16–17).

Unfortunately, man chose to disobey God.

*So when the woman saw that the tree was good for food, that it was pleasant to the eyes, and a tree desirable to make one wise, she took of its fruit and ate. She also gave to her husband with her, and he ate* (Genesis 3:6).

Man's rebellion brought death and corrupted God's perfect creation. Because of man's sin, death became a part of this world and has been ever since.

But God had a plan. He had a plan to defeat death. A plan that would give us the opportunity to be with Him forever in heaven after we die. He sent His Son, the Lord Jesus Christ, to walk this earth as a man. Jesus Christ lived a sinless life so that He could take the punishment that is rightfully ours. He did this when He was crucified, dying on a cross. Then three days later He came back to life, forever defeating death and purchasing for all who trust in Him a life in heaven forever.

*For God so loved the world that He gave His only begotten Son, that whoever believes in Him should not perish but have everlasting life* (John 3:16).

But how can we have everlasting life? We are not worthy of it! God's Word tells us that we are all sinners and therefore deserve to die.

*. . . for all have sinned and fall short of the glory of God* (Romans 3:23).

*. . . the wages of sin is death, but the gift of God is eternal life in Christ Jesus our Lord* (Romans 6:23).

Eventually, our bodies will die, but we can defeat death. Not physical death, but spiritual death. You see, physical death is not the end. We all have souls that will live on into eternity. The choice we have to make is where we will spend eternity, with God or apart from Him forever.

Each of us will die physically and then be judged. But if we have trusted in Christ, we have nothing to fear. Jesus already paid the price for our sins!

*And as it is appointed for men to die once, but after this the judgment, so Christ was offered once to bear the sins of many. . .* (Hebrews 9:27–28).

God promises that all who, repenting of their sins, trust in Jesus, will be saved.

*. . . that if you confess with your mouth the Lord Jesus and believe in your heart that God has raised Him from the dead, you will be saved* (Romans 10:9).

Where will you spend eternity — with God or apart from Him?

# GLOSSARY

**Abduction** — movement away from the midline of the body (For example, you abduct your thigh if you kick a ball with the side of your foot while standing.)

**Acetabulum** — the cup-like depression in the pelvic girdle in which the rounded head of the femur moves; the hip socket

**Acromegaly** — a disorder in which the pituitary gland of an adult produces an excess of growth hormone (GH). This can result in physical abnormalities such as swelling of the hands and feet, enlargement of brow ridges and protrusion of the forehead, among other changes. Excessive growth hormone in a child would produce giantism, not acromegaly.

**Actin** — one of two myofilament types involved in muscle contraction. The thin myofilaments are actin.

**Action potential** — series of events that results in a change in the membrane potential from negative to positive and back again. It is also called a nerve impulse.

**Adduction** — movement toward the midline of the body (For example, you adduct both of your thighs if you squeeze them together.)

**ADP (adenosine diphosphate)** — one of the molecules involved in energy production in a cell. When ATP (adenosine triphosphate) loses a phosphate group, energy is released and ADP is produced. The addition of a phosphate group to ADP again produces ATP (like "charging a battery").

**Aerobic respiration** — an oxygen-requiring series of reactions by which a cell produces energy

**Afferent** — meaning "bringing toward." The sensory division of the peripheral nervous system is called the afferent division, because it brings sensory input to the central nervous system.

**All-or-none law** — the principle that in a given muscle fiber either all the sarcomeres contract or none of them do.

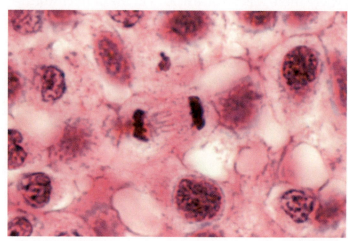

*Mitosis (anaphase-telophase) in the granulosa layer of an ovarian secondary follicle*

**All-or-none phenomena** — as pertains to the nervous system, there is either a full action potential or there is no action potential at all

**Alveoli** — the plural of alveolus.

**Alveolus** — microscopic air sac that is the endpoint of the respiratory system. It is the site of gas exchange between the air and the blood and the alveolar capillaries.

**Amino acid** — molecules that are the building blocks of proteins. There are 20 different amino acids used to make proteins in the human body.

**Amnesia** — loss of memory

**Anaerobic respiration** — a series of reactions by which a cell produces energy without using oxygen

**Anaphase** — the shortest phase of mitosis. During anaphase daughter chromosomes are pulled apart and move to opposite sides of the cell.

**Anatomical position** — standing with the palms facing forward. This position serves as a reference for describing the orientation of the body's parts. There are other positions with names too. For instance, prone means lying face down, and supine means lying face up. (With reference to the anatomical position, please see anterior, distal, inferior, lateral, medial, posterior, pronate, proximal, superior, and supinate.)

# GLOSSARY

**Anatomy** — the study of the body's parts and how they are put together

**Anterior** — the front of the body (For example, the nose is on the anterior part of the head.)

**Aorta** — the largest artery in the body. The aorta carries blood from the left ventricle into the systemic circulation (out to the body's tissues).

**Aortic valve** — the valve between the left ventricle and the aorta. Blood flows from the left ventricle through the aortic valve into the aorta.

**Appendicular skeleton** — the portion of the skeleton consisting of the upper and lower limbs as well as the bones that connect them to the axial skeleton

**Arch of foot** — either of several arch-shaped curves in the sole of the foot that help the foot bear the body's weight most efficiently and adjust to the changes needed to walk most efficiently; arches are formed by the relative positions of the foot and ankle bones, and they are held in place by many tendons and ligaments.

**Arterioles** — the smallest arteries. Arterioles lead into the capillaries.

**Artery** — vessel that carries blood away from the heart.

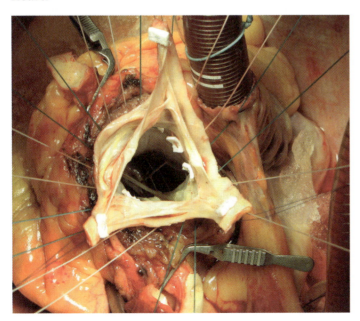

*Surgery on an aortic valve to help protect it from possibly rupturing.*

**Arthritis** — a joint disorder that involves inflammation of one or more joints

**Arthroscopy** — a surgical procedure in which a telescope-like instrument is used to look inside a joint and even to introduce long-handled precision instruments to repair tears in the soft tissues

**Association areas** — cortical areas of the brain that receive input from many other regions of the brain. These areas are responsible for integration of the input and determining appropriate outputs.

**ATP (adenosine triphosphate)** — one of the molecules that stores readily available energy in the cell. When a phosphate group is added to ADP (adenosine diphosphate), ATP is produced. When a phosphate group is released from ATP, ADP is produced and energy is released (like "discharging a battery")

**Atrioventricular bundle (Bundle of His)** — the portion of the cardiac conduction system that conducts the signal from the AV node to the interventricular septum.

**Atrioventricular node (AV node)** — the portion of the cardiac conduction system that is the first step in sending the signal to the ventricles. In the AV node, the signal is delayed approximately 0.1 second to allow the atria to complete their contraction before the ventricles contract.

**Autonomic nervous system** — division of the peripheral nervous system consisting of motor fibers carrying impulses to smooth muscle, cardiac, muscle, and glands. Also called the involuntary nervous system.

**Autonomic reflex** — reflexes triggering smooth muscle and glands. With an autonomic reflex, there is no consciousness of what happened.

**Autorhythmic cells** — special cardiac cells that can generate electrical impulses without any outside stimulus.

**Axial skeleton** — the portion of the skeleton consisting of the skull, vertebral column, and the ribs

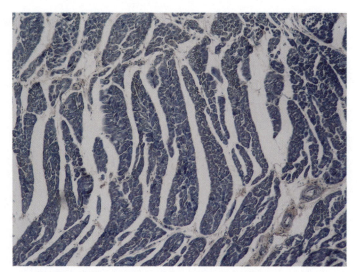
*Cardiac muscle*

**Axon** — the portion of the neuron that carries the nerve impulse away from the cell body

**Axon terminals** — The most distal portion of the axon. The region where neurotransmitters are released.

**Baroreceptor** — special sensory cells found primarily in the aorta and larger arteries. The cells are sensitive to being stretched, thus they are ideal for detecting changes in blood pressure.

**Bipedal** — walking upright on two legs

**Bipolar neuron** — a type of neuron that has only two processes: one axon and one dendrite

**Blood pressure** — the pressure of blood inside a blood vessel.

**Bradycardia** — a heart rate less than 60 beats per minute.

**Brain** — one of the two parts of the central nervous system. The brain is the master control center of the human body.

**Brain stem** — the portion of the brain between the diencephalon and the spinal cord. It consists of the midbrain, the pons, and the medulla oblongata.

**Bronchi** — the main passageways of air into the lungs. The trachea branches to form the right and left bronchi.

**Bronchiole** — smaller branches of the bronchial tree that do not have cartilage in their walls.

**Calcitonin** — a hormone produced by the thyroid gland. It decreases the activity of osteoclasts and reduces the amount of calcium released into the blood.

**Callus** — repair tissue that forms at the site of a bone fracture

**Cancellous bone** — another term for spongy bone

**Capillary** — the smallest kind of blood vessel in the body. The wall of a capillary consists of a single layer of endothelial cells.

**Carbonic anhydrase** — an enzyme that accelerates the conversion of water and carbon dioxide into carbonic acid. It also assists with the conversion of carbonic acid back into water and carbon dioxide.

**Cardiac cycle** — the steps involved in filling the heart's chambers and pumping the blood.

**Cardiac muscle** — the muscle of the heart. One of the three types of muscle. The myocardium is composed primarily of cardiac muscle.

**Cardiac output** — the amount of blood pumped by the heart in one minute.

**Cardiac reserve** — the difference between the cardiac output at rest and cardiac output during maximal exertion.

**Cardiac tamponade** — a serious medical condition is which the pericardial sac has accumulated so much fluid that the heart cannot squeeze properly.

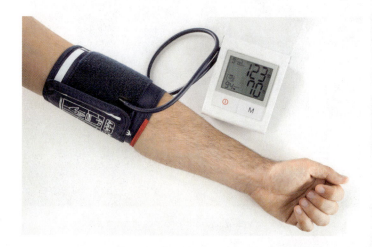

**Cell** — the most basic structural and functional unit of a living organism, such as the human body. A cell generally consists of three parts: the nucleus, the cell membrane, and the cytoplasm.

**Cell membrane** — the cell's wrapper. The cell membrane separates everything outside the cell from everything inside the cell, and it regulates what can and cannot go across. It consists of two layers of molecules called phospholipids. It is therefore called a phospholipid bilayer, and many important molecules are embedded in this "phospholipid sandwich."

**Cellular respiration** — the process inside cells in which nutrients are metabolized into energy.

**Central canal** — a channel in an osteon containing blood vessels and nerves. It is sometimes called a Haversian canal.

**Central nervous system (CNS)** — the portion of the nervous system composed of the brain and spinal cord

**Centrioles** — a pair of L-shaped cellular organelles involved in organizing microtubules to guide chromosomal movements during mitosis.

**Cerebellum** — the posterior portion of the brain. It is involved with balance and position sense.

**Cerebrospinal fluid** — the fluid found in the subarachnoid space surrounding the brain

**Chemical synapse** — a type of synapse that functions by the release and uptake of chemical messengers, called neurotransmitters. This is the most common type of synapse.

**Chemoreceptor** — special cells sensitive to changes in the levels of certain chemicals or substances in the body.

**Chondrocyte** — a mature cartilage cell

**Chordae tendineae** — bands of fibrous tissue that connect the papillary muscles in the ventricles to the tricuspid and mitral valves. These bands help prevent the valves from being pushed backward into the atria during ventricular systole.

**Chromatid** — one copy of a duplicated chromosome

**Chromosome** — tightly packed portions of DNA found in the nucleus of cells. Chromosomes are generally only visible during cell division.

**Cilia** — tiny hairlike structures on cells. These hairs help sweep mucus and debris out of the respiratory system.

**Collagen** — the primary structural protein of connective tissue

**Compact bone** — the dense outer layer of most bones

**Compound fracture** — fracture in which the broken bone protrudes through the skin

**Concentration gradient** — the tendency of molecules to move from areas of higher concentration to areas of lower concentration

**Cones** — photoreceptors in the retina responsible for color vision. There are three types of cones: blue, green, and red.

**Connective tissue** — tissue that helps provide a framework for the body

**Continuous conduction** — the conduction of an action potential along an unmyelinated axon. This involves step-by-step transmission along the length of the axon.

**Contractility** — how hard the cardiac muscle can contract when it is stretched to a certain point.

**Cranial nerves** — nerves that emerge directly from the brain and pass through holes in the cranium. There are 12 pairs of cranial nerves.

**Cristae** — a fold in the inner membrane of a mitochondrion

**Cytoplasm** — the fluid inside a cell plus all of the organelles, except for the nucleus

**Cytoskeleton** — the network of microtubules and microfilaments inside a cell

**Cytosol** — the fluid portion of the cytoplasm

**Decussate** — crossing over to the opposite side

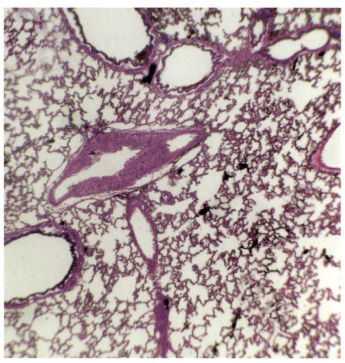

*Normal Lung Tissue. The larger holes are blood vessels. The smaller spaces are alveoli.*

**Dendrite** —- the site where the neurons receive inputs. The signals are then carried toward the cell body.

**Deoxyhemoglobin** — hemoglobin that is not bound to any oxygen molecules.

**Depolarization** — a decrease in the membrane potential, where the membrane potential becomes less and less negative, and then positive

**Dermatome** — The region of the body that provides sensory input to a particular spinal nerve (or segment)

**Diaphysis** — the shaft or midsection of a long bone

**Diencephalon** — the portion of the brain between the cerebrum and the brainstem. It consists of the thalamus and the hypothalamus.

**Distal** — located farther from the center of the body than something else (For example, the hand is distal to the elbow.)

**Diastole** — the period of time when a heart chamber is relaxing.

**Diastolic blood pressure** — the lowest pressure in the arterial system during left ventricular diastole

**DNA (deoxyribonucleic acid)** — a molecule that contains all the genetic information that is needed for the development and function of a living organism, such as the human body

**DNA polymerase** — an enzyme that assembles DNA by linking nucleotides together

**Dorsal respiratory group (DRG)** — a group of specialized cells located in the medulla. Signals are sent from the DRG to the diaphragm and intercostal muscles to stimulate them to contract. This triggers inspiration.

**Dorsal root** — posterior root of a spinal nerve that contains only sensory axons bringing input from sensory receptors throughout the body

**Double-helix** — the structure formed by DNA. It looks like a twisted ladder.

**Efferent** — meaning "carrying out." The motor division of the peripheral nervous system is called the efferent division, because it takes nerve impulses away from the central nervous system.

**Ejection fraction** — the percentage of the volume of blood in the left ventricle ejected with each beat. It is normally 60–70%.

**Electrical synapse** — a type of synapse where the action potential is transmitted directly to the next cell

**End diastolic volume** — the amount of blood in the ventricle when it is full (at the end of diastole).

**End systolic volume** — the amount of blood remaining in the ventricle after it contracts.

**Endocardium** — the inner layer of the heart.

**Endocytosis** — a process of bringing material—usually large molecules or other things too large to be simply transported across the cell membrane—into a cell. The cell membrane folds itself around the needed material and then pinches off, forming a new vesicle inside the cell.

**Epicardium** — the outermost layer of the heart.

**Epiglottis** — a flap made of cartilage located at the entrance of the larynx. When swallowing, the epiglottis closes and prevents food from going into the trachea.

**Epiphyseal plate** — the growth plate in a bone

**Epiphysis** — the rounded end (joint end) of a long bone

**Epithelial tissue** — the tissue that lines body cavities or covers surfaces

**Erythrocyte** — a mature red blood cell

**Evolution** — the belief that all life, including the human body, developed on its own as a result of chemical reactions over million of years

**Exhalation** — the flow of air out of the lungs. It is also called "expiration."

**Exocytosis** — a process of releasing material from inside the cell. A vesicle inside the cell merges with the cell membrane, releasing cellular products into the extracellular fluid.

**Expiration** — the flow of air out of the lungs. It is also called "exhalation."

**Extension** — movement that increases the angle between body parts (For example, you extend your elbow to straighten your arm.)

**External respiration** — the process of moving air from the environment into and out of the lungs

**Extracellular fluid** — the fluid outside a cell

**Fibroblast** — a cell that makes collagen and the extracellular matrix.

**Flexion** — movement that decreases the angle between body parts (For example, you flex your elbow when you bend your arm.)

**Foramen magnum** — the large opening in the bottom of the skull through which the brainstem continues into the spinal cord

**Fracture** — a broken bone

**Gene** — a segment of DNA that codes for a specific protein

**General senses** — senses that do not require a specific sensory organ. Input for the general senses comes from basic sensory receptors located throughout the body.

**Genome** — the complete set of genetic material of an organism. The human genome contains all the information needed to build the human body. The human genome has over 3 billion DNA base pairs.

**Gigantism** — a condition resulting from excessive production of growth hormone (GH) during childhood, while bones are still growing in length. Also called giantism, it can result in persons reaching excessive heights, often well over 7 feet.

**Glandular epithelium** — specialized form of epithelial tissue that forms the body's many glands

**Golgi apparatus** — an organelle responsible for the packaging and transport of many substances in a cell

**Gout** — a form of arthritis due to excessive amounts of uric acid in the joint capsule

**Graded potential** — type of membrane potentials generated in dendrites and cell bodies. This potential varies with the strength of the stimulus.

**Gray matter** — regions of the central nervous system consisting mostly of neuron cell bodies and nonmyelinated axons

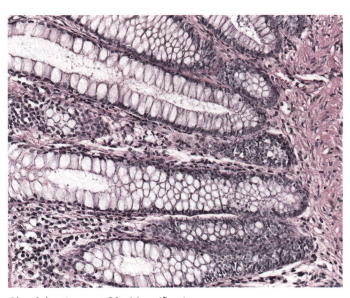

*Glandular tissue at 20x Magnification*

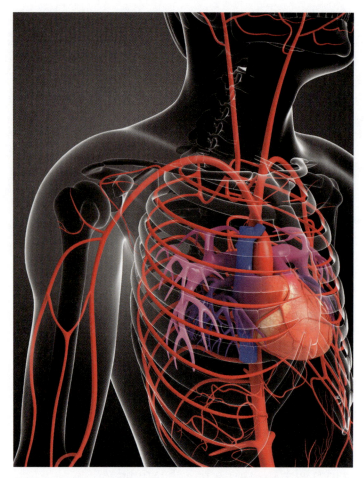

**Ground substance** — the material that fills the space between cells in connective tissue. Also called matrix.

**Growth hormone (GH)** — one of the hormones secreted by the pituitary gland. Among other things, GH helps regulate bone growth.

**Heart murmur** — an abnormal gurgling or rushing sound that occurs as the heart beats. While abnormal, a murmur is not necessarily an indication of heart disease. Many heart murmurs are benign (harmless).

**Heart rate** — the speed at which the heart is beating. It is most often reported as the number of beats per minute.

**Hematoma** — a mass of clotted blood; a bruise

**Hematopoesis** — the process of producing new blood cells

**Hemoglobin** — an iron-containing protein found in red blood cells. It is involved in the transport of oxygen to the tissues.

**Hilum** — located on the medial surface of the lung. It is where the bronchi and blood vessels enter the lung.

**Homeostasis** — maintaining various processes and conditions within appropriate limits. For example, blood sugar, body temperature, and blood pressure need to be neither too high nor too low. Many homeostatic mechanisms maintain equilibrium among the body's systems.

**Homunculus** —– means "little man." It is a mapping of the cortical regions of the brain based on the various parts of the body they either control (motor function) or receive input from (sensory function).

**Hydrophilic** — attracted to water (literally meaning "water-loving")

**Hydrophobic** — something that avoids water, like oil; (literally "water-fearing")

**Hyper-polarization** — an increase in membrane potential beyond the resting membrane potential

**hypertension** — a blood pressure consistently above 140/90; frequently called "high blood pressure."

**Inferior** — below (For example, the inferior vena cava is a large vein that brings blood from the lower part of the body back to the heart.)

**Insulin** — a hormone produced by the pancreas that helps regulate the amount of sugar in the blood

**Integration** — one of the functions of the nervous system. This is the recognition, analysis, and processing of various sensory inputs that results in an appropriate response.

**Internal respiration** — the process of exchanging oxygen and carbon dioxide between the environment and the body's cells. It involves gas exchange between the air and the blood in the alveolar capillaries and the gas exchange between the blood in the systemic capillaries and the body's cells.

**Interneurons** — neurons located in the CNS between the sensory and motor neurons. They are also called association neurons.

**Interphase** — the time in the cell cycle not directly involved with duplicating the cell

**Intracellular fluid** — fluid inside a cell; the cell's organelles and many important molecules are found in the intracellular fluid

**Keratin** — a structural protein found in hair and nails

**Lacunae** — spaces in compact bone where osteocytes are found

**Lamellae** — the rings of extracellular matrix found in compact bone

**Larynx** — the portion of the airway that connects the laryngopharynx to the trachea. It is often called the "voice box."

**Lateral** — to the side (For example, in the anatomical position, the radius is lateral to the ulna. Now you see the importance of the anatomical position. This would not be true in some other positions!)

**Lateralization** — functions performed by only one of the cerebral hemispheres and not both. These functions are said to be lateralized.

**Left ventricle** — the chamber of the heart responsible for pumping blood into the systemic circulation.

**Ligament** — dense connective tissue that binds bone to bone

**Lipid** — an organic molecule that does not dissolve in water (hydrophobic). Fats are the most common type of lipids.

**Lower respiratory system** — the trachea, the bronchi, and the lungs.

**Lumen** — the inner space of a blood vessel. The blood flows in the lumen of the vessel.

**Lysosome** — intracellular vesicle containing enzymes that can digest many kinds of molecules and debris

**Matrix** — the material that fills the space between cells in connective tissue. Also called ground substance.

**Medial** — toward the middle (For example, in the anatomical position, the big toe is medial to the little toe.)

**Meninges** — the three layers of connective tissue covering the brain and spinal cord. They are the dura mater, arachnoid mater, and the pia mater.

**Messenger RNA (mRNA)** — a type of RNA that is made during protein production. When a section of DNA is read (decoded), messenger RNA is produced.

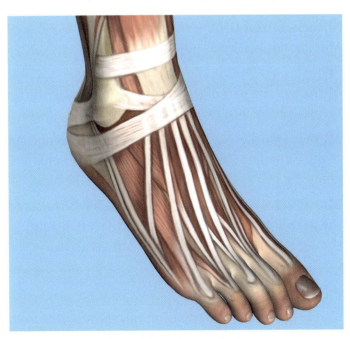

Ligaments and tendons of the ankle and foot

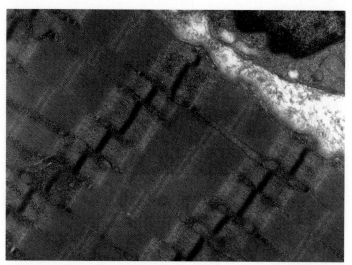
*Transmission electron microscope image of a thin longitudinal section cut through an area of human skeletal muscle tissue. Image shows several myofibrils, each with distinct banding pattern of individual sarcomeres.*

The mRNA then attaches to a ribosome where the information is read by molecules of transfer RNA (tRNA), and a protein is produced.

**Metaphase** — second phase of mitosis when the chromosomes reach the center of the cell

**Minute ventilation** — the amount of air moved into and out of the lungs in one minute.

**Mitochondria** — organelles inside a cell that generate and store energy

**Mitochondrial DNA** — the DNA of a mitochondrion. Mitochondrial DNA is inherited only from the mother.

**Mitosis** — the part of the cell cycle involved with dividing a cell into two daughter cells

**Mitotic spindle** — a structure inside a cell composed of microtubules. This array of microtubules helps guide the cell's chromosomes during cell division.

**Mitral valve** — the valve between the left atrium and the left ventricle. Blood flows from the left atrium through the mitral valve into the left ventricle.

**Mixed nerve** — nerves that possess both motor and sensory fibers

**Motor division** — this division carries impulses from the CNS out to the body. It is sometimes called the efferent (meaning "carrying out") division because it carries nerve impulses "away from" the CNS.

**Motor neuron** — neurons that transmit impulses away from the central nervous system

**Multipolar neurons** — the most common type of neuron. It has several processes (more than three) consisting of one axon and multiple dendrites

**Muscle atrophy** — loss of muscle mass due to disease or disuse

**Muscle fiber** — muscle cell

**Myelin** — a fatty substance that surrounds and electrically insulates axons of neurons

**Myocardium** — the middle layer of the heart. It is composed primarily of cardiac muscle.

**Myofibril** — rod-like structure made of myofilaments extending through the length of a muscle fiber.

**Myofilament** — actin or myosin

**Myosin** — one of two myofilament types involved in muscle contraction. The thick myofilaments are myosin.

**Muscle tissue** — tissue responsible for movement

**Nerves** — a structure composed of many things: bundles of axons, blood vessels, connective tissue, and lymphatic vessels.

**Nervous tissue** — tissue that is the primary component of the nervous system. The nervous system regulates and controls bodily functions.

**Neuroglia** — one of the two types of nervous tissue. These cells support and protect neurons.

**Neurons** — cells that transmit electrical signals. These cells are designed to respond to some type of stimulus

**Neurotransmitter** — the molecules that carry the signals across the synaptic cleft

**Nuclear membrane** — the membrane the surrounds the nucleus of the cell

**Nucleotides** — the molecules that are the building blocks of DNA and RNA. The nucleotides adenine, cytosine, guanine, and thymine are found in DNA.

The nucleotides adenine, cytosine, guanine, and uracil are found in RNA.

**Nucleus** — the control center of the cell. The nucleus contains DNA.

**Odorant** — a stimulatory chemical that triggers a smell response

**Onhalation** — the flow of air into the lungs. It is also called "inspiration."

**Onspiration** — the flow of air into the lungs. It is also called "inhalation."

**Organ** — a group of tissues that have a particular function

**Organelle** — a structure within a cell that has a specific function

**Osteoarthritis** — a form of arthritis caused by deterioration of the joint cartilage

**Osteoblast** — a cell that builds new bone

**Osteoclast** — a cell that breaks down bone

**Osteocyte** — a mature bone cell

**Osteon** — the basic unit of compact bone

**Osteoporosis** — a bone disease that is characterized by a loss of bone mass

**Oxyhemoglobin** — hemoglobin that is bound to one or more oxygen molecules.

**Parasympathetic division** — the portion of the nervous system geared to support the rest and recuperation activities of the body

**Pericarditis** — inflammation of the pericardium.

**Pericardium** — the double-walled sac surrounding the heart.

**Periosteum** — the membrane that cover the outer portion of bone

**Peripheral nervous system** — the portion of the nervous system outside of the central nervous system. It consists of the cranial nerves that extend from the brain, and the spinal nerves that extend from the spinal cord.

**Pharynx** — the portion of the respiratory system beginning at the rear of the nasal cavity and extending down to the larynx. It is commonly called the "throat."

**Phospholipid** — the primary component of cell membranes. It is composed of a hydrophilic head and two hydrophobic tails.

**Phospholipid bilayer** — a term to describe the plasma membrane because it is composed of two layers of phospholipids

**Physiology** — the study of how the parts of the body function

**Plasma membrane** — another name for the cell membrane X

**Pleura** — the double-walled membrane that covers the lungs. It is composed of the visceral pleura and the parietal pleura.

**Pneumotaxic area** — an area of the pons that coordinates the switch between inspiration and expiration. It is also known as the Pontine respiratory group (PRG).

**Pontine respiratory group (PRG)** — an area of the pons that coordinates the switch between inspiration and expiration. It is also known as the pneumatic area.

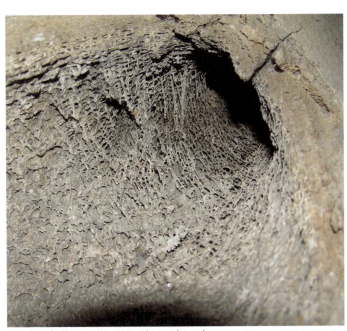

*Inside of a bone showing the trabecular structure*

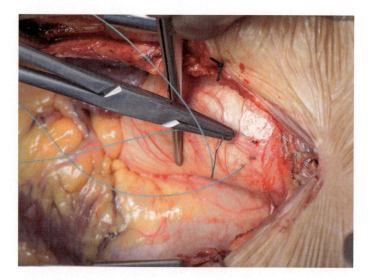

**Posterior** — the back side of the body (For example, the vertebral bones are on the posterior part of the torso.)

**Preload** — the amount that cardiac muscle is stretched by the blood in it before contracting.

**Pronate** — to turn the hand (or the arm) so that the palm faces backward or downward; (also, to turn the foot so that the weight rests on the medial part)

**Prophase** — first phase of mitosis when the DNA condenses into chromosomes and centrioles separate by moving along newly formed microtubules

**Protein** — an organic compound made up of amino acids

**Proximal** — located nearer to the center of the body than something else (For example, the knee is proximal to the foot.)

**PTH (parathyroid hormone)** — a hormone secreted by the parathyroid glands. PTH stimulates osteoclasts to break down bone, thus increasing the amount of calcium in the blood.

**pulmonary valve** — the valve between the right ventricle and the pulmonary artery. Blood flows from the right ventricle through the pulmonary valve into the pulmonary artery.

**pulse pressure** — the difference between the systolic blood pressure and the diastolic blood pressure.

**Repolarization** — the return of the resting membrane potential to its normal level after depolarization

**Resting membrane potential** — the electrical potential across the neuron membrane. It is typically around -70mV.

**Rheumatoid arthritis** — a form of arthritis due to the body's immune system attacking joint structures

**Ribosome** — intracellular organelle where proteins are made

**Rickets** — a bone disease in children caused by Vitamin D deficiency

**right ventricle** — the chamber of the heart responsible for pumping blood into the pulmonary circulation.

**Rigor mortis** — stiffening of the body after death

**RNA (ribonucleic acid)** — single-stranded molecule similar to DNA; it is involved in executing the instructions for protein synthesis found in DNA. See transfer RNA and messenger RNA.

**Rods** — photoreceptors in the retina responsible for vision in dim light

**Rotation** — movement around an axis, like turning your head from side to side.

**Rough endoplasmic reticulum** — series of tubes and membranes connected to the nuclear membrane. Rough endoplasmic reticulum is covered with ribosomes and is involved with protein production.

**Saltatory conduction** — the type of nerve conduction that occurs along a myelinated axon. Here action potentials occur only at the gaps in the myelin sheath.

**Sarcomere** — the simplest contractile unit of a muscle cell; each muscle cell (fiber) contains many sarcomeres.

**Sensory division** — this division carries impulses from the skin and muscles as well as from the major organs in the body to the central nervous system. It is sometimes called the afferent (meaning "bringing toward") division because it carries nerve impulses "to" or "toward" the CNS.

**Sensory neurons** — neurons that carry impulses triggered by sensory receptors into the central nervous system

**Sesamoid bone** — a bone embedded in a tendon or a muscle; the patella (kneecap) is the largest one.

**Sinoatrial node (SA node)** — a group of autorhythmic cells in the upper portion of the right atrium. The SA node is the heart's main pacemaker.

**Skeletal muscle** — striated muscle that is attached to the bone of the skeleton

**Sleep** — a state in which an individual achieves a degree of unconsciousness from which he or she can be aroused

**Smooth endoplasmic reticulum** — a series of tubes and membranes connected to the nuclear membrane. Smooth endoplasmic reticulum is not covered with ribosomes. It is involved with the production of fats and certain hormones.

**Smooth muscle** — non-striated muscle found in the wall of most hollow organs of the body

**Soma** — the cell body of the neuron

**Somatic nervous system** — the division of the nervous system that sends signals to muscles that we can consciously control

**Somatic reflex** — a reflex that results in contraction of skeletal muscle

**Special senses** — senses dependent on special types of receptors and are confined in organs specifically designed for them

**Spinal cord** — one of the major portions of the central nervous system. It is located in the vertebral canal.

**Spongy bone** — a type of bone that is less dense that compact bone. It is found primarily at the end of soft long bones.

**Sprain** — a tear or partial tear in the ligaments in the joint

**Stimulus** — a change in the environment that triggers a neuron or a receptor

**Striated muscle** — muscle tissue in which the orderly repeating arrangement of sarcomeres make it look striped. Skeletal muscle is striated; smooth muscle is not.

**Stroke volume** — the amount of blood pumped with each beat.

**Sub threshold** — a depolarization not reaching the threshold level. It will not trigger a nerve impulse.

**Superior** — above (For example, the superior vena cava is a large vein that brings blood from the head, neck, and arms—the upper parts of the body—back to the heart.)

**Supinate** — to turn the hand (or the arm) so that the palm faces forward or upward (also, to turn the foot so that the weight rests on the lateral part)

**Surfactant** — a detergent-like substance found in alveolar fluid. It helps keep the alveoli from collapsing.

**Synapse** — the area where a neuron communicates with another neuron or an effector cell, such as a muscle cell.

**Synovial fluid** — lubricating fluid produced by the synovial membrane inside joints that move

**Systole** — the period of time when a heart chamber is contracting.

**Systolic blood pressure** — the highest pressure reached in the arterial system during left ventricular systole.

**Tachycardia** — a heart rate greater than 100 beats per minute.

**Tastant** — a stimulatory chemical that triggers a taste response

**Telophase** — the last phase of mitosis in which the nuclear membranes re-form and the cell completes its division into two cells

**Threshold** — the level of membrane depolarization at which an action potential is triggered

**Tissue** — a group of cells that perform similar or related functions. There are four basic tissue types: epithelial, muscle, connective, and nervous.

**Trabeculae** — the functional units of spongy bone

**Transfer RNA (tRNA)** — a form of RNA that is responsible for transporting amino acids during protein production

**Tricuspid valve** — the valve between the right atrium and the right ventricle. Blood flows from the right atrium through the tricuspid valve into the right ventricle.

**Tunica externa** — the outermost layer of a blood vessel. It is made of collage and elastic fibers.

**Tunica intima** — the innermost layer of a blood vessel. It is composed of a smooth layer of tissue called the endothelium.

**Tunica media** — the middle layer of a blood vessel. It consists mainly of smooth muscle and elastic fibers.

**Unipolar neuron** — type of neuron that has only one process extending from the cell body

**Upper respiratory system** — the part of the respiratory system above the level of the chest. It includes the nose, nasal cavity, the sinuses, the pharynx, and the larynx.

**Vasoconstriction** — when the lumen of a blood vessel gets smaller. This is the result of contraction of the smooth muscle in the blood vessel. This increases the pressure in the lumen.

**Vasodilation** — when the lumen of a blood vessel gets larger. This is the result of relaxation of the smooth muscle in the blood vessel. This decreases the pressure in the lumen.

**Vein** — vessel that carries blood back to the heart.

**Ventilation** — the movement of air into and out of the lungs by means of inspiration and expiration.

**Ventral respiratory group (VRG)** — a group of specialized cells located in the medulla. Signals are sent from the VRG to the diaphragm and intercostal muscles to stimulate them. The VRG can stimulate both inspiration and forced expiration.

**Ventral root** — anterior root of a spinal nerve that contains axons of motor neurons carrying nerve signals from the CNS out to muscles and glands

**Venule** — the smallest veins. They carry blood from the capillaries to the larger veins.

**Vesicle** — an organelle inside the cell consisting of fluid enclosed in a lipid bilayer (similar to the plasma membrane)

**Vitamin D** — a molecule that aids the absorption of calcium, phosphate, and iron from the food we eat

**White matter** — regions of the central nervous system consisting mostly of myelinated axons

## Photo and Illustration Credits

Shutterstock.com; iStock.com; Bill Looney; Dreamstime.com; NASA; Pubic Domain; Science Photo Library; Wikimedia Commons: Images from Wikimedia Commons are used under the CC-BY-3.0 (or previous) license, CCA 2.5, CC-BY-CA 3.0, CC BY-SA 4.0, CC-BY-SA 2.1 JP, or GNU Free documentation License Version, 1.2; Itayba; Blausen.com staff; Nephron; Bolzer et al; Rice University; Patricia Curcio; Andrea Mazza; Ganimedes; Rollroboter; Daniel Ullrich; CardioNetworks; CDC/ James Gathany; OpenStax College; E.Faccio P.Saccheri; Patrick J. Lynch; Holly Fischer; Robert Bear and David Rintoul; Helixitta; Andrii Cherninskyi; Patrick J. Lynch; Michael Hawke MD; Welleschik; Mark Fairchild

# INDEX

Acetabulum .................................................................... 76, 302
acid reflux disease ............................................................. 173
acromegaly ..................................................................... 53, 302
action potential ........ 4, 207, 221-230, 281, 283, 302, 305-306, 314
Adam's apple ........................................................ 172, 175, 250
ADP ............................................................. 19-20, 89-90, 302-303
aerobic respiration ......................................................... 89-90, 302
afferent ........................................................ 203, 210, 283, 302, 313
all-or-none law ................................................................ 88, 92, 302
All-or-none phenomena ............................................................ 302
alveolar duct .......................................................................... 176
alveoli .... 163, 166, 176, 179-181, 186-190, 192-193, 302, 306, 314
alveolus ........................................................... 179-180, 189, 302
amino acid ....................................................................... 24-25, 302
amnesia ........................................................................... 255, 302
amplitude .......................................................................... 287-289
Amyotrophic lateral sclerosis ............................................... 260
anaerobic respiration .............................................................. 90, 302
anaphase ............................................................................. 29, 302
anatomical position ......................................................... 39, 302, 309
    anterior ............................................................... 39, 302-303
    distal ................................................................. 39, 302, 304, 306
    inferior ............................................................. 39, 302, 304, 306
    lateral .............................................................. 39, 302, 304, 306
    medial ............................................................. 39, 302, 304, 306
    posterior .......................................................... 39, 302, 304, 305
    proximal .......................................................... 39, 302, 305, 306
    superior ........................................................... 39, 302, 304, 306
anatomy.... 6, 10-11, 39, 47, 106, 109, 156, 165, 173, 175, 178, 188, 237-239, 257, 274, 281, 285, 287, 293, 303
Angina pectoris .................................................................... 128
Angiotensin II ..................................................................... 153-154
ankle sprain ........................................................................... 77
anterior lobe ......................................................................... 249
anterior white column ............................................................ 259
anterograde amnesia ............................................................ 255
aorta .... 110, 112, 119-120, 122-124, 126-127, 129, 133, 143-145, 150, 152, 154, 157-158, 160, 181, 194, 303-304
    abdominal aorta ................................................... 157, 160
    aortic arch ......................................................... 157-158, 268
    ascending aorta .................................................. 126, 157
    thoracic aorta ..................................................... 157
aortic valve ............................................... 120, 122, 126, 150, 154, 303
appendicular skeleton ..................................................... 63, 73, 303
arachnoid mater .......................................................... 235, 257, 309
Arch of foot .......................................................................... 303
arteriole ............................................................................... 152
artery .... 117, 119, 122-124, 127-129, 133, 135, 138, 142-143, 145, 149, 151, 157-160, 188, 250-252, 303, 312
    axillary artery ..................................................... 158
    brachial artery .................................................... 149, 159
    brachiocephalic trunk .......................................... 158
    bronchial artery .................................................. 158
    common carotid artery ......................................... 158
    common iliac artery ............................................. 160
    external carotid artery ......................................... 160
    femoral artery .................................................... 160
    inferior mesenteric artery ...................................... 160
    internal carotid artery .......................................... 160
    left subclavian artery ........................................... 158
    popliteal artery ................................................... 160
    posterior tibial artery ............................................ 160
    radial artery ....................................................... 149
    renal artery ........................................................ 160
    right subclavian artery .......................................... 158
    superior mesenteric artery .................................... 160
    ulnar artery ....................................................... 159
arthritis ................................... 57, 60-61, 67, 269, 303, 307, 311-312
Arthroscopy ..................................................................... 61, 303
association areas ........................................................ 243, 245, 303
asthma ............................................................... 108, 177, 182, 188
Astrocyte ............................................................................ 252
ATP .......................................... 19-20, 89-90, 93, 221, 225, 302-303
atrial diastole ....................................................................... 124
atrial systole ........................................................................ 124
Atrioventricular bundle (Bundle of His) .................................. 303
Atrioventricular node (AV node) ...................................... 131, 303
auditory association area ...................................................... 243
auricle .......................................................................... 285, 289
autonomic nervous system ................... 4, 205, 259, 273-274, 303
autonomic plexuses ............................................................. 276
autonomic reflex ............................................................. 271, 303
autorhythmic cells ......................................................... 130, 303, 313
axial skeleton ......................................... 63, 68, 72, 74, 76, 303
axon ........... 207-210, 213-214, 216-217, 222-228, 230, 275, 280, 304-305, 310, 313
axon hillock ......................................................................... 208
axon terminals ............................................................... 209, 227, 304
bacteriophages ..................................................................... 168
balance . 6, 17, 40-41, 57, 59, 68, 70, 79, 91, 101, 119, 125, 151-152, 220, 231, 246-247, 249, 266, 274, 279, 284, 290-291, 305
Baroreceptor ....................................................................... 304
basal cells .................................................................... 280, 282
Bell's palsy ......................................................................... 265
bicarbonate ion ................................................................... 193
Bipedal People ..................................................................... 79
Bipolar neuron ..................................................................... 304
blood brain barrier ......................................................... 252-253

blood flow.....119, 127-129, 145-147, 149, 151-153, 155, 158, 185, 250-252, 266, 276-277, 295
blood pressure..11, 91, 110, 123, 129, 132, 142, 149-155, 201, 231, 246-247, 249, 251, 266, 273, 276, 304, 306, 308, 312, 314
bone............... 35, 42-43, 45, 47-61, 65-72, 74-76, 78, 80, 86-87, 89, 92, 95-96, 101, 107, 169-170, 172, 234-235, 257, 268, 280, 285-287, 304-309, 311-314
bones ........... 3, 7-8, 10-11, 33, 35, 38, 42-59, 61-63, 65-71, 73-80, 83-85, 91, 95, 101, 115, 170-171, 186, 234, 257, 269, 286-287, 289, 303, 305, 307, 312-313
    cells ............................................................................49, 51
        osteoblast ............................................... 49-50, 311
        osteoclast ............................................... 49-50, 311
        osteocyte ............................................... 49-50, 311
    gross anatomy .......................................... 47, 178, 237-239, 257
    growth ................................................................................ 52-53, 308
    remodeling ...............................................................................55, 59
    skeletal bones
        hyoid bone ........................................................ 69, 170, 172
        lower limb ............................................................75-76, 100, 160
            calcaneus .................................................................78, 101
            femur ................................................. 47, 60, 64, 76, 302
            fibula ...........................................................................47, 76-77
            metatarsals ............................................................... 78
            pelvic girdle (pelvis)............................63, 75-76, 302
            phalanges.....................................................................75, 78
            tarsals...........................................................................77-78
            tibia .................................................................. 47, 64, 76-77
        skull............ 44-45, 47, 62-63, 65, 68-69, 79, 82, 94, 169, 236-237, 287, 289, 303, 307
            cranium.............. 68, 236, 250, 252-253, 259, 265, 305
            ethmoid bone ..................................................68, 282
            frontal bone ............................................................... 68
            mandible ................................................................... 69
            maxillary bones......................................................... 69
            occipital bone ........................................................... 68
            parietal bone ............................................................. 68
            sphenoid bone........................................................... 68
            temporal bone ..............................................68, 287-288
            zygomatic arches ..................................................... 69
        thoracic cage ................................................................... 72
            ribs ............. 47, 63, 68, 72-73, 96, 99, 185-186, 303
            sternum........................... 72, 74, 111-112, 174, 186
        upper limb............................................. 73-74, 84, 96-97, 158
            carpal bones .....................................................47, 75
            humerus ....................................... 47, 64, 74-75, 96-97
            metacarpal bones................................................... 75
            phalanges.......................................................................75, 78
            radius .............................................. 47, 61, 74-75, 309
            pectoral (shoulder) girdle....................63, 73-74, 97
                clavicle ...................................................................67, 74
                scapula ................................................. 74, 96-97
            ulna....................................................47, 61, 64, 74-75, 309
        vertebral column.. 43, 47, 63, 68, 70-71, 75-76, 260, 277, 303
            cervical.................................64, 67, 70-71, 260, 268-269
            coccyx ....................................................................70, 76
            lumbar................ 70-71, 238, 259-261, 269-270, 277
            sacrum .................................................... 70, 75-76

        thoracic.......................................................................70, 72
    types
        compact bone ..........47-48, 51-52, 58, 305, 309, 311, 313
        flat bones......................................................................46-47
        irregular bones ..........................................................46-47
        long bones ....................................... 46-47, 54, 91, 313
        sesamoid bones ............................................................ 47
        short bones.................................................................46-47
        spongy bone ........................ 49, 51-52, 58, 304, 313-314
bony labyrinth ............................................................................. 286
brachial plexus ...................................................................... 267-268
Bradycardia.......................................................................... 150, 304
brain ................... 4, 9, 11, 13, 35-36, 38, 45, 56, 63, 68-71, 84, 87, 109, 112, 116-119, 146, 152-155, 158, 181, 193-194, 202-203, 205, 207, 209, 211-212, 214, 217, 230-238, 240-245, 247-257, 259-264, 266, 269, 271-272, 275, 282-283, 287, 289, 291, 295, 303-306, 308-309, 311
brain dead............................................................................. 254
brain stem..............................................194, 214, 237, 247, 249, 304
brain waves....................................................................253-254
    alpha waves ............................................................ 254
    beta waves .............................................................. 254
    delta waves ............................................................. 253
    theta waves............................................................. 253
Brain, blood supply
    anterior cerebral artery ......................................250-251
    basilar artery ......................................................250-251
    circle of willis .....................................................250-252
    internal carotid artery ............................................. 250
    middle cerebral artery ............................................ 250
    vertebral artery....................................................... 250
Broca's area .......................................................... 242-243, 245
bronchi .....................112, 166, 174, 176, 178, 189, 304, 308-309
    lobar bronchi .......................................................... 176
    primary bronchus ................................................... 176
    secondary bronchi ................................................. 176
    tertiary bronchi ...................................................... 176
Bronchiole ............................................................................ 304
    respiratory bronchioles ...................................176, 179
    terminal bronchioles ........................................166, 176
Bundle of His .................................................................132, 303
calcitonin................................................................................57, 304
callus ................................................................................. 58-59, 304
cancellous bone .............................................................. 48-49, 304
carbaminohemoglobin............................................................. 193
carbon dioxide ...9, 36, 38, 41, 89-90, 109, 144, 162, 164, 166, 176, 179, 181, 188, 191-195, 231, 253, 304, 309
Carbon Dioxide Transport ................................................... 192
carbonic acid ..................................................................193, 304
Carbonic anhydrase ........................................................193, 304
carbon monoxide poisoning .................................................. 190
cardiac conduction system ....................... 130-132, 134, 139, 303
cardiac cycle .................................................................. 123-124, 304

# INDEX

cardiac muscle .......... 33-34, 84, 110-111, 115-116, 125, 130, 135, 137-138, 207, 273, 304-305, 310, 312
cardiac output ..................... 135-137, 139, 149, 153, 304
Cardiac reserve ..................................................... 137, 304
Cardiac tamponade................................................113, 304
Cardiopulmonary Resuscitation (CPR) .................................. 112
cardiovascular center ............................................... 152-153, 155
Carpal tunnel syndrome ................................................. 269
cataract................................................................. 294
cauda equina .......................................................... 258
cell ................... 11-24, 26-27, 29-30, 33-34, 37, 49-50, 83, 85-90, 106, 115-116, 120, 130, 132, 146, 179, 189-190, 197, 207-225, 227-230, 237-238, 247, 249, 252, 258-259, 261, 265, 274-275, 280-284, 295-296, 302-307, 309-311, 313-315, 325
    cycle
        anaphase ..............................................29, 302
        interphase .............................................26-27, 309
        metaphase .............................................29, 310
        prophase ..............................................27, 312
        telophase .............................................29, 314
    membrane ....... 11, 13-14, 16, 21, 88, 116, 209, 220-223, 225, 229-231, 305-307, 311
cell body ...................... 207-208, 210, 216, 275, 304, 306, 313-314
central canal ................................................... 51-52, 236, 305
central nervous system...... 4, 35, 68, 202-203, 205, 210-212, 214, 216, 233, 235, 237, 239-241, 243, 245, 247, 249, 251, 253, 255, 257, 259, 261, 263, 272, 302, 304-307, 310-311, 313, 315
central sulcus..........................................................238, 240
centrioles......................................... 21, 27, 29, 305, 312
centrosome...................................................20-21
cerebellum......................................... 237, 245, 247, 249, 305
cerebral cortex............152, 194, 238, 240-241, 249, 253, 261, 281
cerebrospinal fluid...................... 212, 235-236, 257, 305
cerebrum ..................... 237-240, 242-243, 245, 248, 306
cerumen.................................................................. 285
cervical nerves..................................................... 258, 267
cervical plexus.......................................................... 267
channel............................16, 128, 219, 223, 227, 230, 305
chemical synapse.................................................228, 305
Chemoreceptor......................................................... 305
chimp DNA ............................................................... 25
Chondrocyte............................................................. 305
Chordae tendineae ..................................... 120-122, 305
choroid...........................................................237, 293-296
choroid plexus ......................................................... 237
chromatid................................................. 26, 29, 305
chromosome ........................................ 24, 26, 29, 305
cilia ......................................... 167, 174-175, 212, 280-281, 305
Ciliary muscles ......................................................... 294
ciliary processes .................................................. 293-294
circumflex artery ....................................................... 127
coccygeal nerves....................................................... 258

cochlea ...................................................................286-287
collagen .................. 23, 34-35, 50, 58, 67, 145, 305, 307
Concentration gradient ........................................219, 305
cones ..........................................................295-297, 305
Congestive heart failure ............................................. 125
connective tissue..................................................... 30, 32, 34-35, 50, 52, 58, 65-67, 72, 78, 85, 87, 113-115, 132, 142, 175, 182, 207, 215-216, 234-235, 293, 305, 308-310
consciousness.................................... 190, 245, 250, 254-256, 303
Continuous conduction ........................................225, 305
contractility.................................... 83, 137-138, 153, 305
conus medullaris ....................................................... 257
cornea...........................................................292-293, 295, 297
corneal abrasion ........................................................ 292
coronary angioplasty .................................................. 129
coronary artery bypass surgery................................... 129
coronary artery disease............................. 128-129, 135, 138
coronary circulation ............................................. 126-127
corpus callosum.................................................. 238-239, 245
cranial nerves ........... 4, 69, 203, 262-264, 266, 275, 283, 305, 311
    abducens (VI) nerve...............................................265, 267
    accessory (XI) nerve .............................................265, 268
    facial (vii) nerve.................................................265, 267
    glossopharyngeal (IX) nerve...................................265, 268
    hypoglossal (XII) nerve .........................................265, 268
    oculomotor (III) nerve ..........................................265, 267
    olfactory (i) nerve ...............................................265, 266
    optic (ii) nerve ....................................................265-266
    trigeminal (v) nerve .............................................265, 267
    trochlear (IV) nerve .............................................265, 267
    vagus (X) nerve...................................................265, 268
    vestibulocochlear (viii) nerve .................................265, 268
cranial vault........................................................ 234-235
craniosacral division.................................................. 275
cranium ..................... 68, 234, 248, 250-251, 257, 263, 305
cribriform plate ...................................................264, 280
cricoid cartilage ..................................................172, 175
cricothyrotomy .......................................................... 175
crista ....................................................................... 291
crista ampullaris ....................................................... 291
Cristae ..................................................................... 305
cupola ..................................................................... 291
cytoplasm ..................... 11, 13-15, 17-19, 21, 27, 51, 207-208, 305
cytoskeleton ......................................................20-21, 305
cytosol ..................................... 14, 17, 219-220, 228, 305
Decussate ..........................................................248, 305
dendrite .............................. 208, 210, 216, 227, 280, 282, 304, 306
Deoxyhemoglobin .............................................. 189-190, 306
depolarization ...............................................222-224, 306, 312-314
dermatome .......................................................269-270, 306
desmosome............................................................... 116

## INDEX

diaphragm .... 37, 110, 112, 157, 160, 174, 178, 185-187, 193-194, 260, 267-268, 306, 315
diaphysis ................................................................. 47, 49, 53-54, 306
diastole ................................ 118, 123-124, 127, 137, 150-151, 306
diastolic blood pressure .......................................... 150, 306, 312
diencephalon ............................................. 237, 245, 247, 304, 306
DNA ............................................ 14, 20-27, 29, 73, 305-307, 309-312
DNA polymerase ................................................................. 306
dorsal horn ................................................................. 259, 271
dorsalis pedis ................................................................. 149
Dorsal respiratory group (DRG) ....................................... 193, 306
dorsal root ............................................ 206, 259-261, 269, 271, 306
dorsal root ganglion ............................................. 206, 259, 261
Double-helix ................................................................. 23, 306
dura mater ............................................................. 234-235, 257, 309
ear infection ................................................................. 170
efferent ................................................................. 203, 210, 306, 310
ejection fraction ............................................. 125-126, 136, 306
electrical gradient ................................................................. 219
Electrical synapse ................................................................. 228, 306
electroencephalogram ................................................................. 253
electrolytes ................................................................. 14, 106
End diastolic volume ................................................................. 137, 306
endocardium ............................................. 114-115, 144, 306
endocytosis ................................................................. 16, 306
endolymph ................................................................. 286, 291
endoneurium ................................................................. 216
endothelium ................................................................. 12, 144, 146, 314
End systolic volume ................................................................. 137, 306
Epicardium ................................................................. 114, 306
epiglottis ................................................................. 172-173, 307
epinephrine ............................................. 138-139, 153, 230
epineurium ................................................................. 216
epiphyseal plate ................................................................. 53-55, 307
epiphysis ................................................................. 48, 53-54, 307
epithelial tissue ................................................................. 30-32, 207, 307
Erythrocyte ................................................................. 307
Eustachian tube ................................................................. 170, 285-286
evolution .............. 9-10, 23, 25, 41, 71, 79, 196-197, 256, 295, 307
exhalation ................................................................. 178, 185-186, 307
exocytosis ................................................................. 16, 18, 228, 307
expiration .................... 175, 177-178, 185-187, 194, 307, 311, 315
external auditory canal ................................................................. 285, 289
external ear ................................................................. 285, 289
external nares ................................................................. 166
extracellular fluid ................................................................. 15, 17, 219-220, 307
fascicles ................................................................. 85, 216
fetal hemoglobin ................................................................. 192
fibroblasts ................................................................. 34, 58
fight or flight ................................................................. 205, 276-277

fissure ................................................................. 238
flocculonodular lobe ................................................................. 249
Foramen magnum ................................................................. 69, 307
fovea centralis ................................................................. 294
fracture ................................................................. 50, 58-59, 61, 304-305, 307
Fracture types
    complete ................................................................. 59, 61
    incomplete ................................................................. 59
    simple ................................................................. 59, 61
    compound ................................................................. 59
frontal association area ................................................................. 243
frontal lobe ................................................................. 238, 243
gas exchange ............... 166, 176, 179, 181, 188-189, 192, 302, 309
gene ................................................................. 14, 22-23, 25, 307
general senses ................................................................. 279, 307
genome ................................................................. 21, 307
Gigantism ................................................................. 53, 307
Glandular epithelium ................................................................. 31, 307
glaucoma ................................................................. 293
glycocalyx ................................................................. 17
goblet cells ................................................................. 174
Golgi apparatus ................................................................. 18, 206, 208, 307
Gospel ................................................................. 4, 300-301
Gout ................................................................. 67, 307
graded potential ................................................................. 227, 229, 307
gray matter ............................................ 238, 258-259, 271, 275, 307
ground substance ......................... 32, 34-35, 50-51, 207, 308-309
growth hormone (GH) ................................................................. 302, 307-308
gustatory epithelial cell ................................................................. 282-284
gyrus ................................................................. 237, 240, 242
headache ................................................................. 190, 236, 248
hearing .... 4, 69, 170, 209, 243, 245, 247, 266, 278-279, 284, 287-289
heart failure ................................................................. 125, 182
Heart murmur ................................................................. 308
heart rate 111, 130-132, 135-139, 149-153, 155, 205, 231, 247, 249, 273, 276, 304, 308, 314
Heimlich maneuver ................................................................. 174-175
helicotrema ................................................................. 287, 289
Hematoma ................................................................. 58, 308
Hematopoesis ................................................................. 308
hemoglobin ................................ 13, 117, 189-193, 195, 306, 308, 311
    fully saturated ................................................................. 189
    partially saturated ................................................................. 190
hemorrhagic stroke ................................................................. 251
hemothorax ................................................................. 182
hilum ................................................................. 178, 180-181, 308
homeostasis ............... 3, 6, 40-41, 57, 151, 155, 231, 246, 274, 308
hormone ... 53, 56, 139, 153, 231, 246, 272, 302, 304, 307-309, 312
hydrophilic ................................................................. 15, 308, 311
hydrophobic ................................................................. 15, 308-309, 311

# INDEX

hyoid bone ..................................................... 69, 170, 172
hyperopic ............................................................... 297
hyper-polarization ................................... 224, 229, 308
hypertension ................................................... 154, 308
hypothalamus ................... 38, 237, 245-247, 272, 306
incus ............................................................... 286, 289
inferior lobe ............................................................ 249
inhalation ................................... 173, 178, 185, 188, 311
inspiration ........... 175, 177-178, 185-187, 193-194, 306, 311, 315
Insulin ............................................................... 13, 309
integration ............... 201-202, 205, 211, 243, 271-272, 303, 309
intercalated disc ..................................................... 116
interphase ..................................................... 26-27, 309
intervertebral foramen ........................................... 258
intracellular fluid ....................................... 15, 219, 309
intrinsic conduction system ........................ 130-131, 139
ischemic stroke ....................................................... 251
Joint types
    cartilaginous ........................................... 53, 65, 76
    fibrous ... 35, 48, 50-51, 65-67, 69, 113-114, 132, 258, 269, 305
    synovial joint ........................................................ 65
junk DNA ................................................................... 25
keratin ................................................................ 33, 309
lacunae ............................................................... 51, 309
lamellae .......................................................... 51-52, 309
Laryngitis ................................................................ 173
laryngopharynx ............................................... 169, 172, 309
larynx ...... 38, 165-166, 169, 172-174, 183, 242, 307, 309, 311, 314
lateralized ..................................................... 240, 242-245, 309
lateral sulcus ........................................................... 238
lateral white column .............................................. 259
learning ............... 6, 11, 109, 201, 212, 232, 243, 255-256, 272
left atrium ............................................ 118-120, 122, 127, 310
left bundle branch .................................................. 133
left coronary artery ................................................. 127
left ventricle ........ 118-120, 122, 126-127, 133, 137, 143, 145, 150, 153-154, 188, 303, 306, 309-310
ligament ......................................................... 67, 77, 309
lipid ............................................................ 15-16, 309, 315
longitudinal fissure ................................................. 238
Lower respiratory system ................................. 165, 309
"lub-dub" ................................................................ 121
lumbar nerves ......................................................... 258
lumbar plexus ......................................................... 268
lumen ..................................................... 144-146, 150, 309, 314
Lung Volumes ......................................................... 187
    inspiratory reserve volume ............................. 187
    residual volume ................................................ 187
    tidal volume ............................................... 187-188
    total lung capacity ............................................ 187
    vital capacity ..................................................... 187

Lysosome ........................................................... 18, 309
malleus ............................................................ 286, 289
matrix ............................................ 34-35, 50, 55, 58, 307-309
mean arterial pressure ........................................... 151
Mechanoreceptors ......................................... 273, 279
mediastinum ......................................... 111-113, 178
medulla oblongata ................... 152, 193, 247, 257, 283, 304
membranous labyrinth ................................... 286-287
memory ........................................... 246, 255, 272, 302
meningeal layer .............................................. 234-235
meninges .................................................. 234-236, 257, 309
meningitis ............................................................... 236
Messenger RNA (mRNA) ................................... 25, 309
Metaphase ....................................................... 29, 310
Microglia ................................................................. 212
midbrain .......................................................... 247, 249, 304
middle ear ................................................... 170, 285-286, 289
mind ................................................... 161, 245, 254-256, 299
minute ventilation .................................... 187-188, 310
mitochondria ...... 13, 19-21, 86, 88-89, 91, 110-111, 116, 208, 310
Mitochondrial DNA .......................................... 20, 310
mitosis ..................................... 27-29, 302, 305, 310, 312, 314
Mitotic spindle ................................................ 21, 310
mitral valve ................................................ 120, 122, 310
mixed nerve ............................................. 216, 265-266, 310
monosynaptic reflex .............................................. 271
motor division ................................... 203, 205, 306, 310
motor homunculus ................................................ 240
motor neuron ........................................... 260, 271, 310
motor output ............ 202-203, 205, 257, 260, 266, 271, 273
Movement terms
    abduction .......................................... 64, 96, 302
    adduction ........................................... 64, 96, 302
    extension ...................................... 64, 75, 211, 307
    flexion ................................................. 64, 75, 307
    rotation ........................................ 64, 67, 75, 312
mucous membrane ............................ 167-169, 173-174
mucus ................................................ 167-170, 174-175, 305
multiple sclerosis ............................................ 214, 291
Multipolar neurons ......................................... 210, 310
muscle ................... 3, 7, 11-13, 21, 30, 32-34, 45, 54, 80, 82-97, 99-102, 110-112, 114-116, 119, 123, 125-135, 137-138, 144-146, 150, 157, 177, 181, 185-186, 207-208, 227, 241-242, 245, 248-249, 260, 265, 271-273, 294, 302-305, 310, 312-314
muscles ................................................................ 3, 23, 32, 34-35, 38, 45, 48, 55-56, 67-69, 71, 74, 76, 79-86, 88-103, 110, 116, 121, 135, 145-147, 167, 171, 173, 185-187, 191, 193-194, 201-203, 205, 210, 215, 231, 240-242, 245, 248-249, 259-260, 265-268, 271-274, 276, 290, 294, 305-306, 313, 315
    atrophy ........................................... 93, 262, 310

contraction  7, 32-33, 83-84, 87-88, 92, 115, 121-122, 124, 130-131, 137, 147, 177, 185-186, 193-194, 273, 302-303, 310, 313-314
    cramp .................................................................................. 90
    groups
        chest and abdomen ................................................... 99
            external oblique ................................................ 99
            intercostals ....................................................... 99
            internal oblique ................................................ 99
            pectorals major ............................................... 99
            rectus abdominis ............................................. 99
            transversus abdominis ..................................... 99
        head and face ........................................................ 101
            buccinator ...................................................... 102
            levator labii superioris alaeque nasi................ 102
            masseter............................................................. 99
            mentalis ......................................................... 102
            occipitofrontalis............................................. 101
            orbicularis oculi .................................. 96, 101-102
            orbicularis oris ....................................... 101-102
            platysma......................................................... 102
        lower limb ......................................... 75-76, 100, 160
            biceps femoris ............................................... 100
            gluteus maximus ............................. 95-96, 99-101
            hamstrings ..................................................... 100
            quadriceps..................................................... 100
            rectus femoris ............................................... 100
        upper limb ........................................ 73-74, 84, 96-97, 158
            anterior flexors................................................. 98
            biceps brachii ................................................... 97
            deltoid...................................................... 96, 269
            infraspinatus ..................................................... 96
            posterior extensors ......................................... 98
            subscapularis ................................................... 96
            supraspinatus ................................................... 96
            teres minor................................................ 96, 269
            trapezius ................................................. 96-97, 268
            triceps brachii .................................................. 97
    growth ....................................................................... 91
    pulled ......................................................................... 92
    structure.............................................................. 85, 88
        muscle cell ........ 12-13, 83, 85-87, 89, 116, 130, 229, 310, 313-314
        muscle fiber............................... 85-88, 92, 302, 310
        myofibril............................................................ 86, 310
        myofilament................................... 86-88, 302, 310
            actin ................................. 23, 86-88, 93, 302, 310
            myosin ........................................ 23, 86-88, 93, 31
        sarcomere.................................................... 86-87, 313
    tissue 30, 32-33, 83, 85, 87-88, 91-92, 115, 129, 209, 310, 313
    tone ..................................................................... 7, 33, 92, 251
    types
        cardiac muscle....33-34, 84, 110-111, 115-116, 125, 130, 135, 137-138, 209, 275, 304-305, 310, 312
        skeletal muscle......21, 33, 84-87, 110, 112, 115-116, 130, 209, 274-275, 310, 313
        smooth muscle ... 33-34, 84, 110, 115, 144-146, 177, 209, 273, 275, 296, 303, 313-314
Muscular artery ............................................................. 142

myelin................................................. 212-214, 226, 264, 310, 313
myelination ................................................ 212-215, 217, 226
myocardial infarction ........................................................ 128
myocardial ischemia......................................................... 128
myocardium.......................114-115, 126-129, 133-135, 304, 310
myopic................................................................................ 297
nasal cavity..........................165-169, 175, 264, 280, 311, 314
nasal conchae ................................................................... 168
nasopharynx.......................................................... 169-171, 285
nerve................ 4, 11-13, 32, 35, 48, 51, 75, 83-84, 87-88, 92, 130, 139, 142, 152-153, 167, 193-194, 203, 207-210, 212-219, 221-223, 225-231, 233, 258-260, 263-272, 275, 280-283, 287, 289-296, 302, 304, 306, 310, 313, 315
nerve cell ........................ 12, 83, 207, 215, 218, 222, 259, 282-283
nerves .....4, 35, 38, 56, 63, 69, 72, 87, 112, 175, 203, 205, 215-216, 231, 240, 248, 257-258, 260-269, 275-276, 283, 305, 310-311
    axillary nerve................................................................ 267
    femoral nerve .............................................................. 268
    median nerve........................................................ 267, 269
    obturator nerve ......................................................... 268
    phrenic nerve ....................................................... 267-268
    radial nerve .................................................................. 267
    sciatic nerve ................................................................ 268
    ulnar nerve ........................................................... 267-268
neuroglia .......................... 4, 206-207, 211, 216, 237-238, 252, 310
neuron ...207-212, 215-219, 221-222, 224, 227-230, 240, 247, 249, 258, 260, 271-272, 274-275, 280-282, 304, 307, 310, 312-314
neurotransmitter ............................... 219, 228-231, 283, 310
nociceptor ........................................................................ 273
nodes of Ranvier......................................................213, 226
nonmyelinated ........................................... 213-214, 238, 307
non-rapid eye movement sleep ...................................... 254
norepinephrine................................................ 138-139, 153, 231
nose .... 37-38, 69, 102, 165-169, 171, 185, 207, 210, 264, 280, 294, 303, 314
nuclear membrane.................................. 17, 21, 25, 310, 312-313
nucleotides ..............................................22, 25, 29, 306, 310-311
    adenine...............................................................310-311
    cytosine ............................................................... 310-311
    guanine ............................................................... 310-311
    thymine ..................................................................... 310
nucleus .... 11, 13-15, 17-22, 24-27, 29, 86, 207-208, 246-247, 283, 305, 310-311
occipital lobe ..............................................................238, 243
odorant ....................................................................... 281, 311
odorant molecule............................................................ 281
olfaction .................................................................... 264, 280
olfactory epithelium ................................................. 280-282
olfactory sensory neuron ............................................... 281
optic disc .......................................................................... 294
organ .. 3, 11, 36-39, 41, 44-45, 55-56, 84, 109, 126, 149, 201, 203, 233-234, 252, 274, 276, 279, 284, 287, 307, 311
organelle.......................................................21, 87, 307, 311-312, 315

| | |
|---|---|
| organ of Corti | 287 |
| oropharynx | 169-171 |
| osmoreceptor | 272 |
| osteoarthritis | 57, 311 |
| osteon | 51-52, 305, 311 |
| Osteoporosis | 57, 311 |
| otitis media | 170 |
| otolithic membrane | 291 |
| otoliths | 291 |
| Oxygen Transport | 189 |
| Oxyhemoglobin | 189-190, 311 |
| pacemaker | 130-132, 139, 313 |
| papillae | 282 |
|     circumvallate papillae | 282 |
|     filiform papillae | 282 |
|     foliate papillae | 282 |
|     fungiform papillae | 282 |
| parasympathetic nervous system | 139, 205, 277 |
| parietal lobe | 238, 242 |
| parieto-occipital sulcus | 238 |
| peduncles | 247, 249 |
| pericardial effusion | 113 |
| Pericarditis | 113, 311 |
| pericardium | 113-114, 182, 311 |
|     fibrous pericardium | 113-114 |
|     parietal pericardium | 114 |
|     visceral pericardium | 114 |
| perilymph | 286, 289-290 |
| perineurium | 216 |
| periosteal layer | 234 |
| periosteum | 47-48, 51, 54, 58, 234, 311 |
| peripheral edema | 125 |
| peripheral nervous system | 4, 35, 202-203, 205, 207, 211-212, 215-216, 230, 240, 262-263, 265, 267, 269, 271, 273-275, 277, 302-303, 306, 311 |
| pharynx | 38, 165-166, 169-170, 172-175, 311, 314 |
| phospholipid | 15, 305, 311 |
| phospholipid bilayer | 15, 305, 311 |
| photoreceptor | 273, 295-296 |
| Physiology | 3, 6, 11, 40, 109, 148-149, 151, 153, 155-156, 282, 311 |
| pia mater | 235, 257, 309 |
| pigmented layer | 295-296 |
| pitch | 173, 287-289 |
| placenta | 181, 192 |
| Plant cell | 13 |
| pleura | 181-182, 186, 311 |
|     parietal pleura | 181, 186, 311 |
|     visceral pleura | 181, 186, 311 |
| pleural effusion | 182 |
| pleural fluid | 182 |
| Pleurisy | 182 |
| pneumonia | 166, 180, 182 |
| Pneumotaxic area | 194, 311 |
| pneumothorax | 182 |
| polarized | 220-221 |
| polysynaptic reflex | 272 |
| pons | 194, 247, 249, 304, 311 |
| Pontine respiratory group (PRG) | 194, 311 |
| posterior white column | 259 |
| postganglionic neuron | 275 |
| postsynaptic neuron | 227, 229-230 |
| precentral gyrus | 240 |
| preganglionic neuron | 275 |
| Preload | 137, 312 |
| premotor cortex | 241-243 |
| presynaptic neuron | 227-228, 230 |
| primary motor cortex | 240-243, 248 |
| primary somatosensory cortex | 242-243, 245 |
| processes | 10, 29, 41, 51, 71, 152, 188, 195, 197, 201, 208, 210, 212, 214, 230-231, 243, 249, 252, 276, 293-294, 304, 308, 310 |
| Prophase | 27, 312 |
| Proprioceptors | 273, 279 |
| protein | 14, 16-17, 19, 22-25, 33, 35, 50, 86, 91, 93, 189, 212, 220, 230, 305, 307-310, 312, 314 |
| PTH | 56-57, 312 |
| pubic symphysis | 65, 76 |
| pulmonary circulation | 117-119, 143, 156, 312 |
| pulmonary edema | 125 |
| pulmonary valve | 122, 312 |
| pulse | 108, 149-151, 250, 312 |
| pulse pressure | 150-151, 312 |
| P wave | 134 |
| pyramidal tracts | 240, 248, 260-261 |
| QRS complex | 135 |
| rapid eye movement sleep | 254 |
| receptor region | 228 |
| Renin | 153 |
| Repolarization | 222, 224-225, 312 |
| respiration | 72, 89-90, 176, 178, 193, 195, 231, 266, 273, 302, 305, 307, 309 |
|     cellular respiration | 176, 305 |
|     control of respiration | 176, 193 |
|     external respiration | 176, 307 |
|     internal respiration | 176, 309 |
| rest and digest | 277 |
| resting membrane potential | 4, 220-222, 224, 308, 312 |
| reticular activating system | 249 |
| reticular formation | 249 |
| retina | 209, 264, 273, 279, 293-297, 305, 312 |
| retrograde amnesia | 255 |
| rheumatoid arthritis | 57, 312 |
| ribosome | 18-19, 25, 310, 312 |

rickets ............................................................................54, 312
right atrium .........................118-120, 127, 131, 159, 188, 313-314
right bundle branch ......................................................... 133
right coronary artery ....................................................... 127
right ventricle .......118-120, 122, 127, 132-133, 143, 188, 312, 314
Rigor mortis .................................................................93, 312
risk factor .....................................................................129, 269
RNA ............................................19-21, 25, 309-312, 314
Rods ......................................................................61, 295-296, 312
rough endoplasmic reticulum ........................ 18, 21-22, 208, 312
saccule ............................................................................ 287, 291
sacral nerves .................................................................. 258
sacral plexus .................................................................. 268
Saltatory conduction ...................................................... 313
scala media ................................................................... 287
scala tympani ............................................................... 287, 290
scala vestibuli ............................................................... 287, 289
Schwann cell .................................................................. 214
sclera ............................................................................. 293
scurvy ............................................................................ 145
semicircular canals ....................................................286-287, 291
semilunar valve ............................................................. 122
sensory division ........................................................... 203, 302, 313
sensory function ...........................................................201, 308
sensory neuron ............................................................ 218, 271-272, 281
shingles .......................................................................... 269
shock .....................................................................48, 71, 155, 235
    cardiogenic shock ............................................... 155
    hypovolemic shock .............................................. 155
    vascular shock ..................................................... 155
shoulder dislocation ...................................................... 74
Sinoatrial node (SA node) ............................................. 313
sinus ...................................................................69, 131, 166, 169
sinuses .............. 69-70, 163, 165-166, 168-169, 173, 235, 284, 314
    ethmoid sinus ...................................................... 169
    frontal sinus ........................................................ 169
    maxillary sinus .................................................... 169
    sphenoid sinus ................................................... 169
skeletal muscle ............21, 33, 84-87, 110, 112, 115-116, 130, 207, 272-273, 310, 313
skeletal system ...............................................3, 8, 11, 38, 44-45, 56
sleep ....... 9, 34, 61, 73, 171, 184, 200, 246-249, 253-255, 269, 313
smell .........................4, 68, 183, 251, 263, 272, 278-282, 284, 311
smoking .........................................................57, 71, 129, 180, 183, 251
Smooth endoplasmic reticulum .......................................... 87, 313
smooth muscle .. 33-34, 84, 110, 115, 144-146, 177, 207, 271, 273, 294, 303, 313-314
sneezing ........................................................................ 167
sodium-potassium pump ............................................. 221, 225
soft palate ................................................................... 169-170
somatic nervous system ........................ 203, 205, 273-274, 313

Somatic reflex ..............................................................271, 313
somatosensory association cortex ..................................... 243
somatosensory homunculus ............................................ 242
Special Senses..4, 278-279, 281, 283, 285, 287, 289, 291, 293, 295, 297, 313
spinal cord .......................... 4, 35, 38, 63, 68-72, 87, 202-203, 205, 207-208, 211-212, 214, 217, 230, 233-237, 240, 245, 247-248, 257-262, 266, 268-272, 275, 304-305, 307, 309, 311, 313
spinal cord segment ...................................................... 258
spinocerebellar tract ..................................................... 261
spinothalamic tract ....................................................... 261
spiral organ ................................................................... 287
stapes ..........................................................................286, 289
stenosis ......................................................................... 122
sternum ...................................................... 72, 74, 111-112, 174, 186
steroids ......................................................................... 91, 269
stimulus..83, 130, 195, 207, 218, 222-223, 226-228, 230, 270, 272, 282, 303, 307, 310, 313
Striated muscle .............................................................. 313
stroke ..............................................................135-139, 209, 251, 265, 291, 313
stroke volume .............................................................. 135-139, 313
subarachnoid space ...................................................... 235-236, 257, 305
subclavian steal syndrome ............................................ 250
subdural space ............................................................. 235
Sub threshold ................................................................ 313
sulcus ............................................................................ 238, 240
surfactant ..................................................................... 180-181, 314
sympathetic division .....................................................205, 275
sympathetic nervous system ..................... 139, 205, 231, 276-277
symphysis ...................................................................... 65, 76
synapse ......4, 227-228, 230, 271-272, 275, 280, 283, 305-306, 314
synaptic plasticity ......................................................... 255
synaptic vesicles ........................................................... 228
Synovial joint types ...................................................... 65-66
    ball and socket ............................................... 66-67, 74, 76
    condyloid ............................................................. 67
    hinge ............................................................. 66, 74-75, 172
    pivot .................................................................. 67, 75
    plane .................................................................. 66-67
    saddle .................................................................. 66
systemic circulation................... 117-119, 143, 156-157, 303, 309
systole ........................ 118, 121, 123-125, 137, 150-151, 305, 314
systolic blood pressure .................................................. 150, 312, 314
Tachycardia ...................................................................150, 314
tastant ...........................................................................283, 314
Telophase ..................................................................... 29, 314
temporal lobe ............................................................... 238, 243
tension headache .......................................................... 248
terminal ganglia ........................................................... 275
thalamus ........................................... 237, 245-246, 261, 283, 306
thermoreceptor ............................................................. 272

| | |
|---|---|
| thoracic nerves | 258 |
| thoracolumbar division | 275 |
| threshold | 222-223, 229-230, 281, 313-314 |
| thyroid cartilage | 172, 175 |
| tic douloureux | 264 |
| tissue | 4, 11-12, 30-35, 45, 50-52, 55, 58, 65-67, 72, 78, 83-85, 87-88, 91-92, 111, 113-115, 129, 132, 142, 144, 146, 171, 175-176, 181-182, 191, 206-207, 209, 211-213, 215-218, 234-235, 251-252, 257, 273, 293, 304-310, 313-314 |
| tongue | 23, 69, 170-171, 173, 265-266, 279, 282-284 |
| tonsillectomy | 171 |
| tonsils | 38, 171 |
| trabeculae | 52, 58, 314 |
| trachea | 37-38, 112, 165-166, 172, 174-176, 304, 307, 309 |
| tracheotomy | 175 |
| tracts | 240, 248, 259-261 |
| transfer RNA (tRNA) | 25, 310, 314 |
| tricuspid valve | 120-121, 131, 314 |
| Trigeminal neuralgia | 264 |
| Tunica externa | 145, 314 |
| Tunica intima | 144, 147, 314 |
| tunica media | 144-145, 314 |
| turbinates | 168 |
| T wave | 135 |
| tympanic membrane | 170, 285-286, 289 |
| Types of Tastes | 283 |
|    bitter | 283 |
|    oleogustus | 284 |
|    salty | 283 |
|    sour | 283-284 |
|    sweet | 283-284 |
|    umami | 283 |
| Unipolar neuron | 314 |
| upper respiratory system | 165-166, 314 |
| utricle | 287, 291 |
| vasoconstriction | 144, 153-154, 314 |
| vasodilation | 144, 153, 155, 314 |
| vein | 117-118, 129, 142, 159, 308, 313-314 |
|    basilic vein | 159 |
|    brachial vein | 159 |
|    celiac trunk | 160 |
|    cephalic vein | 159 |
|    common iliac vein | 160 |
|    external jugular vein | 158 |
|    femoral vein | 160 |
|    inferior vena cava | 118, 160, 308 |
|    internal jugular vein | 158 |
|    saphenous vein | 160 |
|    superior vena cava | 118, 131, 159, 181, 313 |
|    tibial vein | 160 |
| ventilation | 176, 178, 187-188, 231, 310, 315 |
| ventral horn | 260 |
| Ventral respiratory group (VRG) | 193, 315 |
| ventral root | 260, 272, 315 |
| ventricles | 118-124, 126-127, 131-133, 135, 138, 145, 212, 235-236, 303, 305 |
| ventricular diastole | 118, 124, 306 |
| ventricular systole | 118, 121, 124, 305, 314 |
| Venule | 315 |
| vertebra | 70-72, 257-258 |
|    intervertebral disc | 71 |
|    vertebral arch | 71 |
|    vertebral body | 70-71 |
|    vertebral foramen | 71 |
| vertigo | 291 |
| vesicle | 16, 18, 306-307, 309, 315 |
| visual association area | 243 |
| vitamin D | 7, 38, 49, 54, 56-57, 312, 315 |
| Vocal cord | 173 |
|    false vocal cord | 173 |
|    true vocal cord | 173 |
| voice | 69, 165, 169, 171-173, 309 |
| voice box | 69, 165, 169, 171-172, 309 |
| Wallerian degeneration | 217 |
| Wernicke's area | 243 |
| white matter | 238-239, 245, 249, 258-259, 315 |

# DR. TOMMY MITCHELL

Dr. Tommy Mitchell graduated with a BA with highest honors from the University of Tennessee-Knoxville in 1980 with a major in cell biology. For his superior scholarship during his undergraduate study, he was elected to Phi Beta Kappa Society (the oldest and one of the most respected honor societies in America). He subsequently attended Vanderbilt University School of Medicine, where he received his medical degree in 1984.

Dr. Mitchell completed his residency at Vanderbilt University Affiliated Hospitals in 1987. He was Board Certified in Internal Medicine. In 1991, he was elected a Fellow of the American College of Physicians (F.A.C.P.). Tommy had a thriving medical practice in his hometown of Gallatin, Tennessee for 20 years, but, in late 2006, he withdrew from medical practice to join Answers in Genesis where he served as a full time speaker, writer, and researcher.

As a scientist, physician, and father, Dr. Mitchell had a burden to provide solid answers from the Bible to equip people to stand in the face of personal tragedy and popular evolutionary misinformation. Using communication skills developed over many years of medical practice, he was able to connect with people at all educational levels and unveil the truth that could change their lives.

Dr. Mitchell was married to his wife Elizabeth (herself a retired obstetrician) for over 30 years, and they have three daughters. He passed into the presence of the Lord on September 17, 2019.